Kildare County Libraries

D0303408

RETHINKING PSYCHOLOGY

Also by Brian M. Hughes
Conceptual and Historical Issues in Psychology (2012, Prentice-Hall)

Rethinking Psychology

Good Science, Bad Science, Pseudoscience

Brian M. Hughes
National University of Ireland, Galway

macmillan education palgrave

© Brian M. Hughes 2016

All rights reserved. No reproduction, copy or transmission of this publication may be made without written permission.

No portion of this publication may be reproduced, copied or transmitted save with written permission or in accordance with the provisions of the Copyright, Designs and Patents Act 1988, or under the terms of any licence permitting limited copying issued by the Copyright Licensing Agency, Saffron House, 6–10 Kirby Street, London EC1N 8TS.

Any person who does any unauthorized act in relation to this publication may be liable to criminal prosecution and civil claims for damages.

The author has asserted his right to be identified as the author of this work in accordance with the Copyright, Designs and Patents Act 1988.

First published 2016 by
PALGRAVE

Palgrave in the UK is an imprint of Macmillan Publishers Limited, registered in England, company number 785998, of 4 Crinan Street, London, N1 9XW.

Palgrave Macmillan in the US is a division of St Martin's Press LLC, 175 Fifth Avenue, New York, NY 10010.

Palgrave is a global imprint of the above companies and is represented throughout the world.

Palgrave® and Macmillan® are registered trademarks in the United States, the United Kingdom, Europe and other countries.

ISBN 978–1–137–30397–4 hardback
ISBN 978–1–137–30394–3 paperback

This book is printed on paper suitable for recycling and made from fully managed and sustained forest sources. Logging, pulping and manufacturing processes are expected to conform to the environmental regulations of the country of origin.

A catalogue record for this book is available from the British Library.

A catalog record for this book is available from the Library of Congress.

Printed in China

Contents

Acknowledgements

I am grateful to all at Palgrave for their work on this book. I pay particular tribute to Paul Stevens, Isabel Berwick, and Cathy Scott for their dedication, support, and superb advice along the way.

Many of the ideas in this book were road-tested in various teaching and presentation contexts over the past decade. In this regard, I thank the many students who shared learning opportunities with me in my classes on pseudoscience, research methods, science communication, and social issues over the years, as well as the very fine PhD researchers I have had the privilege of working with. In the latter regard, I thank Siobhán Howard, Ann-Marie Creaven, Eimear Lee, Niamh Higgins, Aoife O'Donovan, Tracey Quinn, Diarmuid Verrier, Sinéad Conneely, Éanna O'Leary, Sinéad Lydon, Páraic Ó Súilleabháin, and Amanda Sesker.

On a personal note, thanks to my children, Louis and Annie, for their many observations and tips, including their ideas for the title ("*All About Psychology*" was on the shortlist). Particular gratitude goes, with affection, to Marguerite. Bashfulness prevents me from articulating here the full extent of my appreciation, but I trust that the appropriate gist can be derived.

B. M. H.

Preface

I took my first Psychology courses at university twenty-five years ago, and I was far from alone in doing so. In fact, hundreds of other students joined me in enrolling. Given such numbers, only the largest teaching venues could be used for our classes. These weekly gatherings, where throngs of enthusiasts congregated in the cause of a common interest, certainly made for quite a heady atmosphere. But despite the near feverish levels of fandom in the air, the content of what was taught often came as a surprise. For many students, the reaction was more shock than awe.

We arrived expecting to explore the ins and outs of subliminal advertising, or perhaps even the rudiments of romantic prowess. Instead, we ended up drawing flowchart diagrams said to depict various types of human cognition. We thought we might learn how to analyse people's body language or handwriting, or maybe even their dreams. We actually learned how to statistically analyse large numerical datasets, using multi-storeyed mathematical formulae to identify the significance of correlations and mean differences. And while we certainly expected to hear something about the brain, little did we know we would spend long hours exploring the anatomy, physiology, neurology, and biochemistry of both the peripheral and central nervous systems, in rabbits and in sea slugs as well as in humans.

We conducted experiments (of a kind), and maintained handwritten 'laboratory notebooks' in which we carefully detailed the findings of our class-wide investigations. In later years we were sent out to conduct our own research projects. I used this opportunity to pursue animal research. I examined social behaviour in garden beetles one year and in laboratory rats the other. My motivation stemmed more from convenience than from a fascination with the non-human condition: by doing my projects on animals, I avoided the many timetabling difficulties that arise when attempting to schedule appointments with human beings.

That is not to say that our curricula were devoid of crowd-pleasing elements. The classes on child development were particularly well received, provoking as they did our nurturing impulses. But even then, what we were shown was infused with empiricism rather than sentiment. At one stage we watched a film of a study where experimenters allowed a baby to crawl on all fours in the direction of a precipice, gaily unaware of its looming demise (although prevented from *actually* falling by an invisible glass floor). Our social psychology subject matter was also built on research, with everything from altruism to morality subjected to controlled experimentation.

Many students found the scientific bent invigorating, convincing, and powerful. However, for others it presented personal challenges. Some found it frankly objectionable. Their prior interest had been in the therapeutic potential of studying the human mind, and they had no intention now of allowing biological or statistical abstractions to divert them from their vocational destinies. Displaying an impressive capacity for compartmentalization, many deftly navigated all their degree requirements before building successful careers as psychologists, all the while harbouring attitudes towards science that were at best ambivalent and at worst scornful. Some of these dissidents even ended up in professional academia. For them, the scientific method was not an all-informing philosophy from which to draw inspiration; rather, it was an ever-lingering irritation and an awkward reminder of their own discomfort with rigour.

In the intervening quarter of a century, I have seen the cycle consistently repeat itself, as though it were an ineradicable meme embedded in the heart of the discipline. Psychology's subject matter and teaching methods have become more sophisticated, and the research methodologies employed by students today are far advanced beyond those taught previously. However, as the pool of scientifically literate psychologists has expanded, so too has the subset who are pseudoscientifically minded.

The fact that scientists can have a poor grasp of what exactly is scientific about their area is not exclusive to psychology. Nonetheless, the tension between science and pseudoscience is acutely present in this field. The mysteries of consciousness, behaviour, intelligence, and emotion consistently draw more than their fair share of pseudoscientific attention. Some of it comes from psychologists themselves. But the good news is that psychology not only offers case studies in confusion, it also provides a lens through which to view the resulting turmoil. Psychology is, after all, the (scientific) study of human thoughts, feelings, and behaviours – as such, it is the area of research devoted to telling us how people make sense of the information that is available to them. In other words, psychology itself informs us about the nature of science, and the problem of pseudoscience.

This book explores the interplay between psychology, science, and pseudoscience, and in so doing aims to stand as both record and remedy. It is organized around three main sections. The opening section comprises four chapters that consider the overall scope of the problem. In turn, these chapters consider the nature of science, the nature of pseudoscience, the nature of psychology, and the nature of reasoning. The second section examines practical examples, containing three chapters that look at the way pseudoscientific reasoning has been applied to psychological subject matter. The first of these focuses on topics that most observers would recognize as lying beyond the fringes of science. The subsequent two chapters are perhaps more contentious: these focus on examples of pseudoscientific reasoning that have emerged *within* ordinary mainstream psychology. The third and final section attempts to locate the overall discussion within a social context, comprising three chapters that examine the biases and influences that lead to pseudoscientific thinking. These chapters consider the way socio-political value systems affect scientific research, the impact of spiritual and optimistic beliefs, and the broader social milieu within which the quality of psychological science really matters.

Psychology continues to be a popular subject in today's universities. More than that, it is a science that attracts considerable interest from across contemporary society. The empirical rigour that has allowed researchers to reveal some of the most interesting things we have ever known about human beings remains integral to what psychology strives for. But with all progress comes resistance. By disrupting the homeostatic balance between sense and nonsense that characterizes human knowledge at any given moment, the sheer productivity of scientific psychology sends ripples of newly-energized pseudoscience emanating in all directions. The challenges this presents for psychology – and for society more broadly – are many, and they warrant serious attention. This book seeks to provide this, and in so doing to provoke self-reflection, understanding, and action.

Brian Hughes
Galway, July 2015

Part I

Psychology and Pseudoscience in Theory

Chapter 1

What Is Science and Why Is It Useful?

Stereotypes of science

Science is a word that seldom means quite what its users think. That is not to say it is an obscure or unpopular term. In fact, it is one of the English language's more commonly uttered words. Speakers of English are quite likely to encounter it, or one of its derivatives, every day of their lives. The word is considered useful by so many people that it has become conspicuous in modern culture. But despite this mass popularity (or maybe *because* of it), the word *science* has become burdened by layers of nuanced connotations. For some people *science* is simply the name of a subject taught in schools and universities, alongside *history*, *geography*, *business studies*, and the rest. This is *science* as a category-heading, part of the logistics of education delivery. For other users the word *science* signifies the activities performed in a particular occupation: that of the professional scientist; *science* is considered 'what scientists do' in much the same way that *quantity surveying* is 'what quantity surveyors do'. For yet other speakers, *science* is a more sociologically loaded, amorphous concept. These users refer to *science* as a kind of political or cultural belief system, similar to how religion and communism are belief systems. The metaphor works on several levels: each movement is advocated by proponents who present a set of mutually supportive and ideologically informed beliefs about why the world is the way it is; each movement's adherents undertake to spread and defend these beliefs as a matter of group loyalty, likely construing this duty as an important moral obligation.

All of these views of the word *science* contain elements of truth. It is certainly true, for example, that courses called '*Science*' are taught in schools and universities. It is also true that professional scientists conduct an activity that is commonly referred to as *science*. And it is thirdly true that the scientific method relies on a shared and all-encompassing philosophy, which followers ardently defend. And yet, notwithstanding these elements of truth, all of these views are flawed and incomplete. They are, in effect, stereotypes.

Stereotypes frequently hint at underlying realities. However, what makes them stereotypical is that they rely on weak generalizations, which is what

3

makes them insidiously misleading. Indeed, many stereotypes are developed specifically in order to mislead, with the aim of serving the ulterior motives of those who promulgate them. A typical stereotype will take one small detail (which may actually be true but only some of the time), and present it as if it were the fundamental and eternal essence of whatever target concept is under discussion. This is what happens when people use the word *science* to refer to an academic subject, an occupation, or a belief system. *Science* is a far subtler concept than any of these things. In essence, *science is a process by which objectively defensible knowledge is produced and handled by humans.* This is why philosophers often say that science is not so much a type of information that human society knows, but rather is the *way that human society truly 'knows' anything at all.* In simpler terms, then, science is a *way of knowing.*

Considering *science* as an academic subject matter, an occupation, or a belief system refers only to its superficial aspects. It is a bit like describing *jazz* as a category of shelf seen in a music store. While it is true that most music stores will indeed have a shelf labelled 'Jazz', this doesn't reveal *what* jazz is. Such shelves are simply a superficial feature of the world as it is affected *by* jazz. In the same way, the presence of science in schools, in workplaces, and in culture tells us a little bit about the impact of science. However, it doesn't reveal much about what science actually *is*, and so using the word *science* as a synonym for these things is semantically quite opaque.

Stereotypes of science are themselves multifaceted and highly colourful notions. Consider the common portrayal of scientists as people who work in laboratories. Usually, they are also wearing white coats, and fiddling with test tubes and microscopes. That such visual cues are endemic to the cultural perception of scientists has been confirmed in several research studies based on a very simple methodology: where members of the public are asked to describe – or even to draw a picture of – a scientist. These studies show that the relevant stereotypes are acquired early in life. Cultural historian David Wade Chambers (1983) found archetypal laboratory paraphernalia to be a common feature of drawings of scientists by children as young as four years of age. A more worrying aspect of children's stereotypes is that they are typically drawn in a way that assumes male gender: in Chambers's original study, both boys and girls depicted scientists as sporting large amounts of facial hair. In fact, when Chambers set about developing his now famous Draw-A-Scientist Test (designed to help researchers investigate children's stereotypes by establishing the age at which they first produce distinctive imagery), he found it useful to include the presence of beards, moustaches, or 'abnormally long sideburns' as positively scored items.

But at least children's stereotypes of scientists appear, in the main, to be non-threatening. For adults it is different. It seems that their standard depiction is of the scientist as clumsy or nefarious molester of nature. After having studied nearly a thousand 20th-century horror movies, sociologist

Andrew Tudor (1989) found the role of 'mad scientist' to be third only to that of 'supernatural antihero' and 'mentally disturbed villain' in providing the required personifications of horror. In fact, as many as one-quarter of all horror movies depicted a renegade scientist as their primary disaster-inducing plot element. Similarly, literary historian Roslynn Haynes (2003) studied depictions of scientists in English-language literature (and not just 'horror' novels) and identified seven classic scientist roles, virtually all of them unflattering. According to Haynes, scientists are most commonly depicted as: evil alchemists; foolish figures; inhuman researchers; reckless adventurers; mad or bad villains; or helpless investigators unable to control the outcomes of their work. Only a small minority are depicted positively in some kind of noble hero or saviour role. Overall, according to Haynes, the 'master narrative' of the scientist in Western literary fiction is of 'an evil maniac and a dangerous man' (p. 244).

Even when scientists are not depicted as lab-coat wearing horror villains, their activities are nonetheless portrayed as being out of touch with the norms of mainstream social behaviour. In both fictional and non-fiction contexts, a particularly common portrayal is that of scientist as 'geek': a (usually male) person whose high intelligence is matched only by deplorable social skills, a poor fashion sense, and obscure interests (Mendrick & Francis, 2012). A corollary of these attributes is that scientists are accused of using language in ways that ordinary people struggle to understand, making science an activity that demands all the more suspicion. Consider this example of a typical research finding relating to educational psychology:

> High-quality learning environments are a necessary precondition for facilitation and enhancement of the ongoing learning process.

According to UK-based advocacy group the Plain English Campaign (2012), such a statement would probably be clearer if rephrased as:

> Children need good schools if they are to learn properly.

Scientific jargon can cause confusion without having to employ actual sentences. Sometimes a single word is sufficient to bamboozle, and many of the longest words in the English language emanate from fields of science. The lengthiest to appear in any of the major dictionaries of English – *pneumonoultramicroscopicsilicovolcanoconiosis* – comes from respiratory medicine and refers to the consequences of inhaling superfine dust particles (Oxford Dictionaries, 2012). It is the technical term for a type of illness common in miners (the miners themselves had the good sense to call it *collier's lung*). However, even with 45 letters, pneumonoultramicroscopic-silicovolcanoconiosis is relatively concise compared to some other words that do not make it into authoritative dictionaries. For example, the more modest *Student's Dictionary & Gazetteer* (The Dictionary Project, 2011) features the name of a chemical compound that has 1,909 letters, which is reputed to be the longest English-language word ever published in print. But scientists can actually go much, much further. According to the standard system of chemical nomenclature developed by the International Union of Pure and Applied Chemistry (Panico, Richer, & Powell, 1993), the correct chemical name for the protein encoded by the TTN gene in humans has nearly 200,000 letters.

The idea that scientists have a particular dress-sense or are unable to speak coherently might seem mildly amusing in a certain context. However, it should also be seen as something of a creeping problem for scientists themselves, and quite possibly for wider society too. For one thing, the wide acceptance of stereotypes of science suggests an equally widespread resistance among the general population to genuinely appreciating what it is that science really involves. Insofar as they serve to cast science in a negative or ridiculous light, scientific caricatures impede understanding in a way that can lead to the outright rejection of the idea that science is a worthwhile cultural activity.

A second unwelcome consequence of this vivid stereotyping is that it facilitates the imitation of science by imposters. The majority of actual scientists do not in fact wear white laboratory coats. Historically, the use of white coats by scientists stems from the same reason that butchers (and some barbers) wear white coats: this was the clothing worn by people who worked in blood-stained environments, where light colours helped quickly distinguish hygienically clean garments from ones that required laundering. Nonetheless, because of popular stereotypes, people who wish to falsely portray themselves as scientists (or, equally likely, who wish to falsely project their products or services as having scientific merit) often make white coats part of their standard garb. Imitators can expect to succeed before most audiences by supplementing their white coat with a display of technically sophisticated equipment, and a patter consisting of long words bound together in complex sentences.

Likewise, although scientific statements can often be laboriously constructed and difficult to understand, it doesn't necessarily follow that *all* scientific language will be like that. In fact, in many contexts, scientific language falls comically short of the stereotype. Chemist Paul W. May (2008) has identified dozens of technical terms used in chemistry that sound ridiculous rather than scientific, although their exact resonance is likely to be determined by the listener's sense of humour. Examples include 'moronic acid' (derived from the mora tree), 'curious chloride' (contains curium), 'uranate' (anions of uranium oxide), 'diurea' (two parts urea), and 'commic acid' (a constituent of the tree *Commiphora pyracanthoides*), as well as the chemical abbreviations DEAD (diethyl azodicarboxylate), dUMP (2'-deoxyuridine-5'-monophosphate),

DAMN (diaminomaleonitrile), SEX (sodium ethyl xanthate), and PORN (poly-L-ornithine). Biologists, especially entomologists, are also prone to this sort of thing. One Australian wasp was named 'Aha ha' because when the entomologist who found it exclaimed 'Aha!' at its discovery, his sceptical colleague sardonically replied 'Ha!' (Evans, 1983). Similarly, a particular beetle in the *Agra* genus proved so difficult to locate that the biologist who eventually found it gave it the name *Agra vation* (Spicer, 2006).

So in reality, scientists are just as liable to concoct amusing terms for things as anyone else who has a sense of humour. By contrast, people who wish to *imitate* science are likely to avoid such language at all costs. They will generally try to keep their discourse sounding as unfunny as possible in order to conform to the stuffiness demanded by the stereotype of science.

For example, while a field such as homeopathy is very clearly not at all scientific (and we will examine the reasons why later), its journals typically include complex and technical sounding writing designed to emit the thematic vibrations of science, such as the following:

> Our clinical protocol consists in administering a single remedy, starting with 6Q and continuing on a scale of dilutions, from 6 to 9, 12, 18, 24, 30, and, sometimes, 60Q, generally for 45–60 days for each potency. If there is a subsequent phase, the prescription proceeds with a single dose of an high dilution of Hahnemann's centesimal scale (cH), in a '"scale of potencies' (200cH–1M–10M). Acute cases are usually treated with centesimal dilutions in lower potencies.
> (from 6 to 30 cH; Rossi, Bartoli, Bianchi, Endrizzi, & Da Frè, 2012)

Such verbiage combines long multi-clausal sentences with a heavy smattering of numbers, several orthographically styled abbreviations (e.g. '6Q' or 'cH'), some language borrowed from mainstream pharmacology (e.g. 'potency'), and a number of polysyllabic terms that have no common meaning outside the immediate context (e.g. 'centesimal dilutions'). However, despite these characteristics, the text is not, in itself, in any real way scientific.

In summary, there is more to science than appearances or sounds. In fact, while the stereotypical image of white-coated, male, laboratory-dwelling, jargonese-speaking introverts might provide a convenient reference point for popular discourse, it has very little to do with the actual concept of *science*, at least in the true sense of the term. To this extent, it is worth recalling that possessing the superficial aspects of science does not make a person an actual scientist. Donning the accoutrements of science does not render your claims, your thinking, or your work any more scientific than it would otherwise be. It is useful to remember that this principle is not restricted to charlatans or quacks. It applies just as much to psychologists.

'Science' as a way of knowing

Strictly speaking, *science* is the process or state of knowing. It is derived from the Latin word *scientia*, which means 'to know'. The ending '*-ence*' is common in English as a way of denoting a quality. As such, *reminiscence* is a process of reminiscing, *reticence* is the state of being reticent, *adherence* is a process of adhering, *impertinence* is the state of being impertinent, and so on. Thus, *science* is a process or state of knowing.

For many people, this seems a remarkably simple explanation, if not indeed an overly simple one. However, the not-so-hidden complexity here relates to the idea of 'knowing'. When people say they 'know' something, they often mean that they '*think* they know' something or that they '*assume* they know' something. But 'knowing' is different to 'believing' or 'thinking' or 'assuming'. 'Knowing' refers to *being aware of something that is true while also being directly aware of the very fact that it is true*. Therefore, to really know something, you must have a basis for being sure that it is actually true. If you do not have such a basis, then you might suspect, believe, assume, suppose, judge, accept, conceive of, regard, imagine, presume, or think it, but you do not 'know' it.

To be fair, philosophers have spent centuries struggling to pin down their own thoughts on the question of knowing. In classical philosophy, Plato sought to explain that knowledge needs to be evidence-based, or justified in some other way, in order to be considered real knowledge. For example, beliefs that are correct by coincidence alone do not constitute knowledge *per se*. If you believe that it is raining in Hong Kong and if, by coincidence, it actually is, this does not mean that you truly *know* it is raining in Hong Kong (Schick Jr. & Vaughn, 2008). Thus, there is more to 'knowing' than simply believing things that are actually true; you must also be directly aware of the very fact that they are true. Plato's approach has become associated with the idea of 'justified true belief' or JTB. The argument is that for something to constitute knowledge, all three requirements – to be justified, to be true, and to be believed – need to be met.

Some philosophers have argued that the JTB conditions are insufficient. In other words, they argue that it is possible for the conditions to be met by information that nonetheless fails to constitute actual *knowledge*. The most prominent challenge has been put by American philosopher Edmund Gettier. He points to the possibility of holding conjunctive beliefs. This is when two beliefs are combined into a single either/or statement, such as 'I believe that John drives a blue car or that Mary is in Barcelona' (Gettier, 1963). In such a case the first element ('John drives a blue car') might be *justified but untrue* whereas the second element ('Mary is in Barcelona') might be *unjustified but true*. However, because the elements are separated by the conjunction 'or', the statement overall becomes *both* justified *and* true, despite the fact that its second element – the bit that might be construed as knowledge – is essentially a guess. Other philosophers have dismissed such criticisms as being founded on implausible assumptions, especially the way Gettier presents beliefs as capable of being *justified* but simultaneously *untrue*. As such, the JTB account remains the most robust way of explaining the idea of knowledge.

The types of information we can know will fall into different categories. Some information will be *true by definition*. This means that it refers to something that has an already agreed meaning and so cannot be disputed without perverting the language that is used to express it. For example, you can accurately claim to know that a spaniel is a type of dog simply on the basis of the fact that the word 'spaniel' is *defined* as a 'type of dog'. Likewise, you can claim to know that 'all dogs are mammals' because being a mammal is central to the definition of 'dog'. This type of information is referred to as an *analytic proposition* or an *analytic truth* (in contrast to the term *synthetic truth*, which refers to information that requires evidentiary confirmation). Strictly speaking, there is no way of establishing the truth of analytic propositions without referring, in a linguistic sense, to the semantics of the terms used. Even so, such knowledge is not spontaneous. While these points are true by definition, they will not arrive in your mind unless you encounter the information from some external source. Other types of information which are true by definition *can* be generated by your own thoughts, assuming you have been shown an appropriate system for doing so. For example, knowing that $1 + 1 = 2$ allows you to figure out for yourself that $2 - 1 = 1$. Likewise, you can figure out for yourself that $32,134,556 - 1 = 32,134,555$ (which, unlike $2 - 1 = 1$, you are less likely to have ever been told directly). In other words, you can 'know' these conclusions without having to consult an external source or to look for evidence to corroborate your knowledge. That type of information is known as an *axiom*. Both analytic truths and axioms can be classified as types of *truism*, which is the term used to describe claims that are true without question.

The complications arise when we feel we 'know' something with subjective certainty without its actually being true by definition. For example, we probably feel we 'know' that Paris is the capital of France. This belief can be established as being justified by consulting a sufficiently reliable source (or the preponderance of sources, such as all the encyclopaedias in the world). A similar approach can be taken to verifying our belief that the sun will rise tomorrow (although our 'knowledge' of this is more likely to be derived from compelling prior experience than from encyclopaedias). Information like this is just about as close to being 'true by definition' as it is possible to be without *actually* being true by definition. For whatever reason, these points just 'feel' obvious to us. Without constituting either an *analytic truth* or an *axiom*, overwhelmingly obvious information of this kind can still be said to constitute *truisms*, at least in a loose sense of the term.

Of course, the idea of treating something as 'true' simply because it 'feels true' is fraught with peril. The obvious risk is that something that 'feels true' might actually be false. Let us take as an example the claim that poverty precipitates criminality. For many observers such a notion 'feels true', even to the point of being 'obvious'. The reason such a claim is so frequently presented is the fact that the two relevant conditions – poverty and criminality – are often seen together: habitual criminals often come from poor backgrounds and the poorest neighbourhoods are often those most affected by crime (or, more accurately, by specific types of crime). However, when two conditions emerge together, it does not always mean that one has caused the other in the way that happens to be most commonly alleged. A variety of alternative scenarios will always be possible. Firstly, it might be that the causality described is actually being construed backwards. In other words, rather than poverty causing criminality, maybe it is criminality that causes poverty. Secondly, instead of one causing the other, it could be that both conditions are themselves simultaneously caused by some other, third, condition. For example, for a particular person, poverty *and* criminality might emerge together as a result of becoming a chronic heroin user. Thirdly, things might be more complicated than simply one condition causing the other: instead, one condition might indirectly elevate the likelihood of a number of mediating factors that, given the appropriate circumstances, might then in turn elevate the likelihood of the other condition. In other words, poverty may increase the risk of other things – such as stigma, isolation, a disconnection from the concerns of mainstream society – that in turn, in people who are predisposed in some way (for example, who have poor literacy skills), contribute to criminality. Or fourthly, maybe it is all one big coincidence. Maybe poverty and criminality have nothing to do with each other. After all, plenty of criminals come from wealthy backgrounds, and lots of poor neighbourhoods are crime-free.

What we are discussing here, of course, is the incomplete overlap between *correlation* and *causation*. In the context of university research methodology courses, especially in the social sciences, the idea that 'correlation does not imply causation' has become something of an embedded mantra. Students are encouraged to avoid assuming that X causes Y simply because X is correlated with Y. However, while this is sound, students (and others) can be over-cautious in how they generalize the principle. Instead of treating correlational associations as potentially weak, they treat them as if they were necessarily misleading. In other words, perfectly plausible propositions are dismissed as bogus simply on the basis of being supported by a correlation. In past decades, tobacco companies attempted to discredit medical testimony about the link between smoking and cancer by labelling the evidence as 'correlational'. Nowadays the same approach is commonly used to dismiss claims about how the use of fossil fuels might be causing global warming. *It's all correlational; ergo, there is no evidence; ergo, it isn't true.*

However, despite the fact that 'correlations' have been given a bad name, the presence of a correlation is often a very reasonable basis for suspecting the presence of causation. This is because, while correlation doesn't imply causation, causation *does* imply correlation: if X causes Y, then where there is X you will also find Y. The very fact that causation does imply correlation is why correlations generate so much confusion. The four poverty–criminality scenarios described above represent each of the alternative, non-causal interpretations of a correlation, namely: reverse causality; confounding; dependence on synergies; and coincidence. These possibilities are so well rehearsed in the minds of students and scholars alike that the fifth possible scenario – straightforward causality – is often forgotten. When you think about it, the confusion here partly arises because listeners interpret the link between 'correlation' and 'causation' to be causal when it is in fact correlational. In other words, causality causes (and is therefore correlated with) correlation, but correlation does not cause causality.

The incomplete overlap between correlation and causation is deceptively confusing. When explained at first, most listeners feel that the principle is sufficiently clear that they will never fall into the trap of misinterpreting a correlation again. However, treating correlations as indicators of causal events is part and parcel of navigating our way through everyday life. Answering a telephone when it rings, stepping away from a red-hot fire, and punishing a pet dog for soiling the carpet while you were out of the room, all require an ability to infer causality from observed correlations. An inability to do this would be a life-shortening handicap. Nonetheless, while you might be absolutely confident that your telephone is ringing because somebody has called your number and is on the line waiting for you to pick up, you might be *less* sure that the rise in temperature recorded in the Earth's atmosphere over the last century is the result of human-induced emissions of greenhouse gases. You can clarify the former by answering your phone, but what can be done to clarify the latter? In the end, while something might be consensually agreed to be 'obvious', it still might not be true. Unless it can be tested conclusively, the matter will remain uncertain.

For centuries, people had very few ways of resolving such uncertainties. What was passed around as knowledge was often originally generated on the basis of someone's subjective judgement, where a person decided what 'felt obvious' to them after having given the matter some thought. Where uncertainties persisted, attempts to resolve them involved the consulting of authorities. Some authorities were figures inhabiting the most influential echelons of society, such as religious gurus, famous philosophers, or royalty. In this regard, many historians particularly emphasize the role of organized religions in inculcating the masses with their teachings on nature. For centuries, religions succeeded in convincing the population at large that their scriptures constituted true 'knowledge'. Adherents accepted as factual teachings on such matters as the movement of celestial bodies, the origins of biological species, or the causes of storms. It is certainly the case that religious teachings have proved persuasive to people around the world since prehistoric times. However, perhaps a more important, albeit mundane, manner in which authority has shaped human knowledge has been the way people habitually turn to those around them for information. Much of what we 'know' about the world has been told to us by our friends, our neighbours, our parents, or our teachers. Consulting people we believe to be reliable sources, and treating their advice as 'knowledge', is an extremely common practice, and, in terms of child development in particular, an important survival skill.

In summary, for societies throughout their histories and for individuals during their own lifetimes, the things that we claim we 'know' are frequently assembled by relatively informal means. Some information is true by definition, other information intuitively 'feels' true, while other information is passed on by third parties on the basis of their assumed reliability as sources. However, strictly speaking, none of these processes is able to produce authentically new information that can be defended as being true in an objective way. Nor can any of these processes resolve uncertainties where a multiplicity of alternative interpretations is available. Put simply, semantics, intuition, and authority are incapable of yielding actual 'knowledge', in that they cannot produce new information that we can *know* to be definitely true. Actual knowledge-generation requires a systematic form of investigation and reasoning that avoids the pitfalls of these informal systems. And this is what 'science' really is.

Science as a formal process of knowledge-generation

Science involves the use of data to resolve uncertainties and thereby produce objectively defensible knowledge. Such knowledge is likely to be constructed from smaller bits and pieces, in the manner of bringing together already established assertions ('premises') with newly identified information ('observations') in order to create a logically persuasive combination of points (an 'argument') in support of a predictive explanation (a 'hypothesis') that, if confirmed, generates an overall assertion (the 'conclusion'). The ideal of science is to eliminate the role of subjective judgement in knowledge-generation, so that error can be avoided. While the use of observation is important, the way in which pieces of information are combined is also critical. Simply stringing a series of facts together and then declaring a conclusion would not be scientific. Every step needs to be presented in a logically reasonable way. While there have been many different methods of logic described over the centuries, three main processes of reasoning have emerged as being the most important. These are deduction, induction, and abduction.

Deduction is where premises of successively lessening generality are combined to produce an irrefutably true conclusion (Hughes, 2012). For example, if you know that all dogs are mammals and that your friend has a pet dog, then you are using deduction when you conclude that your friend's pet is a mammal. Induction, on the other hand, is where specific pieces of information derived from a larger group are considered together, a pattern is identified, and a conclusion about the larger group is drawn on the basis that the pattern will continue (Hughes, 2012). In this case, if you know that your friend has 25 pets and that at least 24 of them are dogs, then you are using induction if you conclude that the 25th pet is also a dog. It can immediately be noted that while a deduction is always true (in that all dogs are indeed mammals), an induction is not necessarily so (in that your friend's 25th pet might be an iguana). We can also note that deductions are a form of re-stating what is already known, while inductions are a form of generalization.

Abduction is the process of considering a given outcome along with some possible preconditions, and then combining this information in a way that concludes that the outcome is likely to have been caused by those preconditions (Hughes, 2012). For example, if you know that your friend has just bought a ball of wool, and you also know that cat-owners like to give their cats balls of wool to play with, then you are using abduction if you conclude that your friend must now also have a pet cat. Abductions are never *necessarily* true, because there can always be alternative explanations for the outcome under consideration (perhaps your friend has taken up knitting). In addition, it is not possible to be sure that all of the relevant premises are known to you (perhaps, without you being aware of it, iguanas also enjoy playing with balls of wool). However, unlike deduction or induction, abductive reasoning can suggest new ways of explaining things. In other words, abduction can help us to generate theories, the substance of which can then be investigated.

Deduction, induction, and abduction each involve putting two (or more) pieces of old knowledge together in the hope of producing a third, new piece of knowledge. With deduction, the two pieces of knowledge are related to each other but are of successively less generality. As such, it is often said that deduction involves moving from the broad ('all dogs are mammals') to the specific ('my friend's pet is a mammal'). Deductions have the advantage of always being true. With induction, the pieces of knowledge relate to known subsets of cases with the aim of drawing conclusions about the unknown *full* set of cases from which the subsets are drawn. Induction, then, involves moving from the specific ('some of my friend's pets are dogs') to the broad ('all of my friend's pets are dogs'). Inductions are not always true, but have a certain likelihood (that is, a likelihood above zero) of being true. Abduction, meanwhile, involves putting together observed outcomes with suspected causes. Abductions have no particular truth-value in and of themselves. They may be completely true, they may be completely false, or they may be somehow partially correct within these extremes. Nonetheless, abductive reasoning is the underlying process of science because it represents the way theories about the world are framed. The mechanics of scientific research involve searching for reliable information that can be used to test the accuracy of abductive propositions. In other words, scientific research often involves the combining of newly gathered evidence with abductions, in order to generate inductions.

The idea of using evidence to shed light on the reliability of propositions relates to the principle of empiricism, which asserts that knowledge is best produced by observation. However, while philosophers who describe themselves as empiricists have historically united around the point that observation is critical for knowledge production, it would be misleading to suggest that such a view is divorced from rationalism. Empiricism can only lead to knowledge when observations are combined and interpreted logically by, for example, using induction or deduction.

The emergence of 'modern' science

Such themes have been ever-present throughout the history of human thought, and therefore throughout the history of science. Nonetheless, the traditional narrative used in telling the historical story of science tends to include reference to an epoch-defining transition referred to as the 'scientific revolution' (Koyré, 1939). This is said to comprise a very long period, spanning the 16th, 17th, and 18th centuries, during which the prevailing attitudes of educated people in Western societies moved from an almost exclusive reliance on authority-based knowledge to the almost complete valorization of empirically informed and rational knowledge. The transition occurred on many levels. On one level, the standard scholarly explanation of the very nature of the universe changed fundamentally, from that originating with the Ancient Greek philosopher Aristotle (384–322 BCE) to that proposed by the 17th-century English physicist Isaac Newton (1642–1726 CE; DeWitt, 2004). In the Aristotelian worldview, the universe was a finite space with the Earth at its centre. The stars were embedded in a spherical shell which marked the universe's outer periphery. Both the stars and the planets (which comprised those visible to the naked eye, such as Mercury and Venus, along with the Sun and the Moon) were composed of the same substance, which was entirely different to that of which the Earth was composed. In stark contrast, the Newtonian worldview saw the Earth as one of a number of similar planets orbiting a common sun, which itself was one of a multitude of stars extending across the universe in endless space.

The very fact that this transition of thought extended over a number of centuries illustrates both its excruciating gradualness and the amorphous nature of the stage-boundaries that punctuated it. In fact, the oft-presented depiction of the transition as one of a radical and sudden changing of minds in the Western world is as misleading as it is common. It is certainly true that when astrophysical views popular in the 17th century are compared with those that became popular 200 years later, differences in the fundamentals are vivid. However, similarly vivid differences will be identified, without ever being referred to as revolutionary, in just about any domain of human culture to which equivalent comparisons are applied. For example, few commentators describe the differences in clothes worn in the 17th and in the 19th centuries as constituting evidence of a 'fashion revolution'.

Two other features of the popular history might also be considered misconceptions. Firstly, it is often suggested that the Aristotelian view was simply naïve (in that it was so blatantly incorrect as to not bear serious scrutiny), egocentrical (in that it arbitrarily presumed the place inhabited by humans to be the centre of the universe), and biased in favour of the teachings of Abrahamic religions (which also assumed human, and thus Earthly, centrality). However, the Aristotelian worldview was not as naïve as it first appears. In the era before the invention of the telescope, there was no reason to suspect that planets seen as minute specks of light in the night sky might in any way be similar to the ground upon which we walked every day. Similarly, there was no particular reason to assume that the stars were at all like our own daytime sun, or indeed that they were positioned at varying distances from the Earth. Indeed, what distinguished the stars from the planets was the very fact that, as far as the naked eye could tell, they were arranged into permanent constellations that moved only gradually and in unison, all of which strongly implied they were part of a single unified entity. Overall, there were some very good reasons to adopt the Aristotelian approach to such matters. It might be worth remembering that the Aristotelian view was accurate in a number of respects sufficiently subtle as to suggest the presence of great intellectual insightfulness. For example (and again contrary to stereotype), pre-Newtonian scholars correctly understood that the Earth was spherical rather than flat. Similarly, the accusations of egocentricity and Christian bias are inconsistent with contemporary writings and with the Aristotelian worldview's Ancient Greek provenance.

The second significant misconception in the popular history of the scientific revolution is to assume that the pre-Newtonian scholars were entirely reliant on authority and always allowed it to trump observation, and, correspondingly, that the post-Newtonian scholars could be described using exactly the opposite terms. While the formal concept of 'empiricism' is often associated with a number of philosophers who became prominent during the 17th century, the use of empiricism as a process to support (if not indeed to validate) assertions dates right back to the Ancient Greek philosophers, including, prominently, Aristotle himself. In fact, Aristotle was convinced of the sphericity of the Earth purely on empirical grounds, and used the empirical evidence to argue for sphericity in his writings. Therefore, while Newton (and some predecessors, most notably the Prussian astronomer Nicolaus Copernicus [1473–1543] and his Italian successor Galileo Galilei [1564–1642]) broke new ground in explaining the shape and dynamics of the universe, this owed as much to the invention of the telescope as it did to radical shifts in philosophical approaches to knowledge. In other words, it was as much a matter of technology as it was of epistemology.

It can be argued that the most significant changes brought about by the (Western) scientific revolution were not to the theoretical approaches applied to the generation of knowledge, but to the resulting explanatory frameworks used by scholars to describe the world around them. For example, in the Aristotelian view, all objects had essences and purposes that accounted for, and indeed determined, their behaviours. However, in the Newtonian view, objects behaved because of forces exerted upon them by other objects, with all objects in the universe interacting as if parts in a grand machine. Figures such as René Descartes (1596–1650) generalized this principle to biological organisms, including to people. To the extent that such views assumed the universe and its entities to be operating by themselves without external or spiritual guidance, this shift towards a mechanistic view of nature was a much greater challenge to religious orthodoxies than was the emergence of heliocentricity. The mechanistic approach also fostered the view that natural phenomena were amenable to quantitative measurement, and so facilitated the expansion of mathematical approaches to the study of nature.

Of course, while it might be overly simple to depict the scientific revolution as precipitating a sudden and complete change in the way science was done, it would be equally limited to assert that it had no impact at all. A number of distinct developments that have come to characterize science as it exists today were introduced during this period. One important example is the emergence of experimentation. In the 17th century, the English philosopher Francis Bacon (1561–1625) was among the first to argue that reasoning alone was highly limited in its capacity to generate new knowledge, because it was subject to what we nowadays refer to as *confirmation bias*. In other words, unless competing possibilities are considered and compared in an objective manner, observers are liable to interpret the world around them in ways that conform to their prior expectations. Bacon extended arguments in favour of induction by proposing systematic ways in which to set up the comparisons necessary to test predictions. These proposals eventually led to the convention of experimental research, and, as such, Bacon is sometimes referred to as the 'father of experimental science' (Urbach, 1987). An experiment is a method of research involving the systematic comparison of different arrangements or procedures in order to test a hypothesis. A critical aspect of experimentation is that the researcher deliberately seeks to manipulate the relevant features of these arrangements or procedures in order to facilitate inferences about how such manipulations affect outcomes.

In a broad sense, experimentation in scientific research is simply a variation of experimentation in everyday life. For example, imagine you are in the habit of commuting to work via Route A. Imagine then that someone proposes that Route B would be quicker. You can test the merits of this person's hypothesis by establishing an experimental protocol. In order to establish whether Route B is indeed faster than Route A, you can attempt both routes on a number of occasions and compare the resulting journey times. Your 'experiment' would work better if it met some important requirements. For instance, if you gathered your data on very many occasions, your conclusion would be more reliable than if you gathered data on very few occasions. Similarly, if you gathered your data at consistent times of day so that you ended up comparing like with like, then your conclusion would be better than if you gathered your data at inconsistent times of day (for example, only recording Route A journeys in the morning and Route B journeys in the evening). The best approach would be to spread your measures across the day, but to ensure that for each measure for Route A you collected a matching measure for Route B at the same time of day. Eventually, you could extend your knowledge base by using a similar approach to factor in things like weather conditions and public holidays. If you are worried that your prior habits will taint the fairness of the comparison, you could recruit somebody else to collect the data for you, without telling them the reason why. Ultimately, the more equivalent your data about Routes A and B, the larger your overall sample of data, and the more objectively it is gathered, the more confidently can you form a conclusion. Note that the process of interpretation here will be inductive: you will be drawing a general inference about journey times from your particular set of observations. Note also that you need not use an experimental approach to address this dilemma at all. You could use an authority-based approach instead, by simply asking somebody who already knows the answer.

The above example essentially involves a field experiment, in that it requires you to gather data in public settings. As such, its conclusions are inextricably linked to the particular setting being studied. Knowing all about Route A and Route B is unlikely to directly teach us much about Route C. In other contexts, your everyday experimentation might involve exerting much more control over the features of situations, thereby opening a greater range of possibilities as well as the ability to test more nuanced hypotheses. For example, you might experiment with your exercise regimen (e.g. manipulate the type and amount of exercise you perform in order to assess its impact on your sense of well-being), or with your preparation of meals (e.g. manipulate the various ingredients of your daily sandwich in order to assess the impact on its taste), or even with your hairstyle (e.g. manipulate the appearance of your hair in order to assess its impact on other people's reactions). In each case, the same enhancements would be relevant. The more data you gather, and the more comparable those data are, the better. The fact that you have more personal control over the relevant factors here would itself be an enhancement, because it reduces the extent to which outside factors (including unknown ones) need to be borne in mind. This particular principle has led logically to a preference for fully controlled experiments over field-based ones, and so, in scientific contexts, to the development of research laboratories.

During the scientific revolution, the widespread adoption of experimental approaches allowed scholars to move beyond mere superficial observation of the universe, and towards a more interactive form of science. It facilitated the formal testing of hypotheses, the resolving of uncertainties in theories, and the ability to choose between competing views of the world. As a by-product, scholars came to believe that science was best conducted if it championed principles that enhanced the quality of experimentation, such as quantification, precision, objectivity, sampling, and the depiction of complex situations in terms of whatever fundamental variables could be isolated and measured. In that sense, then, one of the most important legacies of the long and winding scientific revolution is that it coalesced centuries of thought into a set of procedures now known as the 'scientific method'.

Some complexities in the scientific method

There is no one single 'scientific method'. However, there is a certain uniformity in the attempts made by scholars to summarize it. In some discussions, the scientific method is broadly described as the overall process of doing science. A more common characterization sets it out as a series of stages, with the following sequence being quite typical: (1) establish the topic for research, ideally in terms of an appropriately phrased question; (2) conduct a review of previous scientific research on the matter; (3) formulate a relevant hypothesis that reflects the next step beyond past research that is needed to elucidate the overall research question, and upon which an experimental or other study can be based; (4) design a research procedure to test the hypothesis; (5) conduct the study according to that procedure; (6) analyse the results of the study to see if the hypothesis is supported or not supported; (7) report the findings. If the hypothesis is not supported, then return to step (3) and proceed again with a different hypothesis. If the hypothesis *is* supported, then consider it a new part of the accumulated research findings on the matter, and return to step (2), or, alternatively, re-run the study in order to confirm the results by replication.

As well as these steps, the scientific method embodies a number of shared metaphysical, methodological, and theoretical assumptions (Valentine, 1992). In terms of metaphysics, the most important assumption is that of *determinism*. This is the idea that all events in the universe have causes, that nothing happens without a cause, and that causes temporally precede their consequences. Such a view belies the claim that things can happen just by themselves. Historically, it was quite common for people to hold views that breached the assumption of determinism. For example, in medieval times, people believed that some animals (such as rodents) could appear out of thin air, especially in damp spaces. If a house became infested by mice, it was uncontroversial for residents to assume that the animals simply appeared there by themselves, perhaps because the house was allowed to become damp. The solution then, after ridding the house of the mice, was to try to make the house less damp. Today, most people (scientists and non-scientists alike) largely accept the principle of determinism insofar as they would likely conclude that mice cannot simply materialize; therefore, if they find mice in their house, they would probably infer that they intruded from outside. The solution to this problem would be to seal the house by blocking up cracks in the walls and floors, or perhaps to install some kind of mouse repellent. Consistent with the mechanistic approach of science, determinism will always assume that for all current events there are prior causes to be investigated; a corollary of this is that scrutinizing current relevant information should enable future events to be predicted. Overall, science tries not to be gullible. When faced with claims that something inexplicable has occurred, it is generally hostile to such rationalizations as, 'It just happened!' (or, in more contemporary parlance, 'Not everything operates according to a simple linear cause–effect system of forward-moving time!').

Methodologically, modern science assumes that the best research will emphasize empiricism, precision, and the favouring of experimentation over non-experimentation wherever the choice can be made. The emphasis on measurement precision brings with it a focus on quantification and computation, while the emphasis on experimentation encourages a pursuit of experimental control and objectivity. An overarching methodological assumption relates to the valuing of a sceptical outlook. *Scepticism* is the view that it is reasonable to question and investigate any assertion purporting to represent factual information. In this sense, accepting claims on the basis of authority (or, more subtly, on the basis of the reputation of the claimant) is unscientific. The sceptical outlook embedded in modern science contrasts sharply with the approach seen prior to the scientific revolution, when it was common to rely on scriptural teaching in order to formulate starting positions for research endeavours. For centuries, many scholarly writings about the nature of human behaviour were based on the religiously informed view that human beings were created in the image of a deity (and so, for example, were fundamentally different to animals). As such, while scholars employed rationality and observation to help describe the human condition, much of what was argued rested on a scriptural premise that itself was not objectively demonstrable. It is worth noting that scepticism does not *compel* scientists to disbelieve the claims of all authority figures. Scepticism simply legitimizes the questioning of such claims, and promotes an expectation that all claims to knowledge will ultimately require verifiable evidence in order to be considered fair.

The modern scientific method also rests on some theoretical assumptions. One of these is the assumption of *parsimony*. Parsimony is a logical extension of scepticism as applied to theoretical reasoning. It effectively means that when faced with two or more assertions of fact, the one that rests on the fewest uncorroborated premises is considered the most convincing. For example, if a colleague earnestly tells you they saw a tiger walking through the main shopping street in town the other day, the *simplest* explanation would be that the claim was correct. However, such an explanation would rest on a number of unverified premises (such as that tigers are present in the vicinity of town, and that they have access to the main shopping street). A more parsimonious explanation would be that your colleague is mistaken. Another would be that your colleague is lying. Both of these possibilities introduce additional complicating elements, such as the premise that people are sometimes mistaken or the premise that they occasionally lie, as well as the premise that in this particular instance there are plausible reasons for your colleague to either make a mistake or tell a lie. These explanations are parsimonious because such premises are well corroborated (people are indeed capable of both error and dishonesty). By contrast, the first explanation – that your colleague is telling the truth about tigers – is less parsimonious because its premises are uncorroborated (tigers do not roam loose in towns). As such, the most parsimonious explanation is not always the 'simplest' one, or the one with the fewest elements. Indeed, sometimes a very elaborate theory can prove to be the most parsimonious available if, in the main, it is bolstered by corroborated assumptions.

A final theoretical assumption relates to the concept of *falsification*, and particularly to the view that falsification is superior to verification. In the early 20th century, many philosophers had come to the belief that science proceeds best when it combines logical reasoning with empirical observation in order to verify hypotheses. This perspective became known as 'logical positivism' (after the 'logic' of logical reasoning and the philosophical term 'positivism', which refers to the notion that authentic knowledge requires sensory experience and objective verification). Logical positivism was originally championed by a group of European scholars who were influenced by the writings of such thinkers as the Austrian philosophers Ernst Mach (1838–1916) and Ludwig Wittgenstein (1889–1951), and the English philosopher Bertrand Russell (1872–1970). It became particularly associated with a group working at the University of Vienna, under the chairmanship of Moritz Schlick (1882–1936). However, its central proposition was eventually challenged on the basis that verification cannot establish truth in a conclusive way. Logical positivism's most prominent critic was the Austrian–British philosopher Karl Popper (1902–1994). To explain the point, Popper famously offered the example of the proposition, 'All swans are white' (Popper, 1934). He argued that finding millions of swans to be white all over the world would never completely *prove* the proposition, because there would always remain the possibility that some as yet undiscovered exception to the trend remained to be found in the future. However, by contrast, the sighting of a single black swan would completely *disprove* the proposition. Therefore, falsification (which requires one swan) is more powerful than verification (which is not achieved even by millions of swans). Popper's argument suggested that the common approach to science needed refinement: instead of trying to develop knowledge by *confirming* theories, science should proceed by vigilantly endeavouring to *falsify* them (or, put another way, by 'testing' them). In other words, scientists should look for counterexamples rather than for examples, and should try to prove themselves wrong instead of right.

As such, if someone asserts that the Earth is flat, the best way to assess the claim is to look for evidence that it is round. If a theory proposes that men are more intelligent than women, it is best elucidated by cases of women who are more intelligent than men. And if it is claimed that childhood vaccinations cause autism, the best way of assessing this is to look for autistic children who have never been vaccinated (of whom you will find innumerable examples) and for vaccinated children who have never developed autism (of whom you will find millions more). Drawing attention to vaccinated children who *have* autism (in other words, focusing on verification) will not establish anything beyond the possibility of coincidences.

Of course, despite the by now mainstream focus on null-hypothesis significance testing in psychology, falsification is not by any means foolproof. For example, the theory of gravitation is widely believed not least because it is consistently corroborated by the fact that buildings remain standing. However, according to strict Popperism, the fact that buildings remain standing does not demonstrate the merits of gravitational theory: the theory of gravity remains unfalsified, which means we should not consider it a reliable piece of knowledge. Taking this approach, we should not risk living in (or even entering) buildings, because the engineering principles applied to their construction must be deemed scientifically parlous. Critics of Popper argue that the exclusion of verification as a means of generating knowledge is just not logically defensible. In essence, theories that are true can never be shown to be true because they can never be falsified (because they are not false!). For good measure, critics also argue that Popper's own writings breach the principle of falsification with reckless abandon, by repeatedly citing case studies in science to support his argument and ignoring the need to consider alternative case studies that might challenge it (Stove, 1991). And there are numerous pragmatic difficulties in attempting to falsify hypotheses in real-life research (Earp & Trafimow, 2015). In short, while Popper's views on falsification are informative, the principle is not without its limitations. Despite this, the Popperian view remains a popular basis on which to question the reasonableness of scientists, and therefore of science itself more broadly. We will find ourselves returning to this issue of anti-science contrarianism in Chapter 10.

Summary: Why is science useful?

It may be the case, in Western societies at least, that the first time citizens hear of the word 'science' is when it is uttered as part of the phrase, 'science fiction'. Science is very much part of everyday culture, although it frequently ends up being wildly misunderstood. Far from being a cultish profession characterized by a narrow range of subject matter and some confusing jargon, science is actually a relatively simple process of corroborating assertions of fact. It combines deduction, induction, and abduction, draws on the convincingness of experimentation, and champions such values as determinism, accuracy, precision, scepticism, objectivity, parsimony, and falsification. The centuries-long scientific revolution has conferred countless benefits to society, and with every passing decade exponentially spawns technological advances that enhance our systems of information, transport, energy, medicine, public health, and food production. The approach of science can be applied easily to fields like physics, chemistry, and medicine, but it can also be applied to such everyday tasks as playing the lottery, packing a suitcase, navigating through traffic, choosing investments, predicting the weather, and deciding how to vote. For that matter, it can be applied to the subject matter of psychology.

The emergence of science has taken many generations of refinement, and remains an ongoing process. Nonetheless, the fact that it is commonly misconstrued has created opportunities for those who wish to convey the *appearance* of science without necessarily submitting to the rigours of having to think scientifically. For as long as there have been champions of science, there have been imitators of science alongside them. Humanistic domains such as psychology have been to the fore in attracting their attention. Such simulation of scientific style bereft of scientific substance is known as 'pseudoscience', and it is to this that we now turn.

Chapter 2

What Is Pseudoscience and Why Is It Popular?

Clarity of Crystals

Crystals have been part of the human world for centuries. While the sight of gemstones glistening in sunlight has long caught the eye, most crystals are in fact mundane substances, appearing across nature in different shapes and sizes. In simple terms, crystals are clusters of solidified chemicals in which atoms have lined up in regular, repeated patterns. They include objects as familiar (and as small) as individual grains of salt or single snowflakes, as well as some as dramatic (and as large) as the Devils Tower rock formation in Wyoming or the Giant's Causeway in Northern Ireland. And notwithstanding the fact that crystals are most commonly acclaimed because of their appearance to the naked eye, it is their microstructure that truly distinguishes them. It is their appearance when viewed through a microscope that counts, the orderly patterned nature of how their molecules are constellated that defines them as crystalline. Their often complex-looking outer appearance is effectively the culmination of many simple patterns compounded with exponential multiplicity: while outwardly dazzling in their detail, naturally occurring crystals can nearly always be mathematically reduced to conglomerates of modestly polyhedral shapes. At this level, crystals become mathematically enthralling.

The study of crystalline microstructures is a field of science in its own right, and sometimes a theatre for dramatically controversial debate. One famous case arose when Israeli chemist Daniel Shechtman published a paper on the non-periodicity of crystal structures (Shechtman, Blech, Gratias, & Cahn, 1984) that completely overturned prior understanding of the nature of crystal formation. Previously, it was believed that the subatomic patterns in all crystals repeated themselves within a limited range of possible rotational symmetries that required patterns to recur, at some point, across a structure (namely, 2-fold, 3-fold, 4-fold, and 6-fold symmetries). However, Shechtman's work suggested some crystals had patterns that, in a mathematical sense, were orderly but *non*-periodic, and displayed other types of symmetry, such as 5-fold, that could not be repeated. He called these entities 'quasicrystals'. While seemingly obscure to non-specialists, Shechtman's paper initiated a significant scientific row in which his American fellow chemist Linus Pauling became his fiercest critic. Pauling insisted that crystal structures could only contain repeated patterns, and that quasicrystals could not exist.

Criticism in science is not necessarily remarkable. However, Linus Pauling had received a Nobel Prize for Chemistry in 1954, which clearly testified to his chemical expertise. Moreover, he went on to win a *second* Nobel Prize – for his high-profile efforts in peace activism – in 1962. As only the second person in history to become a Nobel Laureate twice-over, he was considered a near-unassailable authority on any matter on which he chose to offer an opinion. Pauling witheringly dismissed Shechtman's work with the memorable sound bite, 'There is no such thing as quasicrystals, only quasi-scientists' (Lannin & Ek, 2011). However, as we know, the scientific method attaches little value to the personal nature of comments (which might otherwise have undermined Shechtman) or to the academic standing of commentators (which might otherwise have elevated Pauling). Ultimately what matters is the weight of the available empirical evidence. And on this basis, Shechtman's findings were found to be quite correct. While quasicrystals are rare – only 100 solids are known to form them, compared with over 400,000 solids that form conventional periodic crystals – they certainly do exist. In the end, Shechtman's discovery of quasicrystals was considered so pivotal to the science, that he was awarded his own Nobel Prize for chemistry in 2011.

Quite apart from such decade-spanning elements of soap opera, the field of crystal-related science (more properly referred to as 'crystallography') has also given us knowledge that is as profound as it is productive. For example, we now know that the very planet upon which we walk is composed primarily of silicate crystals. Many of note are embedded deep inside the earth, the most famous being carbon allotropes, better known to most people as 'diamonds'. Diamonds are not only valued as jewels but also have the attribute of being the hardest substance known to exist, making them critical components of industrial manufacturing processes that require cutting-blades and saws. Other crystals have similarly superlative properties that make them central to our everyday lives. Quartz crystals vibrate at such a precise frequency when carrying an electric current that they are used in the timing mechanisms of millions, if not billions, of watches and clocks worldwide. Other crystals, in liquid form, have varying optical qualities in the presence of electrical fields that allow them to be exploited in the manufacture of digital watches and calculators and, more recently, computer monitors and televisions. Crystals are also the basis for semiconductors, which themselves are the foundation of modern electronics and telecommunications. It would not be an exaggeration to say that we live in a crystal-dependent world.

Crystals are endemic to nature and have exerted a far-reaching impact on humanity. Moreover, the empirical study of crystals has produced some of the most significant discoveries in the history of science, as well as direct benefits that have improved the lives of millions of people. In short, crystals are important and profound.

However, to some observers, crystals have properties even more significant than super-precise oscillation, birefringence, or extreme scratch hardness. To these people, the most important applications of crystals lie far beyond the fields of electronic or mechanical engineering. They claim that crystals can enhance human immunity, reduce cholesterol, address problems arising from low self-esteem, and even help promote monogamy. They assert that crystals achieve these effects not through ingestion or injection or any other process of biochemical interaction, but simply by being held close to the human body, perhaps by being worn as a necklace or kept in the pocket of your trousers. These people call themselves 'crystal therapists'. Philosophers of science call them 'pseudoscientists'.

Crystals as prompts to mysticism

By demonstrating symmetry, angularity, light refraction, straight edges, and unusual colours, crystals are unlike other substances typically found in nature. Perhaps for this reason, they have been perceived as noteworthy since the time of the earliest human civilizations. In much the same way that people have long attached spiritual significance to the sun, crystals too have been veiled in mysticism and credited with special powers. Anthropological records suggest the use of crystals for healing purposes dates back at least five thousand years. Crystals were used medicinally in ancient Egypt and Mesopotamia (Raven, 2005), India (Adler & Mukherji, 1995), China (Cohen & Doner, 2006), as well as Greece and Rome (Cuvier, 1830). Many modern practitioners base their use of crystal therapy on concepts drawn from Ayurveda, the Indian (or, more accurately, Hindu) traditional medical system, which itself is many centuries old. Specifically, crystals are placed near parts of the body associated with 'chakras', points where the purported vital energy of the universe is said to intersect with the physical human being. The concept is that different crystals will succeed in 'unblocking' these chakras so that energy can flow freely in ways that heal whatever illness is being treated (McClean, 2013). While it is most typical for therapists to recommend that crystals simply be held near the body, some medical case studies have been reported where healing crystals have been buried into the patient's flesh by an amateur surgeon, presumably all the better from where to target the required chakra (McLemore & Hallengren, 2010).

According to contemporary proponents of crystal therapy (Peschek-Bohmer & Schreiber, 2004), crystals can be used for a wide variety of ailments. Some are directed at orthodox medical problems. For example, the crystal malachite is recommended as a treatment for kidney stones, while aquamarine is purported to enhance immunity. Ruby is supposed to reduce cholesterol. The obscure crystal sard is offered as a treatment for tumours, as well as for ulcers and myomas. Sometimes healing powers are identified not in terms of particular symptoms, but in terms of the different systems of the body: while sapphire is effective for treating the nervous system, quartz is more effective for the musculoskeletal system, and so on. Even the reproductive system can be enhanced:

> During pregnancy the expecting mother should always wear malachite, carnelian and jade, in no matter what form. During labor, carnelian strengthens the mother's reproductive organs. Jade helps stop bleeding during and after birth....
>
> (Peschek-Bohmer & Schreiber, 2004, p. 302)

It is always worth remembering that according to therapists, these effects result from energy flows emanating from the crystals themselves. The medicinal benefits accrue simply as a result of having the recommended crystal somewhere near your body.

All the more intriguing in this context is the idea that some crystals can affect your emotional state, your impulses, and your behaviour. Aquamarine increases happiness; amethyst reduces anger; citrine increases confidence; lapis makes you less shy. Some crystals might even influence the behaviour of other people who interact with you: chrysoprase, a crystal containing silica and nickel, is reputedly capable of ensuring that you enjoy a happy marriage. Once again, you achieve this by carrying it around in your pocket.

Many people refer to the information about crystal energy therapy as constituting a body of 'knowledge'. Proponents will assert that this 'knowledge' is just as legitimate as that espoused by chemists who study the way electrostatic forces dictate the formation and bonding of crystalline substances, by physicists who assess the large-scale properties of materials that differ at the atomic scale, or by biologists who explore the crystal structure of macromolecules in the proteins that make up DNA. It is true that much of what is written about crystal therapy is produced by authority figures, and contains lots of jargon. However, the claims of the crystal therapists are fundamentally different from those of crystal scientists. The main difference is quite a significant one: the purported therapeutic effects of crystals have not been demonstrated – irrefutably and objectively – to exist. To this extent, the body of 'knowledge' underlying crystal therapy could therefore be described as inaccurate, or maybe even fraudulent. But perhaps the more important difference relates to the way crystal therapists *talk* about crystal therapy (and maybe even to the way they *think* about it). Crystal therapists rely heavily on anecdotal reasoning and subjective assessment, and eschew such notions as parsimony, falsification, or experimentation. In other words, the discourse resembles the way knowledge was generated in the centuries before the scientific revolution. For this reason, crystal therapy could be said to be unscientific. The fact that crystal therapists frequently *claim* to be speaking scientifically makes their field more than just unscientific, however. Such claims render the field *pseudoscientific* too.

The demarcation problem

For as long as formal science has existed, there have been competing fields making unscientific claims regarding parallel subject matter. Astronomers, who study the trajectories of bodies in outer space, have coexisted with astrologers, who claim that the stars and planets influence people's personalities and experiences on earth. Metallurgists, who scrutinize the physics and chemistry of metallic elements, have coexisted with alchemists, who try to extract gold and silver from otherwise valueless base metals by employing Hermetic powers. Scientific medicine has coexisted with traditional, complementary, and alternative healing practices. And so on.

In truth, many of these movements have coexisted because of shared overlapping histories, with the newer science-based branch having evolved from the older competitor as regard for the merits of scientific awareness has grown. However, human societies being what they are, competition has also played a part. In seeking to differentiate themselves, older disciplines have been motivated to accentuate their differences and to evolve new forms and emphases of their own. This can be detected in the history of non-scientific medicine (Whorton, 2002). For example, during the 19th century, when most of what we recognize today as 'modern medicine' first emerged, the new approaches embodied understandings of biology that were radically different to what most people had known up to that point. Indeed, much of scientific medicine was seen as bloody, gruesome, and unconsoling. Accordingly, the most successful *non*-scientific competitors were those, such as homoeopathy, that offered entirely *alternative* models of how the human body works and promised few if any side-effects. Later, as scientific medicine saw the widespread introduction of therapies based on pharmacologically produced drugs, the competing practitioners began to emphasize methods of drugless healing, like osteopathy, chiropractic, and naturopathy. And as the 20th century became the 21st, with mainstream science and its associated commercialized technologies a dominating feature in the lives and livelihoods of the masses, movements aimed at challenging the validity of the conventional view of the world began to strike a chord among sceptical citizens. In medicine, this coincided with a rise in so-called 'holistic' approaches to healing, including those with ethnocultural connotations unfamiliar to the non-Western audience, such as acupuncture and Ayurveda. In short, whenever and wherever science threatens to become the dominant influence on common thinking, you can surely expect a counter-science movement to emerge with a competing alternative.

Where the contrasts are outwardly adversarial, distinguishing between science and non-science is usually straightforward. For example, astronomy is scientific because it uses empiricism to resolve uncertainties regarding the contents of the universe beyond the earth's atmosphere. However, astrology is unscientific because it *rejects* empiricism, and freely relies on assertions that are superstitious and traditional. Further, astrology adopts as assumptions claims that, when tested scientifically, turn out to be false. And the universe as described by astrology contains many features that *cannot ever* be tested, and which many people would likely describe as inherently illogical in any other context: astrological prognostication, for example, requires some kind of process wherein time goes backwards instead of forwards. Therefore, on the whole, the 'knowledge' of astronomy can be said to be clearly scientific, whereas that of astrology can be said to be clearly not.

However, in many other cases the distinction between scientific and unscientific can be much less clear-cut. For example, around the world and for many years, quite a number of people have used their telescopes to scour the skies for evidence of extra-terrestrial visitations. Many have claimed to have seen objects – flying vehicles, mainly – that can only have been put there by beings from another planet; some claim to have recorded photographic and video evidence of these alien craft. Adopting a corresponding paradigm, other observers claim to have identified physical structures on faraway planets (such as Martian canals) that further demonstrate the existence of intelligent life beyond earth. Nonetheless, the scientific consensus weighs heavily against such claims. In fact, the vast majority of scientists have rejected the vast majority of this work as being wholly unscientific. And yet, so long as observations are analysed and interpreted objectively and empirically (which most 'ufologists' would ardently claim to be doing), there is nothing inherently unscientific about the practice of observing the skies with telescopes. So why is the bulk of ufology considered to lack scientific credibility?

Before addressing that question, let us consider a second grey-area example. Specifically, let's consider psychology. Psychology uses scientific methods, such as experiments, to study the thoughts, feelings, and behaviours of human beings (and other animals). Thousands of psychologists are committed to the use of science to help explain these phenomena, and psychology is classified as a science in virtually all formal encyclopaedias of science and by all mainstream organizations charged with promoting science. Despite this, there remains widespread scepticism toward the assertion that psychology is, in fact, scientific. Some of this scepticism comes from proponents of other academic disciplines, some comes from within psychology itself. Indeed, in the history of the philosophy of science, the study of human thoughts, feelings, and behaviour has occasionally been used as an example of what *the opposite* of science looks like.

In short, some fields are sciences and others are not, but apart from the clear-cut cases, it is not always obvious which is which. This challenge of separating one from the other is known in traditional philosophy of science as the 'demarcation problem'.

Pseudoscience

In separating fields that are sciences from ones that are not, a number of relevant categories have been identified and labelled. To start with, many branches of academia are straightforwardly 'not sciences'. While they consider themselves producers of knowledge, the knowledge being produced is part of a cultural stream of subjective evaluations, rather than a body of truly factual information about the world. Such fields do not consider themselves to be sciences, nor do they want to be considered as such. Take, for example, literary theory. As a scholarly discipline, literary theory generates an enormous amount of 'knowledge', and a competent literary theorist is required to become familiar not only with a vast base of published works, but also with the mechanisms by which scholarship in literary theory is produced. Nonetheless, literary theory is not a science, and literary theorists do not consider themselves to be scientists. In fact, it would be quite unfair to accuse literary theory of being 'unscientific', as this would involve testing it against standards that it does not aspire to meet. Doing so would constitute a non sequitur. It would be akin to denigrating particle physicists for adopting clichéd metaphors.

Sometimes fields that are clearly 'not sciences' nonetheless get sucked into debates about what is or is not scientific. Religion is a conspicuous example. As we will see in Chapter 9, religion is often considered so much 'not a science' as to be the *opposite* of a science. What counts as true in religious terms depends on faith rather than on evidence. Theological debate can centre on questions of morality, ontology, and whether or not several angels can be in the same place at the same time (such as on the head of a pin). Such debates are not resolvable scientifically. However, sometimes people use religious platforms to support positions in debates that are actually scientific. For example, that most clichéd of medieval religious claims – that the earth lay at the centre of the universe – was intended to be understood as empirically factual rather than as figurative or faith-based, and religious astronomers were genuinely surprised when contrary evidence emerged. Likewise, when spiritual healers or church authorities endorse miraculous cures or prayer-based recoveries, they are making *scientific* claims about the nature of physiological processes involved in human biochemistry. Similarly, when religiously motivated groups cite evidence from empirical studies to support a particular policy position (such as data on the impact of parental sexual orientation on child welfare), they are engaging in scientific rather than theological debate. They are engaging in science itself, and their points are open to be quality-assessed in scientific terms. Accordingly, it is fair for critics to examine the extent to which the overall thrust of argument comprehensively reflects the accumulated scientific literature as a whole, and to call foul on those who fail to acknowledge the totality of scientific evidence when mounting their scientific argument. This is the case even though the field from which protagonists draw their authority – namely, religion – does not wish itself to be a science.

In contrast, several other fields classified as 'not sciences' are very keen indeed in their ambition to *actually be* sciences. Some of these are fields so nascent as to not yet have the empirical methods, epistemological conventions, or agreed base of major theories necessary to allow researchers to be consistently scientific. For example, they may have difficulty in mounting experimental studies or in adequately measuring (or even proving the existence of) core subject matter. When such fields are peopled by scholars with aspirations to achieve orthodox scientific standards, and where a reasonable likelihood of success in such ambitions exists, these can be considered liable to become fully fledged sciences in the future. In philosophy of science terms, such embryonic fields are commonly referred to as *protosciences*.

Although a truism, it is fairly conventional for philosophers of science to point out that before the scientific revolution, *all* fields of inquiry were protosciences. In this sense, then, Isaac Newton was a protoscientist. Of course, it almost makes no sense to say such a thing. It is akin to claiming that all human beings were technophobes before the invention of technology. Clearly there is a difference in kind between the protosciences that predated modern science and those that now post-date it, with the former group surely to be excused of the accusation implicit in the term. One can hardly be labelled a law-breaker for breaking a law that has not yet been passed.

A related observation is to say that astrology, alchemy, and herbalism were protoscientific precursors to the modern fields of astronomy, chemistry, and pharmacology. It is indeed a point of historical fact that the three former movements preceded the latter three in time. However, it is a much greater historiographical challenge to demonstrate a direct causal lineage from old to new. This is not least because astrology, alchemy, and herbalism continue to be practiced today, giving us *prima facie* evidence that not all astrologists, alchemists, and herbalists were transformed into astronomers, chemists, and pharmacologists. Nonetheless, the notion of historical provenance is often invoked in defence of such fields, and so it is important to consider the merit of the claim. In teasing out some finer points, let us again take astrology as our main example; essentially the same logic applies to alchemy and herbalism.

It is undoubtedly true that, after spending their careers studying the night sky, many historical astrologers became doubtful of astrological concepts, and instead found astronomical ones more plausible and fulfilling. However, to say that this pattern is one of astrology transforming into astronomy would be strange. If a subset of Christians decided to become Scientologists, we would not describe this as Christianity's metamorphosis into Scientology. Instead we would say that they had converted *from* Christianity *to* Scientology. The two worldviews remain unaltered; it is the allegiance of (some) adherents that changes. The point is that astrology didn't stop being astrology when astronomy arrived, and to say that astrology is the ancestor of astronomy is historically misleading. Moreover, it distracts us from the most important difference between astrologers old and new. This is that today's practitioners *know* (or at least *should reasonably* know) that their field has been thoroughly debunked in empirical studies, time and time again. Therefore, in order to remain loyal to astrology, contemporary astrologers must knowingly choose to ignore that which has been shown in scientific research. In other words, while the folly of pre-scientific astrologers can be excused on the grounds of ignorance, that of modern astrologers simply cannot be.

The key matter here is the *intent* of practitioners in a given field. It is simply the case that some fields are characterized by intellectual practices that are morally questionable, comprising sins both of commission and of omission. This is what we mean when we use the term 'pseudoscience': pseudosciences are fields which *purport to be sciences*, but which *fail to adhere to the principles and practices of true science*. Unlike literary theorists or Isaac Newton, pseudoscientists are conspicuous in wearing the outward cloak (or white lab coat) of science, while simultaneously eschewing, or even rejecting, its demanding epistemological standards. The underlying motivations are themselves complex. Some pseudoscientists are well aware of scientific principles and practices, but nonetheless choose to dismiss them; others are aware, but choose never to think too much about them; and still others are *un*aware, but in a way that is wilful and ethically problematic.

The term 'pseudoscience' became prominent in the mid-19th century during a milieu of rapid scientific development, and was used primarily in discussions of the so-called demarcation problem. In that context, as a staple in philosophy of science, the term has some gravitas. However, nowadays the term is often projected with much less decorum. It is mostly used to berate practices that deliberately seek to capitalize on public confusion about science, and to warn a vulnerable public about the dangers posed by those willing to exploit them. So when psychology is accused of being pseudoscientific, psychologists should pay close attention – and be somewhat worried – for at least two reasons. Firstly, their field is being criticized on moral grounds. And secondly, the accusation might actually be true.

Differences between science and pseudoscience

The criterion that pseudosciences *purport* to be scientific is a positive claim: 'purporting' can be observed. However, the criterion that pseudosciences *fail to adhere to principles* is essentially a negative claim: it refers to the *absence* of a feature, rather than its presence, and so is more difficult to demonstrate. Therefore, one way to help explain what pseudoscience looks and feels like is to show how and where it contrasts with science.

Being able to tell the difference between science and pseudoscience is not only useful to philosophers who want to discuss the empirical merits of fields like psychology, it is also of practical utility to the general public. This is because pseudoscience is encountered by most people every day.

Pseudoscientific claims are frequently used in consumer advertising or to defend otherwise unconvincing political positions. They are endemic to the pitching of complementary medicines and other forms of health advice. They are regularly used to rationalize folklore, cliché, and convention. And they crop up time and again in casual conversation, especially when people try to explain the world around them in some kind of all-encompassing way. In fact, if science is an objective method of producing defensible knowledge, then just about any discourse intending to incorporate such knowledge faces the risk of being adulterated by pseudoscience. This is why it is beneficial to be able to distinguish the two.

In particular, it can be useful to focus on two overarching domains within which science differs from pseudoscience. The first relates to *differences of substance*: these refer to aspects of science and pseudoscience that pertain to core subject matter, such as whether claims are valid or not valid, accurate or not accurate, and so on. The second domain relates to *differences of style*: these refer to the ways in which proponents of a given field behave when participating in that field, such as whether they promote peer-review of research or prefer to keep their data secret. We will consider three of each here, starting with three differences of substance.

Differences of substance

A very important difference of substance relates to the value placed on falsifiability. As described in Chapter 1, falsification relates to the manner in which an assertion can be objectively disproved. If an assertion cannot be disproved (in other words, if it can only be confirmed), then there is no way of establishing its accuracy; if accuracy cannot be determined, then science cannot progress. Remember that falsification is more powerful than confirmation. While repeated corroborations might sway an observer to conclude that a particular claim might be believable, all that is needed to convince them that the claim is false is one exception. Just one black swan will show us that not all swans are white. Accordingly, although factually incorrect, the claim 'all swans are white' is nonetheless scientific. This is because its incorrectness is *demonstrable*. However, not all claims are demonstrably falsifiable. Typically this is because they are so multifaceted as to support predictions that cover almost any eventuality.

Sometimes claims lead to multiple predictions that directly contradict each other. For example, a given theory of human personality might be premised on the belief that early life trauma is what shapes a person's character, but that it does so in different ways for different people. Such a theory is quite likely to produce contradictory predictions. While some sufferers will be said to repress all memories of a trauma and to reach adulthood unable to remember the terrible thing that has happened to them, other sufferers will be said to react quite differently: rather than repress the memory, they will recall the trauma on a daily basis, such that it comes to dominate their thinking and drive them towards depression. In other words, according to the theory, some people will repress while others will obsess. Either eventuality is plausible. The problem with a theory that allows for two contradictory outcomes is that the theory itself becomes useless. If you encounter a traumatized child, for example, it will be impossible to offer a clear prediction about what will happen: she may forget all about the trauma, or she may think about it every day for the rest of her life. To offer both prognoses for an individual child would not be particularly helpful. Correspondingly, if you try to conduct a scientific study on a sample of such children, it will be impossible to predict how many will end up falling into either category. *Any* ratio of repression-to-obsession will be possible, and so any data gathered can be said to be consistent with the theory. The theory can never be rejected, and so is essentially useless to science. When somebody defends such a theory by describing it as scientific, when it is so clearly not, then in reality they are muddying the waters with pseudoscience.

Such reasoning is not unheard of in real psychology. One example is from a book written by an Ivy League senior research fellow (attached to the Bush Center for Child Development and Social Policy at Yale University), who has received several National Media Awards from the American Psychological Association (Scarf, 2005). The book aims to promote a number of so-called 'power therapies', interventions that are argued to be far more powerful than the traditional talk therapies used in clinical psychology (the particular therapies mentioned in the book are each scientifically controversial, although this is not explicitly acknowledged). In promoting the merits of one psychotherapy for trauma, the author becomes agitated at a client who cannot remember being traumatized as a child:

> I prompted her a third time, but she still couldn't remember any scapegoating incident in particular. This was atypical...here was Claudia, usually so articulate and forthcoming, acting as if she'd taken an eraser and run it over certain regions of her brain—those areas where you'd suppose the relevant memories and associations would be stored. And I couldn't help but wonder if she was 'zoning out' in order to steer clear of certain experiences too distressing or overwhelming to think about or even remember. Mind/body disconnections of this sort are often linked to severe early stress...Claudia's period of wild drinking, wanton sexuality, and cocaine involvement was a further indication that she might have been self-medicating and acting out around an unprocessed history of trauma. (pp. 79–80)

Later, in discussing another client who had no recollection of childhood trauma, the author notes that 'amnesias are common in the wake of cataclysmic experiences' (p. 156). In short, the author's argument is as follows:

if your client cannot remember being traumatized, then this stands as evidence that they actually *were* traumatized. They could simply be repressing the memory, or maybe suffering from post-cataclysmic amnesia. Presumably, although it is not explicitly stated, the author believes that clients who *can* remember such experiences are also likely to have been traumatized. Therefore, in the end, everyone is likely to have been traumatized. Given this heads-I-win/tails-you-lose framework, such a theory is hardly useful for dealing with the nuances of trauma and its impact on mental health.

Apart from direct ambiguity, other claims are non-falsifiable because they produce predictions that are inherently vague. For example, many alternative medicines promise to contribute to physical and mental well-being in a very general way. This generality makes such promises unfalsifiable because it cannot be determined in advance what would constitute evidence of a successful treatment (such as the improvement in a particular blood cell count or the reduction in size of a tumorous growth). This is one of the reasons alternative therapies are widely regarded as being pseudoscientific. By not specifying the type of health improvement expected, claims surrounding such therapies can always be defended as being somewhat true, so long as a patient experiences some kind of positivity at some stage following treatment. For example, maybe a patient will report a period of positive mood. If so, then claims about well-being can be defended. The point is that such claims cannot ever be shown to be *false*. To do so would require a way of demonstrating that absolutely no well-being of any kind was experienced by the patient at all. The very fact that the patient has not died could be said to be a form of well-being. And even if the patient does die, the therapist can always claim that he would have died *sooner*, or more painfully, had the therapy not been administered. In an argument where 'well-being' is not defined, a therapist who promises such an outcome will never be proved wrong.

Similarly, claims can be non-falsifiable when they are immersed in a milieu of complicating factors that are difficult, if not impossible, to track. Consider the fact that many alternative therapies are administered simultaneously with orthodox medical treatments. If a recovery ensues, it could be because of the alternative therapy – but it could also be because of the orthodox treatment. It will be impossible to be sure. Even if a recovery doesn't ensue, it could be argued that the alternative therapy was somehow incompatible with the orthodox treatment and would have worked had it been administered on its own. With all that wiggle room, a person who defends alternative medicine can never be contradicted. This is why such claims are pseudoscientific.

Still other claims are non-falsifiable because they depend on assumptions that are themselves intrinsically unknowable. These claims are non-falsifiable in the pragmatic sense of the term. For example, a claim that some otherwise inexplicable medical recovery was the result of celestial interventions by a deity cannot truly be falsified. This is because the elements of the claims – deities and the mechanisms by which they intervene in our world from beyond nature itself – are, by definition, unexaminable. Other assertions come with get-out clauses that place them beyond objective scrutiny. For example, many telepaths and clairvoyants argue that their powers are dependent on 'positive energies' and so are adversely affected by observer scepticism; whenever a scientist attempts to study telepathy or clairvoyance, there will always be a risk that the very act of observation will serve to suppress the psychic's power to read minds or predict the future. This outcome, known as a 'shyness effect' (Taylor, 1975), has the convenient impact of making such claims pragmatically non-falsifiable.

Science deals only in falsifiable claims. Therefore, it is a matter of logic that non-falsifiable claims are pseudoscientific, and that fields that purport to be sciences but which regularly rely on non-falsifiable claims must be considered pseudosciences. Science valorizes falsifiable assertions as representing the *sine qua non* of intellectual excellence. By contrast, pseudosciences are entirely relaxed about the matter of falsifiability: falsifiable claims are seen as no more important than non-falsifiable ones. This point of substance is probably the most significant dimension along which science can be distinguished from pseudoscience.

A corollary of this relates to another, more subtle, difference of substance between science and pseudoscience: science and pseudoscience differ in their view of the implications of ignorance. To consider this distinction, it might be useful to refer to the satirical term *truthiness*. This word refers to the extent to which a claim can 'feel' true without there being an otherwise sound reason for believing it to be actually true (Merriam-Webster Online, 2006). Truthiness is an example of an appeal to emotion, a logical fallacy wherein a claimant uses sentiment or passion, rather than logic, to try to win an argument. Science eschews the notion of truthiness. However, pseudoscience holds truthiness in high regard, especially as a way of compensating for empirical ignorance. As a result, the following contrast emerges: in science, a claim that is unproven is simply that – unproven – and so has no truthiness; in pseudoscience, failure to disprove a claim is considered to have the effect of subtly *increasing* its truthiness.

This approach is seen in the argument that *X must be true because it has never been shown to be false.* Again, alternative medicine produces many relevant examples. Acupuncturists, chiropractors, and homoeopaths frequently point out that critics have been unable to show that their practices do not work. Rather than presenting evidence in favour of a given therapy, the therapist shifts the burden of proof onto the sceptic, demanding instead evidence in favour of the criticism. For example, acupuncturists note that their critics have been unable to *prove* that inserting needles into the skin has no effect on the flow of cosmic energy (which they call 'qi') in the human body. This is true, but largely because nobody other than acupuncturists know how to detect this energy flow. Physiologists, biochemists, and anatomists have no idea what this energy is, where it comes from, or how to find it in the human body. There is no machine, for example, that can detect the presence of qi. Even two different acupuncturists who examine the same patient are unlikely to agree on where it is (Kalauokalani, Sherman, & Cherkin, 2001). Unsurprisingly, mainstream medical scientists also have no idea how changing the direction of qi using metallic needles could possibly affect, say, cancer cells. So ultimately, yes, it is true that scientists have been unable to say anything definitive in criticizing acupuncture. However, it is just as true that scientists have been unable to say anything definitive *in support* of acupuncture (or of chiropractic or homeopathy). So this leaves us with a set of practices that may *or may not* work. To interpret this state of affairs as implying that the as-yet-not-disproved claim has elevated truthiness seems arbitrary. It seems especially arbitrary when you consider the extent to which research has tested alternative medicine using rigorous methods, and has yet to demonstrate a single replicable therapeutic effect. And yet it is very common to hear proponents of alternative medicine defending these practices by drawing attention to the failure of scientists to prove their *inefficacy*. Rarely do they point out the conceptual vagueness inherent in such fields, or the abundant non-falsifiability of their assertions.

As we will see in Chapter 5, alternative medicine has been subjected to a vast amount of empirical research. However, imagine a proposition that has never been researched at all. British philosopher Bertrand Russell posited the claim that there exists a china teapot in outer space somewhere between here and Mars, beautifully coasting along an elliptical orbit about the Sun (Russell, 1952). The teapot is too small to be seen using even the strongest telescope on earth. He quite reasonably pointed out that it would be absurd to accept such a claim on the basis that scientists have never disproved it (or are unlikely ever to do so). In one fell swoop, Russell's proposition elegantly exposes the fallacy of considering a failure to disprove as having any bearing on truthiness.

Another misinterpretation of ignorance occurs when it is argued that something must be true because no other explanation is available. In 2011, a coroner in Ireland used this reasoning when declaring the death of an elderly man to have resulted from spontaneous human combustion (McDonald, 2011). Spontaneous human combustion is widely regarded by scientists as being highly implausible, as it involves a person suddenly bursting into flames through some as yet undiscovered physiological mechanism. The coroner formally ruled that the man had instantaneously combusted on the simple basis that there was 'no other adequate explanation for the death'. However, such reasoning is spurious. The coroner himself noted that spontaneous human combustion is something for which there is no scientific evidence. Therefore, he was faced with a choice of at least two possible findings, either: (a) the beliefs and theories of the various witnesses before the court were erroneous (in other words, fallible); or (b) a supernatural event occurred. The latter explanation – the one the coroner chose – relied on the premise that supernatural events *can* occur, which is itself undemonstrated, if not false by definition. The former option is far more sensible because it relies on the premise that humans are fallible, a premise we know to be true. Overall, to conclude that the lack of an obvious alternative explanation actually *supported* the claim of spontaneous human combustion was erroneous. At best the court could conclude that, given the available information, the old man's death would have to remain unexplained. Importantly, 'unexplained' is not the same as 'inexplicable'. Generally speaking, when a claim runs counter to the mainstream, the burden of proof should be with the claimant rather than with the critic.

Claims that espouse truthiness on the basis of ignorance are clearly pseudoscientific. This is because they allow claimants to draw subjective conclusions in support of their own theories in ways that are not supported by available data. Such approaches betray a discomfort with uncertainty. Science, on the other hand, is quite comfortable with uncertainty. This is not to say that scientists are unwilling to balance different sources of evidence depending on their strengths. However, the scientific method adopts the view that the unknown is simply unknown, and that uncertainty is best resolved using data.

The question of uncertainty permits an observation on the way science is discussed in public fora. In the cut and thrust of media debates about scientific controversies, pseudoscientists often accuse scientists of endorsing a philosophy entirely opposite to the standard approach to uncertainty. Instead of being comfortable with (if not somewhat addicted to) uncertainty, scientists are regularly accused of being *over*-certain, even dogmatic. However, as we saw in Chapter 1, scientists endorse the principle of parsimony: when faced with a choice between two or more competing explanations, the one based on the fewest uncorroborated assumptions must be seen as most reasonable. In contrast, pseudoscientists casually eschew the principle of parsimony, and are just as happy to endorse a more complex (and thus logically less reasonable) interpretation if it happens to cohere with their prior beliefs. This contrast in attitudes toward the importance of parsimony is the third important difference of substance between science and pseudoscience. An example of this contrast that is particularly relevant to psychology concerns the difference between *reductionism* and *holism*. Simply put, reductionism is the epistemological view that complex systems can be adequately considered in terms of the sum of their constituent parts. Holism is the view that natural systems should only be viewed as wholes, and thus that viewing their parts in isolation will make it impossible for us to understand them. In reality, complex systems are often very difficult to understand in terms of their full complexity, and reductionism offers a parsimonious means by which to build an incremental picture.

Attitudes to the value of falsifiability, the implications of ignorance, and the importance of parsimony highlight very important differences between science and pseudoscience. Each refers to a difference of substance, in that they refer to the meaning of core subject matter. Science and pseudoscience also differ in the way in which they elect to deal with their chosen subject matter in a superficial sense. In other words, science and pseudoscience also differ in matters of style. Scientists and pseudoscientists talk differently, behave differently, and endorse different stylistic approaches to scholarship and education. Many of the most important differences in style relate to the value placed on opinion over fact. These emerge in several ways, including the way scientists and pseudoscientists approach the transparency of their work and its amenability to independent review, the way guru figures or thought leaders are treated, and the overall status attached to anecdotal evidence.

Differences of style

Scientists and pseudoscientists differ conspicuously in their attitudes toward peer review. Over the past two hundred years, scientists have adopted the approach of insisting on peer review as a matter of principle, and scientific work is not considered appropriate for dissemination until it has been thoroughly assessed by knowledgeable experts who do not have a stake in the evaluation. In practical terms, this means that research cannot be published in a credible journal unless it has been assessed for quality by independent reviewers appointed by the editors. Usually reviews are both blind and anonymous: blind in the sense that the names of the researchers are not known to the reviewers, and anonymous in the sense that the names of the reviewers are not known to the researchers. This helps ensure that a particular researcher's status or reputation cannot influence the evaluation of the work. An equivalent approach is used when scientists apply for research grants. In reality, even Nobel Prize-winning scientists can expect to have their submissions bounced now and again, on the basis that reviewers will have judged them to be lacking in some respect. It is true, of course, that all such systems are subject to human failings and that many flawed research papers slip through the peer-review net. Peer review is certainly imperfect, but as a quality control system for curating research, it is probably the least worst method available. Because the scientific community have universally embraced the principle of peer review, any research work suspected of being improperly reviewed can expect to be stigmatized and derided.

In contrast, pseudoscientists take quite a different approach. For them it is quite reasonable to publicize a purported research finding that has not been peer reviewed in any way. In fact, for many, peer review is never conducted and any references to their 'research' will refer instead to in-house studies that have never been formally published. Sometimes research findings are disseminated directly to the public in the form of news releases or media announcements, a pattern that has become known as *science by press conference* (Spyros, 1980). Improperly reviewed studies, or ones not reviewed at all, can enter the public domain without suffering any stigma. This tendency to ride roughshod over the principle of peer review does not in itself guarantee that a particular piece of research must be methodologically flawed. However, given that thousands of mainstream journals publish millions of peer-reviewed papers every year, research that has not been reviewed must be seen as standing at the fringes of scholarship. For researchers to *choose* this position for their work certainly raises questions as to its (and their) trustworthiness.

Science's focus on blind and anonymous reviewing relates to the idea that personal opinion should not outweigh evidence, reflecting the way the scientific revolution promoted empiricism ahead of authority. This theme underlies the two final differences in style between science and pseudoscience. Firstly, pseudosciences are often dominated by key guru-like figures, whose opinions are held in such high esteem as to be treated as bodies of evidence in their own right. And secondly, pseudoscientists are much less sceptical than scientists when it comes to assessing the likely validity of anecdote, and are quite comfortable about treating hearsay testimony as if it were objective research evidence.

Any consideration of the role of guru figures in pseudoscience must acknowledge the fact that mainstream science regularly champions its own famous personalities. It is indisputable that many orthodox scientists achieve renown in their own lifetimes, with a small number ending up celebrities admired by vast popular audiences. It is also true that the individuals concerned can rarely be blamed if their utterances end up misinterpreted by acolytes or by the media at large. Nonetheless, an important distinction between science and pseudoscience relates to the veracity attached to the utterances of gurus. If a statement is considered beyond reproach simply because of the fame of the person who makes it, then it is likely we are encountering pseudoscience.

A defining feature of science is the view that claims must be supported by evidence, and that any person's opinion – even that of a distinguished expert – should be considered open to critical evaluation. When a mainstream scientist becomes famous, it is usually because he or she is uniquely adept at gathering and marshalling empirical evidence in a way that produces convincing accounts of the world around them. Often they will synthesize different types of knowledge to shed light on what were previously regarded as puzzling aspects of nature. As such, scrutiny of their writings is likely to reveal a heavy reliance on the empirical work of predecessors. Moreover, their works will generate falsifiable hypotheses, thereby enabling successors to test the soundness of the overall accounts by seeking to disprove them. In pseudoscience, the role of the guru is quite different. Pseudosciences are frequently dominated by prominent figures who develop complex theories that lack direct empirical support, but which are nonetheless regarded as intact from the outset. Admirers do not gather evidence with the aim of falsifying the account, but instead cite the guru's writings as the basis from which to make claims about the world.

One way to illustrate this contrast is to consider two examples. In contemporary mainstream science, few scientists are more famous than physicist Stephen Hawking of Cambridge University. Much of Hawking's celebrity stems from his success in communicating complex astrophysical concepts to mainstream audiences, especially through his best-selling popular book *A Brief History of Time* (Hawking, 1988). However, Hawking is also deeply admired by specialists with advanced technical appreciation of his expertise. He is particularly famous for developing a series of theorems aiming to explain aspects of gravitation in the universe. One feature of this work is that, rather than providing conclusive answers to questions about the nature of time and space, Hawking's theorems provide a set of propositions that can be tested – by others – through empirical research. In other words, Hawking is famous for providing the type of answer to a question that itself raises further questions. As such, he is playing his part in the historical lineage of scientists who aspire to take incremental steps towards the ultimate clarification of knowledge. The extent to which he is renowned as a leading figure in astrophysics is derived from his resolution of past ambiguities using empiricism and his creation of a platform for new research questions.

In pseudoscience, thought leaders play a different role. The German physician Samuel Hahnemann developed homoeopathy in the late 18th century, but is still revered as an important authority figure. Rather than seeking to resolve past ambiguities in an incremental manner, Hahnemann developed homoeopathy as an entirely new way of treating human illness. Specifically, he formed the view that substances known to cause particular symptoms when unadulterated would be medicinally effective in eliminating those symptoms if administered to patients in highly dilute form. Nowadays, it is generally well known that the dilutions involved in homoeopathic products are extreme. In fact, the standard dilutions far exceed that which would allow the preservation of even one molecule of the original substance being diluted. It is for this reason that homoeopathy is not considered to be biologically plausible in mainstream terms. Hahnemann himself could hardly have known this, given that his work on homoeopathy predated the appreciation of molecular chemistry needed to understand its biochemical implausibility. Nonetheless, modern homoeopaths elect to overlook this limitation, and persist in promulgating Hahnemann's original theory as the basis for their work. Moreover, they continue to cite Hahnemann as the final authority on homoeopathic matters. Rather than being seen as having developed an imaginative theory than can now be tested (and perhaps rejected) on the basis of empirical evidence, Hahnemann is instead held up as a guru figure whose writings are treated as factually valid in their own right. Indeed, modern homoeopaths regularly dismiss attempts to empirically test homoeopathy as constituting hostile attacks on their noble profession. Despite the passing of over two hundred years since he first presented the term *homoeopathy* in print (Dean, 2001), Hahnemann continues to be depicted by devotees as having had the insight to produce an explanation of human illness sufficiently complete as to support 21st-century treatments that the rest of mainstream science considers medically controversial.

In science, famous figures who get things wrong are likely to be caught out eventually. As we saw earlier, despite his fame as a double-Nobel Laureate, American chemist Linus Pauling was exposed as being totally wrong in his assumptions about quasicrystals. The fact that his originally strident views ultimately attracted ridicule (consistent with the anti-authoritarian spirit of peer-review) highlights the self-correcting thrust of mainstream scientific discourse. With pseudoscience, it appears more often to be the case that the views of guru figures are held to be meritorious even in the face of the most extreme criticism and counterevidence. As a result, their utterances and writings – even if very old, as in the case of homeopathy – are cited as the ultimate authority in contemporary disputes. In insisting on the infallibility of primary sources, such fields are the very opposite of self-correcting. This is why pseudosciences are frequently accused of exhibiting a lack of scien-tific development, with dominant theories often dating back decades or, as in the case of homoeopathy, centuries.

The final distinction of style between science and pseudoscience is quite straightforward and relates to the standing accorded to anecdotal testimony. Put simply, anecdotal testimony is information relayed through word-of-mouth channels, most typically referring to a casual observation of an event, and often in an effort to corroborate an assertion. The term *anecdotal evidence* is in common use, but might be seen as something of an oxymoron. This is because anecdotes rarely meet the standards normally expected of 'evidence'. As anyone familiar with TV courtroom dramas will be aware, hearsay testimony is seldom taken seriously in judicial contexts (it is valued even less than 'circumstantial evidence'). This is also the case in science, and for several good reasons. Human recollection, upon which anecdotal testimony relies, is subject to high rates of error; to make matters worse, people greatly underestimate the degree of error typically involved in ordinary remembering. These two effects together mean that most wit-nesses will unknowingly provide error-strewn testimony while conveying inordinate levels of confidence. Anecdotal testimony is also subject to bias: a witness may deliberately choose to colour their version of events in a particular way, or may choose to be selective in what events they report to others. Most importantly, anecdotal testimony can rarely be independently verified. Indeed, if such testimony *were* independently verifiable, then there would be no need for the anecdotal version. In essence, anecdotal testi-mony stands almost as the antonym of scientific evidence. Anecdotes are defined as unscientific precisely because, by their nature, they cannot be scientifically investigated. As such, the conventional view in philosophy of science is that reliance on anecdotes is a conspicuous sign of pseudoscience.

When considering the nature of atoms or the origin of the universe, it is likely that most people will recognize the limited usefulness of anecdotes. Likewise, it is reasonably apparent that anecdotal claims regarding homoeopathy are not equivalent to evidence from scientifically conducted trials. So in these cases, it should be relatively easy for people to detect which fields are pseudoscientific by reflecting on the degree to which anecdotal data are being employed.

However, when we attempt to consider the nature of human behaviour, the role of anecdotal evidence appears less obviously problematic. First of all, human audiences have an instinctive tendency to attach weight to first-hand accounts, especially when these refer to personal stories and experiences. While audiences expect information about atoms or medicine to come from science, they are happy for information about human behaviour to come by word of mouth. And secondly, some of the subject matter of psychology – concepts such as 'thoughts', 'emotions', 'self-esteem', or 'personal identity' – exist only at the level of people's internal cognitive narratives. They cannot otherwise be observed, and it can be difficult to conceive of ways of studying them *without* relying on anecdotal testimony, at least in some senses of the term. As such, to many observers, the standing of anecdotal evidence in psychology is less controversial than it would be in, say, physics. Nonetheless, even in psychology, anecdotal testimony remains no less subject to error, overconfidence, or subjective bias, and continues to lie beyond the scope of independent verification. In other words, it remains scientifically dubious.

The standing of anecdotal testimony continues to be a basis for distinguishing science from pseudoscience. At the extremes, science considers anecdotal testimony to be of very low worth, if not indeed to be completely worthless, while pseudoscience considers anecdotal testimony to be perfectly fine. Somewhere within these extremes – with a history that features introspection, ideography, and qualitative research – lies psychology.

The demarcation problem in philosophy of science is certainly complex, and there are many ways of distinguishing science and pseudoscience. It might be useful to recap on the main distinctions. Therefore, as an overall summary, the differences can be listed, in two categories, as follows:

- *Differences in substance*
  - *The value placed on falsifiability*
  - *Views of the implications of ignorance*
  - *Attitudes toward the importance of parsimony*
- *Differences in style*
  - *Attitudes toward peer review*
  - *The veracity attached to the utterances of gurus*
  - *The standing accorded to anecdotal testimony*

Why is pseudoscience so popular?

Ignoring the merits of falsification, misconstruing ignorance, lacking parsimony, belittling peer review, deferring to authority figures, and relying on anecdotes all heighten the risk of inaccuracy. In essence, therefore, the difference between science and pseudoscience is that while science seeks to maximize built-in accuracy, pseudoscience – while purporting to be scientific – actually has built-in *in*accuracy. This is why pseudoscience is less likely to provide useful knowledge over time. Nonetheless, despite such shortcomings, most pseudosciences maintain a very high level of popular fascination. For example, even though alternative medical therapies are pseudoscientific (and, thus, lacking in medical efficacy), they are hugely popular in industrialized countries, with some estimates suggesting that the global market for alternative medicine is worth some $115 billion (Global Industry Analysts, 2012). Even though astrology has been repeatedly shown to possess no predictive power whatsoever, millions of people consult their daily horoscopes with at least some hope that their destiny will be revealed. Even though dowsing, a practice where diviners attempt to locate ground water (or other buried valuables) by watching slight movements in a stick or rod, has repeatedly been shown not to work, it continues to attract believers in all cultures around the world, as it has done for centuries. In short, even though pseudoscientific practices are pseudoscientific, they are still extremely popular.

It may be valid to recast that point as follows: *precisely because* pseudoscientific practices are pseudoscientific, they are extremely popular. It sometimes appears that the very unlikelihood of a proposition makes it more believable to some. The more frequently sceptics debunk telepathy, clairvoyance, palmistry, UFOs, the Loch Ness Monster, the predictions of Nostradamus, and the curse of King Tutankhamun, the more convinced devotees become of their feasibility. Some audiences are attracted by the prospect of conspiracy: by subscribing to an elaborate claim that mainstream society wishes to reject, a believer can garner the sympathy and solidarity that comes with being part of an unfairly oppressed minority group.

Some inaccurate views are widely held without there being any obvious cachet. According to a large-scale survey conducted by the UK's Royal Statistical Society (Ipsos MORI, 2013), the majority of British citizens hold wildly inaccurate beliefs about some very commonly debated issues: respondents feel that 28 per cent of British adults are single parents, when in fact the correct number is 3 per cent, and that 15 per cent of teenagers become pregnant each year, when the true figure is actually 0.6 per cent. Most people believe the government spends more on foreign development aid than on either pensions or education, when in fact government spending on development aid is just one fifteenth of spending on pensions and education. The majority feel that crime is on the increase, when in fact it is on the decrease and had been for several years. At first glance it would appear that a large number of citizens are either very badly informed about such issues (which seems disappointing given the intense focus on them in daily news coverage), or else are extremely poor at interpreting the available data. However, further consideration of such survey results might suggest a pattern. Most of the errors made by respondents are in the direction of casting both society and government in a negative light, suggesting that the mistakes being made are far from random. Errors appear to be systematic, driven by cultural influences and adhering to a wider sociopolitical narrative.

And so it typically is with pseudoscience. While philosophers of science may feel that the strength of empiricism is its logical rationalism that, by definition, produces objectively accurate knowledge, the fact is that the intended consumers of such knowledge are seldom motivated by a wish for accuracy alone. Knowledge consumers – ordinary people – are quite often living life in a hurry, worried about their place in the world, anxious to be loved, distrustful of authority, conscious of how they fit in with those around them, and frustrated with the status quo. Moreover, they usually carry very many things in their minds at any one time, and rarely have the bandwidth to fully care about the concerns of those who wish to edify them with new knowledge about how the universe really works. When a scientist speaks publicly about the flawed nature of astrology, very few ordinary people will choose to concentrate on the case that is being made. And when a statistician speaks about the underlying trends in the latest crime figures, well, most ordinary people will not even hear them. Most ordinary people acquire their knowledge on crime rates from second-hand sources, interspersed with anecdotally-based opinion, blended together with their own (probably false) prior beliefs, and viewed through the lens of personal value systems.

In this context, pseudoscience prospers over science for one particularly good reason. Pseudoscience, by not constraining itself to a need for accuracy, is far more *adaptable* than science. Pseudoscientific views permeate culture *precisely because* they are not passed through a validity filter. When scientists insist on blind peer review, they immediately fall behind science-by-press-conference in the race for public attention. When science insists on accurate language and pedantically defined terms, it immediately seems less friendly than sentimental cant. By embracing subjectivity, pseudoscience is also free to choose the message it delivers. Scientific medicine attempts to provide the simple facts about health and disease. In contrast, alternative medicine attempts to inspire patients out of illness by couching language in terms of positivity, self-healing, and the flattery of individualized treatment. Scientific medicine tells you how to delay death (thereby reminding you that you are, in fact, going to die); alternative medicine purports to tell you how to live (without being held back by the very vagueness of such a notion). Scientific medicine, obliged to stick to the data, must tell you that there is no ready cure for low back pain, allergies, or terminal cancer. Alternative medicine, playing fast and loose with empiricism, does not have to let the evidence get in the way of a rousing message.

In general, pseudosciences actively discourage difficult forms of thinking and so are easier on the mind. When faced with the challenge of predicting the weather, scientific meteorologists will gather data on x, factor in base-rate prevalence of y, consider the competing probabilities of different contingencies both in isolation and in terms of their possible interactions, and so on. Pseudoscientific weather prognosticators will rely on a proverb, perhaps taking account of bird migrations or sunspots. It is worth noting that most people in everyday life adopt an approach more akin to the latter, choosing to grab an umbrella or to wear shorts based on a combination of what it says on TV and what the sky looks like when they look out the window. In short, pseudoscience is more in tune with how people think in everyday life.

Pseudoscience and psychology

Psychologists have a particular interest in the demarcation between science and pseudoscience. This is at least partly because psychology concerns itself with subject matter that is of intense interest to mainstream popular culture: namely, the way people think, feel, and act, and the reasons underlying human affairs. The subject matter of psychology is part of the mass universal discourse of culture, something in which everyone has a legitimate stake. In contrast, the subject matter of physics is unquestioningly delegated to professional physicists. The behaviour of atoms is seldom debated at the dinner table, whereas the behaviour of humans is discussed at every meal. The very familiarity of psychological concepts – *thoughts, emotions, behaviours, personality, intelligence, attachment, memory, stress,* and so on – belies the fact that each term has very specific meanings in scientific psychology, and so creates a recipe for total confusion. In such chaotic circumstances, the agility of pseudoscience is highly adaptive. Overall, it is always likely that psychological subject matter will be subjected to pseudoscientific attention.

A second reason for psychology to be interested in these issues is that the very demarcation of science and pseudoscience is itself part of the subject matter of psychology. Few other issues provide better case studies about how people handle their cognitions about the world around them, how they are affected by social influences, and how they allow motivations to govern their perceptions. The history of the scientific revolution is akin to the history of human cognitive development. And the problem of how people come to hold beliefs that aren't true is one for which psychology, alone among academic disciplines, might be best placed to offer a solution.

A third reason for psychology to take an interest in pseudoscience is that psychology has long been accused of *being* a pseudoscience. As alluded to earlier, some philosophers of science have chosen psychology as their example of a field that purports to be a science without being scientific. It can be argued that many of these discussions have focused specifically on the subfield of psychoanalysis, which most contemporary psychologists would nowadays consider a fringe pursuit far from the mainstream of scientific psychology. However, many psychoanalysts not only consider their field to be scientific, but also consider it to be very much part of psychology (regardless of what non-psychoanalysts might think). And secondly, as we will see in Chapters 6 and 7, it would be misleading to say that psychoanalysis is the only part of formal psychology to have attracted scientific criticism.

Of course, such an accusation would be unremarkable if it was true. That is to say, if psychology is in fact a pseudoscience then it doesn't really matter if people say so. But psychology is not a pseudoscience. Despite the type of scholarly self-loathing that sees many an academic instruct undergraduates to write essays entitled '*Is Psychology a Science?*', the simple point is that, yes, psychology is a science. At least it is so in epistemological terms. In cultural terms, there may well be many parts of our psychology departments that fall short of the scientific ideal. Indeed, it could be argued that some very prominent areas of psychology have drifted disturbingly close to the demarcation boundary. There are certainly very many psychologists whose scientific literacy is sorely lacking, and who take approaches that for all intents and purposes are pseudoscientific. Having said that, there are also many jobbing physicists about whom such a description could equally be offered, without there being a body of critics standing by to argue that physics must therefore be a pseudoscience.

In short, psychology *is* a science. Let us now look at the reasons why.

Chapter 3

The Scientific Nature of Psychology

Psychology's scientific deficit

Debates about the status of psychology can turn up in the most unexpected places. Occasionally, the issue arouses controversy in the mainstream media. After one academic psychologist wrote a piece for the *Los Angeles Times* complaining about being patronized by biologists (Wilson, 2012), several other writers penned articles ridiculing his reaction and objecting to his sensitivity. According to them, it was perfectly reasonable for a biologist to be dismissive of psychology because after all, as one of them put it, 'psychology isn't science' (Berezow, 2012). A number of common themes were invoked in order to support the point. It was stated that psychology 'does not meet the … basic requirements' for a field to be considered scientific. The most significant of these, it was said, is the need for clear terminology, without which the other prerequisites (such as reproducibility of research findings) cannot ever be met. The example of research into human happiness was identified as an example of a suspect endeavour, along with the observation that happiness cannot be quantified using a ruler or microscope and what makes one person happy will not necessarily make everybody happy. Because of these unassailable points, psychology cannot be considered a science because its 'basic terms are vague and unquantifiable' and disregarding such criticisms would amount to 'an attempt to redefine science' (Berezow, 2012). According to viewpoints like these, psychology suffers from a fatal scientific deficit.

While strident and no doubt heartfelt, these criticisms of psychology are themselves open to challenge. For one thing, the study of happiness is not particularly common in psychology. It is part of what is widely regarded to be a peripheral, and possibly ephemeral, sub-discipline known as *positive psychology* (which we will hear about in Chapter 9). Therefore, it is far from clearly reasonable to consider happiness as representing psychological subject matter as a whole. Secondly, while it might indeed appear difficult to measure happiness in numerical terms, it is certainly clear that some people exhibit more happiness (i.e., are more happy) than others. In other words, it is clear that the concept of happiness occupies a continuum ranging from low to high. The implication of *this* is that happiness does exist in quantities and so is, in essence, quantifiable. The fact that critics of psychology are unable to imagine the method by which it can be quantified is neither here nor there. (To illustrate: we can suspect that such critics would also have considerable difficulty articulating – in their own words – precisely how the total volume of all the Earth's oceans might be measured, without this leading them to the view that the study of marine mass is somehow unscientific.)

But a more important limitation of such critiques is that they focus on the superficial. Just as stereotypes of scientists invoke white coats and jargon rather than falsification and parsimony, these critiques dismiss psychology's scientific pretentions on the basis of appearances rather than of substance. The imagined *difficulty* of measuring happiness is seen as outweighing the overall *rigour* of the epistemology or methods employed.

One way to highlight the point is to imagine applying such a criticism to fields other than psychology. Take, for example, quantum mechanics. In quantum mechanics, the notion of a *wavefunction collapse* is widely regarded as both vague and unquantifiable. Whether or not such collapses even occur is itself unclear (as Schrödinger attempted to demonstrate using his thought experiment involving a cat), despite the fact that they are fundamental to the interpretation of quantum mechanics. However, quantum mechanics is certainly a science. Likewise, consider *neutrinos*, subatomic particles as studied in particle physics. Their exact nature remains utterly unclear. Their mass has never been properly measured, but it is nonetheless agreed that they exist in vast quantities (with 65 billion passing through every square centimetre of the universe each second). Furthermore, it is assumed neutrinos will one day help us to probe previously impenetrable parts of our solar system. These are profound ideas, and yet so little is known about neutrinos that it is not even clear whether they have mass. But nobody claims that particle physics is not a science. Even the more everyday areas of high-school physics throw up examples. The familiar notion of *gravitation* is one. Although extensively described in physics, its nature is poorly defined; physicists do not really know why gravitation happens – it just does. In short, disputed or amorphous terminology does not, in and of itself, prevent scientists from doing science. The claim that psychology cannot be scientific because some of its constructs are vague (or, more reductively, because nobody really knows what happiness is) amounts to a weak critique indeed.

Furthermore, reservations about terminology distract attention from epistemology. For example, criticizing psychology's definition of happiness (or any other variable) fails to consider the fact that science is about parsimony, falsification, and scepticism, and not about vocabulary. A difficulty quantifying a particular construct does not prevent a field from being scientific, whereas a disregard for scepticism, parsimony, or falsification certainly would. As with other scientific disciplines, what makes psychology a science is (partly) its prioritization of such principles. Indeed, by focusing on the example of happiness research (to the exclusion of other subject matter in psychology) and by failing to examine how this critique might apply to other fields (such as quantum mechanics or particle physics), critics commit a fallacy of logic known as *confirmation bias*. In other words, they confine themselves to examples known to support the criticism, while failing to consider examples that might refute it, thereby eschewing the principle of falsification. While intended to highlight scientific shoddiness in others, such criticisms themselves exemplify poor scientific reasoning.

Of course, debunking criticisms of psychology does not separately elevate the contrary view. To think so would represent an example of Russell's teapot fallacy: just because a particular counterargument is poorly based does not in itself make the case that psychology must be a science. An assertion to that effect warrants independent support, ideally from multiple perspectives. Therefore, we will approach this issue by considering three basic questions. Firstly, is psychology considered a science by a consensus of well-informed observers, including non-psychologists? Secondly, does psychology research possess the determining features of science? And thirdly, does psychology meet the basic epistemological assumptions of science, or does it fail to do so in a way that other sciences are able to avoid?

On balance, is psychology considered a science by well-informed observers, including non-psychologists?

In the main, reference to popular sentiment is not the most assured way of establishing the truth or falsity of an assertion. After all, people are fallible. In this vein, it is worth repeating the point that classifying psychology as a science is not the same as stating that all psychologists think and behave scientifically. Many psychologists are far from scientific in their worldviews. Similarly, many physicists are far from scientific in theirs, as are many chemists and biologists. It is sometimes noted that a student can train to be a professional scientist without necessarily attaining an impressive level of scientific literacy. One can press the buttons and run the statistical tests without fully appreciating the epistemological reasons for doing so in a particular way. Even Albert Einstein conceded that professional scientists make pretty awful philosophers of science (Einstein, 1936). Practitioners and scholars of a discipline represent part of its human aspect, and are subject to ordinary human failings. To judge the merits of an academic field on the competence of its contributors can be quite misleading.

Such a comment can sound like a pre-emptive attempt to evade legitimate criticism. After all, when evaluating a proposition, surely it is relevant to note the rationality of its proponents? Well, not really. First of all, it is possible to come to an accurate conclusion without fully grasping a given context. People often correctly predict the weather without having all the information necessary to *guarantee* accuracy, or without any appreciable grasp of meteorology. In fact, people can sometimes make correct weather predictions based on methods that are manifestly nonsensical in context (for example, by examining the entrails of a recently slaughtered chicken), their apparent accuracy resulting from pure luck. In other words, even when an argument is flawed, its conclusion can be correct. Likewise, a particular psychologist (or even many psychologists) can be ridiculous without psychology itself having to be derided.

Secondly, remember that science is not based on an assumption of consistent human rationality. In fact, this is far from the case: science exists as a method for drawing conclusions precisely because it appreciates the innate tendency for *irrationality* in human judgement. It allows scientists to outsource the drawing of conclusions to a practice that can be observed and replicated. In his famous philosophical novel, *Zen and the Art of Motorcycle Maintenance*, Robert Pirsig (1974) puts the point as follows:

> The real purpose of scientific method is to make sure Nature hasn't misled you into thinking you know something you don't actually know. There's not a mechanic or scientist or technician alive who hasn't suffered from that one so much that he's not instinctively on guard. That's the main reason why so much scientific and mechanical information sounds so dull and so cautious. If you get careless or go romanticizing scientific information, giving it a flourish here and there, Nature will soon make a complete fool out of you. (p. 98)

Ultimately, commentators who talk sense might not themselves be sensible, whereas sensible people can, through human fallibility, end up talking nonsense. Given that it would be unfair to expect practitioners of a science to be consistently rational in thought and action, it would be correspondingly unfair to use their behaviour as the basis on which to judge the merit of their field. Sciences are scientific because of how they apply empirical methods to epistemological questions, not because of the views or attitudes of their individual proponents.

Another complication needs to be borne in mind when reflecting on prevailing attitudes towards the scientific standing of psychology. Some conscientiously scientific psychologists seek to *exploit* the popular belief that psychology is not a science in order to generate support for their activities. Sometimes clinical psychologists argue that their clients simply would not trust them if they portrayed their services as puritanically scientific; accordingly, clinicians occasionally exhibit disproportionate interest in clients' pop-psychology stereotypes regarding therapy as a way of establishing rapport before commencing empirically-based interventions. This can give clients an impression that the pop-psychology elements have professional currency.

Such a dynamic has a long history in psychology. In the late 19th century, the early North American university psychology departments were quite happy to embrace misperceptions of psychology in order to garner public support. In order to ensure a steady flow of philanthropic donations intended for 'psychological' research, some of the most prominent departments – including those at Harvard and Stanford – pandered to the view that psychologists specialized in conducting séances and investigating the paranormal. Among the apparatus itemized in the inventory of William James's laboratory at Harvard, the first experimental psychology laboratory in the world, was a Ouija board. The approach proved so lucrative that the words 'psychic' and 'psychological' became virtually interchangeable in popular usage, leading some less-than-enamoured academics to complain that the term *psychology* should be dispensed with from scholarly usage. Overall, this flirtation with popular notions of spiritualism may in fact have accounted for the relatively smooth emergence of psychology into a North American university system that was, in the late 19th century, predominantly religious in its collegial orientation (Coon, 1992).

The theme remains relevant in the modern university psychology department. This is because many students are attracted to the field more by an amorphous curiosity about human nature (and its problems) than by an interest in using scientific principles to resolve uncertainties about objectively collected data. Indeed, a very common first impression among psychology students at university is that their classes are unexpectedly 'scientific' (in the stereotypical sense), with previously unanticipated emphases on human biology, research design, and statistics. In reality, professional bodies for psychology insist on these areas being included in curricula as absolute requirements, and any course of study that omits them, or teaches them poorly, will not be recognized as offering even a basic psychology education (British Psychological Society, 2013). Nonetheless, the experience of psychology curricula is so jarring for some students that they harbour ongoing misgivings throughout their undergraduate years, and sometimes into their postgraduate ones. Some navigate the path with sufficient deftness as to lead to prolific academic careers, characterized by more than a passing scepticism towards the scientific nature of research in psychology (e.g., Gergen, 2001; 2008). Many others retain a somewhat tokenistic scientific approach, lacking in true commitment, such as when they allow cultural values to shape inferences made from casually analysed statistical evidence (Coyne & Tennen, 2010). We will return to these specific problems in Chapter 8.

Nonetheless, while it is a scientific truism that unbolstered human observation is subject to human failing and bias, we can momentarily ask ourselves whether psychology is – by and large – considered a science by the weight of relevant opinion. It is of course possible that this opinion is wrong and that reference to such authorities is ultimately treacherous. However, when marshalling arguments, it is useful to at least set the scene in terms of what it is that the majority of knowledgeable sources have to say. Let us take as initial examples the positions of the major international psychology associations, bodies which will have invested considerable effort in refining fair and comprehensive depictions of what it is that psychology is supposed to involve. For example, the front page of the British Psychological Society website describes psychology as 'the scientific study of people, the mind and behaviour' (British Psychological Society, 2013). The American Psychological Association (2013) contextualizes psychology as follows:

> ... a diverse discipline, grounded in science, but with nearly boundless applications in everyday life. Some psychologists do basic research, developing theories and testing them through carefully honed research methods involving observation, experimentation and analysis. Other psychologists apply the discipline's scientific knowledge to help people, organizations and communities function better.

The German Society for Psychology (Deutsche Gesellschaft für Psychologie) describes its core purpose as 'the promotion and dissemination of scientific psychology' ('*die Förderung und Verbreitung der wissenschaftlichen Psychologie*'; DGP, 2013), while the National Academy of Psychology in India aims to 'foster training for growth of psychology as a science as well as profession' (NAOP, 2013). In France, the two main psychology organizations have joined forces to form a National Committee on Scientific Psychology (Comité National Français de Psychologie Scientifique; Guinot & Schneder, 2011). The Chinese Psychological Society is part of that country's National Academy of Sciences (Institute of Psychology, 2012). The Psychological Society of South Africa is affiliated with the Academy of Science of South Africa (PsySSA, 2013). In Brazil, the Federal Council for Psychology operates a code of conduct obliging psychologists to promote 'freedom of information, psychological science, services, and professional ethics' ('*a universalização do acesso da população às informações, ao conhecimento da ciência psicológica, aos serviços e aos padrões éticos da profissão*'; CFP, 2005, p. 7). Meanwhile, the Russian Psychological Society, first established in 1885, requires members to take an oath 'to protect and develop the glorious noble traditions of Russian psychological science' ('*беречь и развивать славные и благородные традиции российской психологической науки*'; RPS, 2013). In summary, as far as mainstream organizations who represent psychologists around the world are concerned, psychology is very much a science.

Of course, psychologists might have a vested interest in depicting themselves as such. Thus, it is helpful also to consider the views of more generic scientific groups, such as major nongovernmental organizations, state agencies, and producers of globally influential reference texts. Once again, the trend is very clear. In the UK, the Science Council awards Chartered

Scientist status to experienced psychologists (Science Council, 2013). Psychology is one of 24 subject-area sections of the American Association for the Advancement of Science (AAAS, 2013), the body responsible for setting standards for science and technology education in the United States (and for publishing the leading academic journal, *Science*). The relevant United Nations agency charged with promoting science at a global level, UNESCO (the UN Educational, Scientific, and Cultural Organization), operates a classification system 'for fields of science and technology' that gives extensive coverage to psychology, both as a whole field and in terms of its various subdisciplines (UNESCO, 1988). The Oxford English Dictionary (2013) defines psychology as 'the scientific study of the human mind and its functions, especially those affecting behaviour in a given context', while the Encyclopaedia Britannica (2013) goes with 'a scientific discipline that studies psychological and biological processes and behaviour in humans and other animals'. For good measure, the latter is included in Britannica's *Science & Technology* subsection.

Therefore, with reference to the question under discussion – namely, *is psychology considered a science by well informed observers, including non-psychologists?* – there would appear to be a very clear consensus indeed. Virtually no authoritative organization considers psychology *not* to be a science.

Does psychology research possess the determining features of science?

Consulting definitions embodies both anecdotal evidence and argument from authority, and so is not the strongest approach to dealing with this question. However, it is perhaps useful for establishing a general contextual point. Another important contextual point is that psychology appears rare among academic disciplines in placing such an emphasis on the question of its own scientific status. While university psychology students are often asked to write essays with titles like '*Is Psychology a Science?*', equivalent tasks are seldom set for students of, say, biochemistry or geology. It is almost as though academic psychologists bear a lingering doubt (or even a self-loathing) about their standing, regardless of what the professional bodies, science advancement organizations, or encyclopaedias have to say.

Broadly speaking, psychology students are taught to boil down this task to a process of benchmarking: namely, identify the main descriptive features of fully established sciences and assess whether psychology possesses these features. Comparisons are usually made with physics and chemistry, typically held to be archetypal examples of this thing called 'science'. These fields are certainly very well established, both having been the focus of the post-Renaissance scientific revolution. And they possess features considered to be the very core of science, including controlled experimentation, precise measurement, an emphasis on parsimony, and the generation of predictions.

Psychology, it is said by many, fails to exemplify these features in an authentic way. Controlled experimentation is not possible because researchers will never be able to manipulate all pertinent variables in an experiment (for example, when a psychologist wishes to conduct a gender comparison, it is not possible to *allocate* participants to comparison groups – after all, women are women, and men are men). Likewise, psychology has comparative difficulty in achieving precision in measurement: whereas a physicist can obtain the weight of a particle to the nearest microgram, a psychologist will not be able to emulate this type of accuracy when measuring personality or depression. The emphasis on parsimony – the principle of avoiding unsupported assumptions – is also challenged, in that quite a lot of psychological research is entirely *reliant* on such assumptions (such as the assumption that mental representations, like mental imagery or memory, are things that exist). Finally, psychology's powers of prediction appear quite feeble when compared to those seen in physics: physicists can predict the trajectory of a moving object when a known influence is applied under given conditions, but psychologists cannot do the same for a person's behaviour. All told, psychology's shortfall in the areas of controlled experimentation, precise measurement, parsimony, and prediction leads some observers to conclude that psychology falters at being a science.

However, this argument assumes that physics and chemistry are the only benchmarks for making the required comparisons. The main problem with this is that many other benchmarks are available. Indeed, neither physics nor chemistry is a unitary discipline, so their use as comparison archetypes is somewhat oversimplified. The terms *physics* and *chemistry* each refer to clusters of subfields that could reasonably claim to be separate 'sciences' in their own right. Physics contains many subfields that only modestly overlap with each other, such as astrophysics, nuclear science, and optics. Likewise, researchers working in organic chemistry will test very different questions and use very different methods to those working in polymer science, even though both areas fall within the broader domain of chemistry. Another complication is that physics and chemistry are simply two parts of a wider panoply of sciences that has evolved rapidly over recent centuries. Even at the most simplistic level of categorization, there exist several other higher-order domains of science that constitute neither physics nor chemistry, such as astronomy, environmental science, and, of course, biology. The last of these itself comprises a plethora of freestanding sciences, ranging from aerobiology and botany and cell biology all the way to zoology.

The boundaries between one science and another can be fuzzy, but a generally accepted convention is to distinguish fields on the basis of how scientists draw on each other's contributions. By examining patterns of citations in research (in other words, by looking at the way research reports list previously published works as relevant sources), it becomes possible to separate out clusters of research activity that are sufficiently self-contained as to warrant identification as free-standing sciences. When this has been done, bibliometric analysts have tended to identify *hundreds*, rather than a handful, of separate sciences. Even the more conservative calculations put the total number at well over 150 (Leydesdorff & Rafols, 2009). So in response to the question '*Is Psychology a Science?*', comparing psychology to a putatively singular physics or chemistry is quite unsophisticated, if not somewhat erratic. It is akin to being asked 'Is a starfish an animal?' and then basing your response on a comparison between starfish and horses (with the assumption, in turn, that there is only one type of horse).

In fact, when psychology is compared with fields from across the spectrum of science, its claim to scientific status appears relatively strong. Let us again consider the most common comparison criteria, namely: controlled experimentation, precise measurement, parsimony, and predictions. Psychology is far from alone among sciences in deviating from perfection in these standards. Indeed, it can be argued that psychology deviates much less than many other disciplines that are widely regarded, without dispute, to be sciences. For example, many sciences having much more difficulty than psychology in exerting control over experimental variables. When studying weather events such as tornadoes or tsunamis, meteorologists have little or no scope to randomly manipulate conditions in their attempt to examine outcomes. Likewise, palaeontologists cannot conduct controlled experiments when testing hypotheses about, say, the breeding habits of dinosaurs (Hughes, 2012). Except in the most abstract of senses, research in meteorology or palaeontology is rarely experimental at all. Many sciences also have problems with measurement accuracy. At a very basic level, palaeontologists often use carbon dating methods for which there is a considerable margin for error. At a more conceptual level, scientists can struggle to even define that which is to be measured: zoologists have long wrestled with the task of identifying which micro-organism is which, often in ways that have impeded the scientific study of diseases and their treatments (Tappe, Kern, Frosh, & Kern, 2010). Astrophysics lacks parsimony as much as psychology, with major theories reliant upon highly complex and unproven assumptions about the scope, substance, and development of the universe. And with regard to prediction, some sciences have difficulty achieving it (such as when climate scientists argue over the impact of anthropogenic global warming), while others (such as plant taxonomists) do not even attempt it.

The overall point is this: while psychology falls short of physics and chemistry in achieving control, precision, parsimony, and prediction, in reality only very few sciences ever match these benchmarks. To use this to declare psychology not to be a science would force a conclusion that meteorology, palaeontology, zoology, astrophysics, and plant taxonomy, along with dozens of other fields, are not sciences either. It is rare to hear the critics of scientific psychology argue that this is in fact the case.

Does psychology meet the basic epistemological assumptions of science?

Even the benchmarking approach is insufficient to conclusively address the overall question of whether psychology is a science. After all, sometimes what appears to be a family resemblance is really just a coincidence. Ultimately, it is its scientific ethos that separates psychology from everyday speculation about human nature, and which makes it a subject suitable for academic study. Therefore, a final approach to this issue is to consider the philosophical assumptions of science and how psychology meets these assumptions, irrespective (largely) of how other disciplines may or may not do so. In this context, three categories of assumption mentioned in Chapter 1 loom large (Valentine, 1992). These are the metaphysical assumptions of science, its theoretical assumptions, and its methodological assumptions.

Metaphysics is that branch of philosophy concerned with the fundamental nature of being. More concretely, metaphysics is concerned with what things are, and how we know that things are the way we think they are. A metaphysical concern of key relevance to science is the notion of causality, or the conditions under which we know that x causes y. A philosopher will ask whether the fact that x preceded y is sufficient grounds to conclude that x caused y; a scientist will examine whether exposure to second-hand tobacco smoke is causally related to the onset of lung cancer. As discussed in Chapter 1, science proceeds on the basis of an assumption of *determinism*, the idea that all events have causes (insofar as no event occurs in the absence of a cause). Although it is sometimes hard to visualize, one significant implication of determinism is that all events are predictable, at least in principle. If you mix baking soda and vinegar, the acid in the vinegar will react with the sodium bicarbonate in the baking soda to form carbonic acid, which will immediately decompose to produce carbon dioxide. We can make an exact prediction now because we have gathered sufficient information about this type of double-replacement chemical reaction in the past. However, even if we didn't have the information today, the event would still be *theoretically* predictable in the sense that the only thing preventing prediction would be our lack of *access* to the relevant facts, rather than the lack of their existence. In other words, the production of carbon dioxide from baking soda and vinegar is not random, and so the causal elements of the reaction are there to be scrutinized. This is what we mean by predictability in the context of determinism.

If science assumes determinism, and if psychology is a science, then this means that psychology assumes determinism. Herein, it is said, lies the problem for scientific psychology: an assumption of determinism means that psychological events – thoughts, feelings, and behaviours – are predictable. This clashes with a common impression among ordinary people, and one imbued with not inconsiderable spiritual significance, that humans actually possess ultimate independence of mind or, as it is traditionally referred to, 'free will'. The doctrine of free will makes it impossible to ascribe thoughts, feelings, and behaviours to preceding knowable causes, because it implies that people are at all times able to make personal and independent choices and, indeed, to come to personal and independent opinions. Indeed, the doctrine of free will is not just a spiritual belief, it is also a working assumption of democracy. As people intuitively regard their personal autonomy to reflect the possession of free will, many critics have argued that psychology cannot truly make an assumption of determinism and so cannot truly be a science.

However, when we talk about prediction in science, we should remember we are referring to theoretical, rather than actual, prediction. *De facto* prediction is possible only if we have access to all the relevant information, such as is the case with regard to chemical reactions between sodium bicarbonate and acetic acid. However, it may be that the causal elements underlying an event will actually never be known (the problem of known unknowns), and it may even be that the full extent of relevant elements is itself unknowable (the problem of unknown unknowns). In reality, the nature of human thoughts, feelings, and behaviour might be so multifaceted as to defy any real predictability in practical terms. Therefore, the challenge to predictability is *complexity* rather than any exception to deterministic principles.

Included within this complex human nature may well be a default intuition, on the part of all individual humans, that there exists within their psychological selves the kernel of true free will. It is certainly plausible to imagine that humans come pre-programmed with such an assumption (in the same way as they appear to come pre-programmed with instincts for self-preservation and curiosity). However, the fact that people intuitively feel they have free will does not, of course, guarantee that they actually have it. Ultimately, the relevance of determinism in psychology is this: human behaviour can only be truly unpredictable if it is truly random. People who argue for human free will – be they doing so from scientific or spiritual positions – rarely argue that human behaviour is actually *random*. But so long as behaviour is non-random, it lies within the realm of that which can be predicted, at least in principle. As such, psychology's assumption of determinism is reasonable, and the problem of free will does not make it any less so.

The very complexity of the human condition is sometimes itself offered as a reason why psychology cannot be a true science. In epistemological terms, this relates to the various *theoretical* assumptions of science, including those of *systematicity* (the ability to organize scientific knowledge into a coherent system) and *generality* (the ability to create laws and principles that are applicable in different contexts), which are argued to be confounded by complexity. Consider the following assertion from a psychologist contrasting psychology with other sciences by bemoaning the fact that human behaviour cannot be studied scientifically (Goertzen, 2008):

> In other words, the natural sciences lack the tensions which stem primarily from the subject matter of psychology—which includes (amongst other thorny issues) the problem of subjectivity— since the level of complexity of their subject matter, in general, falls short of human mental and socio-cultural life. (p. 832)

Overall, this argument rests on two assumptions: firstly, that psychology is indeed truly complex; and secondly, that scientific methods lack the capacity to elucidate highly complex subject matter. Neither assumption is straightforwardly demonstrable. Firstly, some phenomena in psychology (such as reaction time) are clearly less complex than others (such as happiness). How the overall complexity of the field might be computed or described seems arbitrary. It may be that humans overestimate the complexity of their 'mental and sociocultural lives' simply because they have a front-row perspective on it. Our tendency to ascribe greater significance (for example, greater complexity) to that with which we are familiar is well established in cognitive psychology research, and is known as the availability or recognition heuristic (Gigerenzer & Gaissmaier, 2011). Compounding this tendency, we also know that humans are prone to perceive their own importance in unreasonably self-flattering terms (Bering, 2006). So although we might *think* our lives are highly complex, this might not actually be the case. It could be that human mental and sociocultural life is, relatively speaking, pretty simple in comparison to the structures and processes of other multifaceted systems. The sheer complexity of (say) digestive system chemistry in the marine worm *Turbellaria* might be, in comparative terms, truly breathtaking. Given the sheer number of possible fields of research, how many critics of scientific psychology would be in a position to know that all other subject areas are less complex? Even within the digestive system of a single person exists a greater number of living organisms than the number of human beings who have ever walked the Earth. Given that the entirety of human history has barely involved more than 100 billion people (Haub, 2002), the complexity of the ecology of the ten *trillion* or so gut flora inside *your stomach alone right now* must be acknowledged as being vast. In this context, claims as to the unique complexity of human mental and sociocultural life, and its accordant inaccessibility to scientific research scrutiny, appear quaintly self-flattering.

Secondly, it is something of a miscalculation to suggest that ordinary scientific methods are incapable of dealing with complexity. Complexity in and of itself is not a barrier to science. Indeed, several concepts in science are regularly described as being so complex as to essentially defy any eventual complete comprehension, yet are nonetheless amenable to scientific investigation. One example is the notion of *dark energy*. This form of energy is hypothesized to permeate the entire universe and to underlie its expansion. Beyond such introductory details, dark energy 'is a complete mystery' according to NASA (2013). About the most concrete explanation of what it might be is that it represents 'a property of space', in the sense that empty space is not nothing *per se*. The related concept of *dark matter* is another example. It makes up 85 per cent of the universe, yet is utterly unexplained at present and may never be truly understood. Astrophysicists are agreed that, despite their almost irredeemable levels of complexity, both dark energy and dark matter are core constructs that warrant vigorous ongoing scientific investigation. The implication that this means astrophysics cannot be a true science is unheard of.

As alluded to in the above quotation, some critics hold that 'the problem of subjectivity' represents a special barrier to psychology's scientific status. This is sometimes presented as an example of *reflexivity* (Ashmore, 1989), a problem of interpretation that occurs when there is confounding of causes and effects. The argument is that because conducting psychology research is itself an example of human behaviour, and thus of the subject matter of psychology, there arises an incorrigible problem of self-referencing: psychologists cannot study themselves in a truly objective manner. Given that objectivity is required for science, psychology therefore cannot be a science. One oft-cited quotation outlines this point by describing the putative plot of a science fiction story:

> The master-chemist has finally produced a bubbling green slime in his test tubes, the potential of which is great but the properties of which are mysterious. He sits alone in his laboratory, test tube in hand, brooding about what to do with the bubbling green slime. Then it slowly dawns on him that the bubbling green slime is sitting alone in the test tube brooding about what to do with him. This special nightmare of the chemist is the permanent work-a-day world of the psychologist— the bubbling green slime is always wondering what to do about you. (Bannister, 1966, p. 22)

In short, it is argued that psychology faces a problem with reflexivity that is absent from other sciences. However, the fact that human beings are liable to shape their behaviour to context when participating in psychological research is largely a circumstantial matter, and is substantively unrelated to the point that the researchers themselves are also human. Human beings shape their behaviour to context in all settings, such as in public places as opposed to private ones, but this does not prevent us from examining their thoughts, feeling, and behaviours while doing so. Surely only a very naïve researcher would omit to take cognizance of the settings in which observations are made? In essence, reflexivity is just another layer of complexity, and so is not a true barrier to science. Indeed, the reflexivity issue can (and perhaps should) be seen as an argument *in favour* of scientific psychology, rather than one that undermines its feasibility. This is because reflexivity highlights the shortcomings of trying to draw conclusions about human behaviour from everyday experience. Ordinary human interactions are embedded in an unmonitored contextual milieu, wherein the people around you respond to your presence in the context of their own presuppositions and interpretations of your intentions. When you ask a simple question such as 'What do you think are the effects of violent video games on children?' your respondent will firstly attempt to gauge why exactly it is you want to know such a thing, and then will calibrate their response within the range of what they feel is socially acceptable to say (to you). Your chances of acquiring any kind of reliable insights are at worst remote, and at best incalculable; as such, far better to organize a controlled empirical study of *actual* children, where they can be exposed to *actual* video games.

As well as the metaphysical and theoretical assumptions of science, some of its basic *methodological* assumptions are occasionally asserted to be beyond the reach of psychology. For example, the assumption of *empiricism* – namely, the expectation that the phenomena under scrutiny can actually be observed – is challenged by the way psychology often focuses on that which is unobservable, such as feelings, memories, attitudes, thoughts, and experiences. Once again, this criticism is premised on a misreading of what goes on in other sciences. It is true that in psychology, information about participants' thoughts is gauged inferentially by examining and interpreting their utterances and behaviours: we infer what people are thinking from what they say and do. However, the gathering of information by inferential means is the norm in most sciences, at least to some degree. For example, even the measurement of length or distance requires the use of tools that are in essence inferential. We derive measures of length by comparing things to separate devices (such as rulers) of pre-calibrated lengths, or perhaps by measuring the time taken for signals (such as sound waves) to traverse distances. Such devices appear reliable when the lengths being measured are small, but when sending space probes to Mars, we quickly discover some of the shortcomings in our ability to quantify distance. In truth, all tools for measuring such variables do so by secondary approximation, and thus contain error. The amount of error will certainly vary, and it will always be easier to measure out a metre than to quantify a person's intelligence. However, this just means that the measurement of distance is *easier* than the measurement of intelligence, rather than more scientific. Both approaches are inferential.

It is a common misconception that the various epistemological assumptions of science – the metaphysical assumptions of determinism and predictability, the theoretical assumptions of systematicity and generality, and the methodological assumptions of empiricism and measurement precision – are all absolute requirements for a field to be considered a science. In reality, these assumptions are aspirational, and are rarely fully met by any science. The challenges faced by psychology in meeting them are more pragmatic than epistemological, and any such limitations identified in psychology are greatly outweighed by the extent to which psychology meets the assumptions overall. A determination of whether a field meets the epistemological assumptions of science is one of degree rather than kind (Valentine, 1992) and it is clear that psychology meets all the requirements to some degree. As such, an assessment of its own epistemological assumptions serves to corroborate the view that psychology is in fact a science.

Some technical problems for psychology as a science

To conclude that psychology is in fact a science should not be taken to imply that it faces no limitations as a discipline. Indeed, the fact that psychology *is* a science allows us to scrutinize its limitations in a systematic and quantified way. In recent years, a number of important pragmatic (rather than epistemological) challenges have become apparent, which occasionally provide encouragement to those who feel that psychology *should* not be a science (even though it *is* one). Three important and interlinked examples are worth discussing here: the problem of replication; the problem of improbable findings; and the problem of small sample sizes.

What makes science so attractive to many of its proponents is the way it excites our instinct for curiosity. People who enthuse about science do so because of its capacity to unveil hidden truths, to unravel devilish mysteries, and to unleash the potential of our otherwise confined imaginations. However, valorizing the fruitfulness of scientific discovery often involves overlooking one of science's most important roles. This is the role of ensuring not only that we gather information to help us know new things, but also that we gather information to make sure that the things *we presumed we already knew* are actually true. In other words, science is as much about testing longstanding assertions as it is about testing novel theories. As such, you might expect the activity of actual scientists to be dominated by checking and rechecking the findings of previous research. However, this is far from the case, especially in psychology.

One of the reasons for this is undoubtedly the very fact that, as life choices, journeys of discovery are far more seductive than tasks that comprise a lot of checking and re-checking. It is simply the case that more researchers want to conduct novel studies than want to conduct replications. Given various aspects of human motivation, this is perhaps unsurprising. For centuries, scientific researchers have found fame and adulation through discovery rather than through corroboration. Indeed, the fact that scientific research is predominantly published in journals and not in books reflects this point: journals emerged in the late 17th century as a way for scientists to publish their findings immediately, rather than take the risk of another scientist making the same discovery and publishing it before them. Competitiveness around the notion of scientific priority – the idea that special credit is owed to scientists who make discoveries, rather than to those who corroborate them later – is as apparent today as it was three hundred years ago (Strevens, 2003). In this regard journal editors are as complicit as paper authors. Scientific journals are nowadays run as businesses and find themselves subject to market forces. This encourages a focus on publishing novel research instead of replications, simply because the presentation of new information makes the content of a journal far more eye-catching to readers (and, it follows, to paying customers).

The problem with this focus on novelty at the expense of replication is that it undermines the self-correcting nature of the scientific method. After all, the whole point of replication is to double-check the results of a prior study. If that prior study was flawed or if it somehow produced a freak finding, then a replication (and, indeed, the pattern of multiple replications over time) will eventually reveal this to be the case. However, if the fashion among scientists or the incentive-structure of scientific publication actually *discourages* replication, then it stands to reason that rogue findings will likely go undetected. In one famous case, a group of sceptical scientists found themselves unable to replicate a controversial study on precognition which had appeared in the prominent *Journal of Personality and Social Psychology* (Bem, 2011). When they submitted their own report to the same journal, the editor rejected it on the grounds that, by policy, the *Journal of Personality and Social Psychology* 'does not publish replications' (French, 2012). Pitifully, we must conclude that this high-profile journal is not interested in double-checking material that it itself chooses to disseminate to the public, even in cases where its sheer implausibility appears nakedly extravagant. Exactly how many erroneous findings exist unpoliced in the literature is difficult to quantify. However, if they represent even a small percentage of the 50 million or so journal articles that comprise the cumulative corpus of academic work published to date (Jinha, 2010), then their number will run to at least the thousands, if not tens (or even hundreds) of thousands.

The skewing of scientific publication by non-scientific judgement calls can be revealed statistically. For example, in psychology, around 97 per cent of empirical papers report conclusive findings in support of *a priori* hypotheses, compared to only 3 per cent of papers reporting findings that lead to the rejection of such hypotheses (Fanelli, 2012). This suggests that nearly all the predictions made by psychology researchers turn out to be correct in the end, which simply raises the question of why we need to bother actually *conducting* so much research. Moreover, the pattern of successful prediction appears to be bound by national borders, with researchers based in the United States displaying higher rates of prediction success than those seen in other places (Fanelli & Ioannidis, 2013). Of course it could be that psychologists achieve such reliability by playing it safe and replicating old studies over and over again. However, only a tiny number of these positive findings are based on replications. In fact, according to one analysis, just 1 per cent of psychology articles since the year 1900 has reported on replication studies (Makel, Plucker, & Hegarty, 2012).

So are psychologists just astoundingly prescient in only developing predictions that turn out to be supportable with data? Unfortunately, other patterns suggest this is unlikely to be the case. For example, published research is disproportionately likely to report levels of statistical significance just inside the conventional threshold for positive results. In psychology, statistical conclusions are almost universally derived using a system known as *null hypothesis significance testing*. This involves computing the probability that a given set of observations (the data) will be a fluke under a given set of circumstances (the hypothesis). Deciding whether the data are non-random (i.e., whether they represent signal rather than noise) is based on computing what their probability would be if the *alternative* to the hypothesis (the 'null hypothesis') were true. By convention, if this probability is lower than one-in-twenty (i.e., a probability less than 5 per cent or '$p < 0.05$', a numerical threshold known as the *alpha level*) then the finding is declared to be 'statistically significant' and the initial hypothesis is deemed to be supported. Of course, the way researchers do this is *itself* amenable to testing against randomness. For articles in psychology journals, analysts who have examined the frequency of different probability levels have found a spike in p-values lying just below the alpha level of 0.05, relative to those lying in other ranges (Leggett, Thomas, Loetscher, & Nicholls, 2013). There is no statistical reason for research studies to organically produce data where the relevant probability statistic ends up lying just below 5 per cent. The preponderance of such results suggests that researchers may be habitually pursuing idiosyncratic strategies in order to boost their chances of significance. For example, they may embark on a series of trial-and-error attempts – treating subsets of participants separately, including or excluding variables, using slightly different versions of the required statistical techniques, and so on – so that their p-value gets lower and lower. When it falls below the magical 5 per cent threshold they stop analysing their data and start writing their paper, the overall outcome being an over-representation of marginally significant results.

The focus on publishing positive (rather than null) findings, coupled with the idiosyncratic nature of individual data analyses across studies, has a number of important implications. Firstly, the research that appears in journals is unlikely to fully represent the totality of work conducted. The extent of null findings will be underreported: either journal editors will decline to consider them or else researchers won't bother to submit them (the so-called 'file-drawer' problem). Secondly, the amount of tweaking that takes place when psychologists analyse research data means that reported findings could be distorted to an extent that renders them unreliable.

And thirdly, the consequent uniqueness of individual studies makes them extremely difficult to replicate in practical terms.

The idiosyncratic way in which data can be analysed in order to yield significant findings was satirically highlighted by a group of US psychologists who published a report describing a study on 'musical contrast and chronological rejuvenation' (Simmons, Nelson, & Simonsohn, 2011). In the study, students were assigned to listen to one of two pieces of music, after which they filled out a form that required them to report their age. According to the statistical results, students who listened to the Beatles' 'When I'm Sixty-Four' were found to be statistically significantly *younger* than students who listed to Mr. Scruff's 'Kalimba'. This was despite the fact that they were allocated the music at random. The authors referred to a previous study in which participants who listened to certain music reported *'feeling* younger' after doing so, and made the point that their Beatles study could be seen as a 'conceptual replication' of this earlier finding. Now, of course, the results of the second study could not feasibly be meaningful – not even the Beatles can cause a group of college students to actually *be* younger – and yet the study authors were able to assure readers that the data and procedures reported were genuine. So what happened? Well, the authors went on to reveal that their statistically significant finding was based on a highly selective analysis: in reality more than 30 students were tested but only 20 were included in the study; a third group of students who listened to another track ('Hot Potato' by The Wiggles) was ignored entirely; analyses were conducted periodically throughout data collection, so that the researchers could decide whether they needed to test more participants or whether they could stop where they were; and the finding was significant only when one (and only one) of a plethora of alternative variables was included as a covariate in the statistical test used. In other words, the researchers dabbled with an open-ended series of different combinations of approaches until they encountered one that yielded a statistically significant finding. Their purpose for doing so was explicitly to demonstrate the perils of 'flexibility' in data collection and analysis, and so they should be lauded for providing us with a valuable theoretical paper. However, in orthodox research reports, use of such flexibility is typically never described – despite being the norm rather than the exception.

Perhaps unsurprisingly, when researchers *do* attempt to replicate previous studies, it seems remarkably difficult for them to produce like-for-like results (Pashler & Harris, 2012). This has become widely discussed as representing a replication problem – even a replication crisis – for psychology. In fact, it often seems that replications are successful only when carried out by the *same* research team who conducted the original study, or by researchers who have in some way been in contact with the original investigators (Makel et al., 2012). This may well relate to the need for replicators to properly understand all of the methods and procedures used in the original studies, but it greatly threatens the underlying principle that scientific findings should be independently verifiable. But perhaps the larger problem is the overall cultural bias against replication in psychology: so long as all those replication attempts linger in researchers' file drawers and editors' trash folders, condemned to remain forever unpublished, whatever few replications *are* attempted are rendered essentially uninterpretable.

All this feeds into the second major technical problem for scientific psychology. This concerns the probabilistic nature of statistical inference. The very fact that researchers draw conclusions by eliminating alternative explanations on the balance of probability, virtually *guarantees* that a subset of findings will turn out to be false positives (known in research terms as 'Type I errors'). This will be true even if researchers analyse their data conscientiously without seeking to artificially extract significance, even if they report studies comprehensively in all their detail, and even if they conduct independent and authentic replications. If you have technically established that there is a one-in-twenty chance of your observation being a fluke, then, who knows, maybe your study is that one-in-twenty of the relevant possibilities. Overall, one logical generalization is that there will always be a subset of published research findings that are simply false: some studies will be reporting flukes as if they were facts. Therefore, the replication problem faced by psychology does not stem only from the prevalence of idiosyncratic research designs and convoluted methodologies. It also relates to prevalence of Type I errors.

This technical problem might be considered acceptable in that, when data are analysed using probabilistic methods, the probability of error is something that goes with the territory. Comparisons can be made with legal systems that use the balance of probabilities as their burden of proof: some unfortunate litigants are indeed wrongly faulted, but most of the time the system distributes valid justice and so the pay-off in cost-benefit terms is ultimately handsome. Legal systems usually include remedies for those who are treated unjustly and, insofar as can be established, miscarriages of justice appear to be quite rare. However, in psychology, given that the practice of remediation (through replication) is almost unheard of, its cost-benefit pay-off will hang largely on its Type I error rate. Worryingly, there has recently emerged a concern that the prevalence of false positives is much higher than was previously appreciated.

Sometimes psychologists argue that an alpha level of 5 per cent will ensure that no more than 5 per cent of findings will be false positives (Pashler & Harris, 2012). However, this is not a logical conclusion. Firstly, such an extrapolation would imply an assumption that study results are selected *at random* to appear in the scientific literature. This is of course not true in any sense. Editors and reviewers get to choose what appears and what doesn't. Moreover, given the notion of idiosyncrasy (or 'flexibility') in data treatment, there may well be a tendency for researchers to bias their choice of reportable findings towards those whose statistical significance is on weak ground to begin with. Reporting all conducted tests would allow readers to gauge for themselves whether the result is indeed one finding drawn from twenty attempts; but publishing standards appear to favour a practice of incomplete reporting, where researchers filter out mention of non-significant outcomes. Without knowing the full extent of statistical analysis from which reportable findings are selected, it becomes impossible to know the prevalence of Type I errors.

In the absence of being able to know, we are forced to resort to mathematical projection combined with intelligent speculation. Having taken this approach, at least one prominent health sciences researcher has asserted that 'it can be proven that most claimed research findings are false' (Ioannidis, 2005). More tempered critiques have reckoned that 'between 17% and 25% of marginally significant scientific findings are false' (Johnson, 2013). The scale of the problem in psychology, as distinct from the scientific literature at large, is hard to tell. According to such analyses, the unreliability of a published statistical finding becomes more likely if a field deals with small effect sizes, uses lots of different statistical analyses and flexible research designs, or involves subject matter that is simultaneously studied by different researchers around the world (thereby elevating the risk that one-in-twenty of them will get their Type I error published). All of these features could be said to be true of psychology. On the other hand, such analyses of the scientific literature are themselves imperfect in a number of ways (Goodman & Greenland, 2007). They tend to assume that scientists select research questions at random, rather than on the basis of prior suspicion that there exists a problem worth studying. In other words, they tend to assume that the prior probability of most hypotheses is below 50 per cent, when in fact researchers typically focus on smoking guns (such as the suspicion of a link between exercise levels and heart disease) rather than on random subject matter (such as a link between body mass index and being born on a Tuesday).

One thing that is clear is that the use of small research samples increases the risk of drawing erroneous conclusions, both conceptually (in the sense that your small sample might not be sufficiently similar to the wider population you wish to talk about) and statistically (in the sense that statistical power is contingent on having a sufficient quantum of data in your analysis). This raises the final technical problem facing scientific psychology: the risk that its research studies are simply so small as to be intrinsically unreliable. The observation that psychology studies tend to be underpowered in this way was first offered over half a century ago by American psychologist Jacob Cohen (1962). His assessments focused on computations of statistical power: scores from 0 to 100 per cent that reflect the ability of an analysis to avoid a Type I error. Having reviewed the literature, Cohen established that the average power of psychology research was less than 20 per cent. Thirty years later, Cohen was invited to publish a follow-up paper on how psychology research had improved through the decades. He reported that very little had changed (Cohen, 1992). Today, the problem is particularly illustrated in the field of neuroscience, where sample sizes can be restricted for various practical and ethical reasons. One recent analysis of the available neuroscience literature computed that the average statistical power of all neuroimaging studies published between 2006 and 2009 was no more than 8 per cent (Button, et al., 2013). This means that fewer than one in ten research findings derived from brain imaging techniques, such as fMRI, are based on sufficient data as to allow researchers to confidently avoid false positives. In other words, you could say that 92 per cent of such findings are essentially uninterpretable. This occurs despite the fact that neuroimaging studies are frequently touted as representing the cutting edge of psychology, as well as its most viable future direction as a field.

If most psychology research is underpowered (due to sample sizes insufficient to actually detect the effect that is being investigated), then you might expect that the bulk of published research would have produced *no* findings of substance, rather than a *huge preponderance* of such findings, as discussed above. The fact that the literature is replete with significant results despite being derived from underpowered research further illustrates the level of filtering that must be taking place in the editorial and review processes of scientific publication. It also highlights the slipperiness of research designs and analytic approaches that arise in a world where replication is frowned upon, rather than encouraged. To the extent that these technical problems reflect severe challenges in producing competently effective research, then it is true that psychology faces multiple causes for concern.

But, of course, psychology is still a science.

Ultimately, these issues reflect pragmatic limitations and not conceptual ones. They do not affect the scientific standing of the work being attempted or the ethos used to approach it. Scientific research that is idiosyncratic in design, hard to replicate, blighted by an unknown error rate, or statistically underpowered, is still scientific research, in the same way that the work produced by a poorly trained or modestly talented sculptor still constitutes art. And to be fair, it would be misleading to suggest either (a) that psychology is utterly contaminated by incorrigible limitations at a technical level, or (b) that other sciences face no technical problems at all. Many areas of psychology insist on empirical rigour before, during, and after publication. Indeed, psychology is notable for the level of emphasis it places on methodological and empirical matters. The extent of self-scrutiny that has served to highlight the issues discussed here is rarely matched in the literature on, say, toxicology. Further, the problems relating to obsession with significance, publication biases, and difficulty with replication exist in lots of sciences. For example, as with psychology journals, journals in fields such as biochemistry, materials science, pharmacology, and clinical medicine all seem suspiciously loaded towards the publication of significant results. Around 90 per cent of published papers in these fields report significant findings (Yong, 2012), indicating that the file-drawer problem extends far beyond psychology. Similarly, highlighting the degree to which non-replicable research makes its way into scientific publication, the vast majority of papers retracted from scientific journals due to flaws or fraud are from fields like cell biology, genetics, and clinical medicine, rather than psychology (Marcus & Oransky, 2013). Type I error rates appear to be no higher in psychology than they are in epidemiology or other health sciences (Ioannidis, 2005). The same can be said for the problem of underpowered research, which is common in a variety of biological and other sciences, particularly when the effects being investigated are in any way nuanced (Gelman & Weakliem, 2009).

If anything, that such issues arise at all in psychology is testament to the very fact that its subject matter is directly amenable to scientific scrutiny. Psychology fits right in. Moreover, these limitations and complexities demand *more* commitment to the scientific method, rather than less: the problems arise from the way personal judgements (by editors, researchers, and readers) interfere with scientific practice. The compromised nature of unassisted human reasoning, the contaminating effects of social contexts and biases, and the adulterations of mass communication all distort the way science is handled by humans. It is these distortions that we turn to next.

Chapter 4

The Psychology of Evidentiary Reasoning

The risks posed by aeroplanes (or, even worse, by donkeys)

Many people suffer from a debilitating fear of air travel usually referred to as *aviophobia* (although sometimes more awkwardly called *pteromerha-nophobia*). The fear can manifest itself as extreme anxiety, panic attacks, fainting, nausea, or vomiting, and is usually sufficient to prevent sufferers from availing of air transport for the rest of their lives. Symptoms can arise even at the mere mention of flight. In almost every case people with such a fear will have one overriding anxiety, one scenario that they cannot put out of their minds: they are afraid the plane they are flying in will crash. This makes aviophobia somewhat different to other phobias. People who are afraid of spiders do not genuinely believe that a house spider crawling up the sitting room curtains is going to kill them. People terrorized by public speaking do not feel the activity poses a tangible risk to their health. But people who are afraid of flying really are afraid of *dying*. Typically unworried about being hit by cars or falling in front of trains, their compelling concern is with aviation. They are specifically, and obsessively, scared of dying in a plane crash.

The best statistics we have suggest that, once you step onto a plane, your odds of dying in an accident are around 1 in 11 million. This makes air travel one of the safest forms of transport. You are much more likely to be killed in a car: in a given year, the average person's chances of dying in a road accident are around 1 in 20,000 (Bandolier, 2007). Even if you are unfortunate enough to fly with one of the 39 airlines who have the world's worst accident rates (National Transportation Safety Board, 2014), you are still more than 100 times safer in the air than on the road. The phobia towards air travel appears to be so wildly at odds with the actual dangers posed, it suggests that the most people's assessments of its risks are excessive, and egregiously so.

Sometimes the error loads in the opposite direction; sometimes people can perceive air travel to be disproportionately *safe*, at least in terms of its comparative safety relative to other threats. Ever since a 1987 article in *The*

Times, in which a columnist glibly cited an expert opinion that 'more people in the world are kicked to death by donkeys than die in plane crashes' (Gill, 1987), the belief that donkeys are more dangerous than aeroplanes has gained worldwide notoriety. While the comparison is intended to allay people's fear of taking to the skies, the statistic implied by the sound bite is just not true. Donkey-related death statistics are seldom compiled, but it is certain that few if any human fatalities ever occur due to donkey aggression. On the other hand, even though the odds are slim, the annual number of aviation deaths is certainly above zero. So not only is a morbid fear of flight statistically unjustified, but the associated claims about death-by-donkey are just as poorly calibrated. And yet both beliefs are widely held. The fact they are wholly inaccurate appears not to undermine their plausibility in the minds of thousands, if not millions, of people.

The idea that millions of people can sincerely believe propositions that are manifestly untrue is sobering, but also informative. It helps us understand the true nature of consensus. Just because a claim is widely believed, it doesn't necessarily carry weight. If many people buy a product, it doesn't imply that the product works. If many people feel that something is dangerous, it doesn't mean that it is. And if lots of academic psychologists perceive a particular theory to be sound, they could all be completely wrong. In this chapter, we will consider the reasons why all this is so, by examining the various ways people struggle with evidentiary reasoning. Simply put, the uncoached human mind does a pretty poor job of trying to make sense of evidence. It is burdened by elementary problems with computation and cognition. It compounds error by allowing social influence to skew interpretation. And it is generally oblivious to the distortions caused by the many mass media filters through which raw information passes. In fact, when you consider the range of problems, it is a wonder human minds can make any sense at all.

The problem of relative likelihoods

Human beings are particularly poor at judging relative probabilities. Even with the appropriate statistics to hand, figuring out whether one scenario is more likely than another – such as whether the risk of dying in a plane crash exceeds that of dying in a car accident (or at the hooves of a donkey) – strikes most people as hard. In fact, even when the required comparisons are *as simple as they can logically be*, the majority of people stumble at the challenge.

Mathematically, we can use that description – 'as simple as can logically be' – when a comparison takes the following form: the likelihood of *two events occurring at the same time* compared to that of *either event occurring in isolation*. In this situation, you do not need to know the exact likelihood of either event; it is simply a matter of logic that the probability of both occurring together will always be lower than that of either happening on its own. There are no other factors to consider. And yet people regularly handle this comparison very poorly.

To explain why it is actually quite straightforward, let us consider an example. Imagine any single unusual event. Next, imagine that particular unusual event occurring twice. Does it not seem less likely that the event would occur twice rather than just once? The chances you will buy the winning ticket in a raffle are very remote; the chances you will buy *two* winning tickets are surely *even more* remote. The likelihood you will bump into one long lost relative while on a foreign holiday is low; the likelihood you will bump into two is even lower (unless, of course, there is some connection between them that causes them to take holidays together). The prospect of being hit by a falling frog on a Tuesday, in the month of March, while wearing your favourite blue jumper, is unlikely; the prospect of it happening two years in a row is almost inconceivable. And so on. An unusual event is, by definition, not usual: therefore, while it is unlikely enough to happen once, by logic it continues to be unlikely thereafter and so is certainly hardly expected to happen a second time.

Imagining unusual events occurring twice helps us to appreciate the idea of remote probability. Thus it also helps us appreciate the idea of *comparative* probability. The critical breakthrough in reasoning comes when you realize that the same principle applies when thinking about two *different* events. For example, the probability that you will lose your hat on a particular day when there happens to be a thunderstorm is always less than *either* the probability of you losing your hat on any day (regardless of the weather) *or* the probability of a thunderstorm (which, presumably, will be unrelated to the whereabouts of your hat). The second event occurring *together with* the first is clearly of a lesser likelihood than either one of the events occurring on its own. Putting all this another way, we can reflect on the fact that two things happening together requires a *coincidence*, whereas one thing on its own is merely an *incident*. And even common sense should tell us that *a coincidence is less likely than an incident*.

Now let's try to express the point statistically. Take for example the probability that both members of a married couple will be left-handed (and I will presume that people do not choose their marriage partners on the basis of handedness!). Approximately 10 per cent of the general population are left-handed. Therefore, the probability that any randomly selected person will be left-handed is somewhere around one-in-ten. In statistical terms we would usually refer to this as a probability of 0.10. From this we can say that the probability of a randomly selected *married* person being left-handed will also be 0.10 (on the basis that, just like everyone else, one in ten married people will be lefties). Now, in order to establish the probability that *both* members of a couple are left-handed, you need to put yourself in the position of a left-handed married person (one in ten of the population) and then ask yourself this question: *what is the probability that my spouse will be left-handed?* Remembering that spouses are chosen from the general population, you should see that the probability of having a left-handed spouse will duly be 0.10. So, if every tenth married person is left-handed, and every tenth one of *those* has a left-handed spouse, then the overall proportion of married couples where both are left-handed will be one tenth of one tenth (or one in a hundred), making this a probability of 0.01.

Two useful principles can now be inferred. Firstly, the probability of both spouses being left-handed (0.01) is much lower than the probability of a single married person being left-handed (0.10). Secondly, the arithmetic computation of the probability of two unrelated events co-occurring is achieved very simply: by multiplying the two individual probabilities together (in this case, multiplying 0.10 by 0.10 to give us 0.01). This latter principle is what ensures that the probability of a coincidence will always be lower than that of either incident. Imagine two independent events that have individual probabilities of 99 per cent; in other words, there is a 99-in-100 chance of each occurring. The probability that they will occur together will be 0.99×0.99, which is 0.9801. Although close to 99 per cent, the combined probability is nonetheless lower. This will always be the case: probabilities are fractions, so the multiple of two probabilities will always be a fraction of a fraction, and always therefore a lower number.

The failure to realize that a coincidence is less likely than an incident is known in logic as a *conjunction fallacy* (Tversky & Kahneman, 1982), a term that alludes directly to the 'conjunction' of two events. Conjunction fallacies are very common, and can catch people out even when they are concentrating hard to solve serious problems that they have been professionally trained to handle. In one early study, now a classic (Tversky & Kahneman, 1983), a group of Harvard-based physicians were tested for their ability to avoid conjunction fallacies. To achieve this, they were shown a case summary for a 55-year-old woman with pulmonary embolism, and then asked to rate the probability of a selection of different diagnoses. Within the list of six options, the following two were of interest:

dyspnea and hemiparesis

hemiparesis

Note that the former represents a coincidence while the latter represents one of the relevant incidents. As such, the probability of 'dyspnea and hemiparesis' (together) must be lower than the probability of 'hemiparesis' (on its own). Nonetheless, over 90 per cent of the doctors in the study made a basic error: they rated 'dyspnea and hemiparesis' as being the *more* probable diagnosis of the two. In reality, dyspnea (shortness of breath) is a much more common outcome than hemiparesis (partial paralysis) in patients with pulmonary embolism. This real-world trend may have misled the physicians to an intuitive view that any diagnosis involving dyspnea must be more probable than one of hemiparesis alone. However, as the dyspnea option they were asked to rate *included* hemiparesis, such reasoning was flawed. The arithmetic probability of 'dyspnea and hemiparesis' will be the probability of dyspnea multiplied by that of hemiparesis, and so will always be lower than the probability of hemiparesis in isolation.

One way to gauge the implications of such a finding is to consider the treatment options for dyspnea and hemiparesis. Dyspnea is sometimes treated by using a breathing apparatus to administer oxygen to patients which, in medical terms, is fairly benign. In contrast, hemiparesis might indicate an underlying stroke or brain lesion, and so would call for a much more radical intervention. The doctors' tendency to falsely conclude that hemiparesis was present would have led, in a clinical context, to a needless recommendation that the patient undergo radical treatment. When the doctors who participated in the study were told about their judgements, they expressed dismay at how easily they had made such an elementary mistake. Nonetheless, subsequent research has continued to show that these misjudgements are very prevalent in medical decision-making contexts, even amongst experienced and vigilant professionals (Reyna & Lloyd, 2006).

Conjunction fallacies result from our tendency to dodge the requirement to formally compute probabilities. We are not naturally inclined toward multiplication and division, and so we rely instead on a proposition's superficial elements when attempting to judge its relative likelihood. For example, if a proposition is similar to one we have encountered before, we're likely to use this as the basis for estimating its chances of occurring. Alternatively, we may base our estimates on the extent to which the proposition resembles a distilled version of the subject matter. This was memorably illustrated in a now classic series of studies by Amos Tversky and Daniel Kahneman (1983), two psychologists who went on to dominate the field of logical cognition and reasoning. They presented college students with a description of 'Linda', a fictitious (and now somewhat famous) character who:

...is 31 years old, single, outspoken and very bright. She majored in philosophy. As a student, she was deeply concerned with issues of discrimination and social justice, and also participated in anti-nuclear demonstrations. (p. 297)

Of the 140 or so students who were shown this text, 85 per cent felt it was more probable that 'Linda is a bank teller and is active in the feminist movement' (the coincidence) than, simply, that 'Linda is a bank teller' (the incident). When an additional group of students were given the same description of Linda, a majority agreed with a statement that she is more likely to be a *feminist* bank teller 'because she resembles an active feminist more than she resembles a bank teller'. Only a third of students agreed with the alternative statement offered, that 'some women bank tellers are not feminists, and Linda could be one of them'. In short, two thirds of the research participants not only made illogical determinations about relative likelihoods, but they *knowingly* chose to base their judgments on appearances and stereotypes rather than on logical deduction. They dodged logic in favour of gut instinct. And they were quite happy to tell researchers that they did so.

The difficulty of pure probabilities

The task of estimating a probability is itself inherently difficult because, in the end, there are no certainties. After all, even though the odds are against it, the patient may in fact have hemiparesis, and Linda may indeed be a feminist bank teller. However, other experimental demonstrations show us how we commit logical errors, more or less automatically, even when there is no such room in which to wiggle. Even when there is indeed a 'right' answer to a question about probabilities, we find it very hard to produce that answer.

Take, for example, the task of computing exact probabilities of particular outcomes, given a total number of possible permutations with known relative frequencies. Or, more simply, imagine tossing coins or throwing dice. In a coin toss, the total number of permutations is two (heads or tails), and their relative frequencies are known (they are both equally likely). It is straightforward for most people to correctly answer the question, 'What is the probability of getting "heads" when you toss a coin?' The correct answer, of course, is 0.50 (or, in more common language, 'fifty-fifty'). When throwing a dice things are more complex, although only slightly. This time the total number of permutations is six (each of the six sides of the dice), and again their relative frequencies are known (inasmuch as we know all sides are equally likely to face up). Therefore, it is again straightforward for most people to answer the question, 'What is the probability of getting a "six" when you throw a dice?' The correct answer here is a probability of 'one-in-six' (or, arithmetically, 0.17).

Most people will find it straightforward to answer such questions about coin tosses and dice throws. But this is true *only in the simplest circumstance of a single event.* Consider instead asking people the following question: 'What is the probability, when tossing a coin *twice*, of getting two "heads" in a row?' You will find that the majority of people will immediately become more tentative when attempting to come up with an answer. Indeed, their answers will frequently be wrong. Likewise, if you ask people 'What is the probability of getting two "sixes" when you throw two dice?', the task of providing a confident response will flummox the majority of respondents. (Perhaps you have been considering these questions yourself. Are you confident about your answers?) The procedure to use is exactly as outlined above with regard to conjunctions: when dealing with unrelated events, the probability of two outcomes is the probability of one multiplied by the probability of the other. Therefore, the probability of throwing two 'heads' will be 0.25 (0.50 × 0.50), while the probability of throwing two 'sixes' will be 0.03 (0.17 × 0.17; although an arithmetically more adept answer would be 'one-in-thirty-six'). A way to summarize all this is to point out that, while a single event with known parameters is straightforward enough to judge, adding even just one further event (for example, just one extra coin toss or dice throw) makes the whole business disproportionally more difficult. In fact it escalates the difficulty of estimating probabilities to almost universally devilish proportions. And given that everyday life requires us to judge a future comprising not just two events, but several thousand details ranging from the mundane to the highly pertinent, it is little wonder that we consider even the near future to be essentially unpredictable.

Sometimes, however, we must press on and attempt our forecasts anyway. Consider the following medical conundrum (from Casscells, Schoenberger, & Graboys, 1978; see also Bennett, 1998; Pinker, 1997):

> If a test to detect a disease whose prevalence is one in a thousand has a false positive rate of 5%, what is the chance that a person found to have a positive result actually has the disease, assuming you know nothing about the person's symptoms or signs? (p. 999)

The correct answer for computing such probabilities is provided by applying the following formula:

[Base rate] × [Hit rate]/[Overall rate of positive results]

The base rate refers to the prevalence of the disease, which in this case is 1/1,000 (or 0.001). The hit rate (or test sensitivity) refers to the proportion of sick people who will test positive for the disease (bearing in mind that sometimes truly sick people do not get positive test results). In the present example, no reference is made to test sensitivity and so we can assume that the test is, in fact, perfect. Therefore, the hit rate will be 100 per cent or, in the context of fractions, simply 1/1 (i.e., 1). The overall rate of positive results is a little more complex. This value represents the number of positive results for every 1,000 cases. As well as the number of positive results associated with persons who actually have the disease, these results will also include any *false* positives, reflecting the fact that sometimes people *who are not sick* do actually get positive test results. We are told that the false-positive rate is 5 per cent. Given the base rate, we know that at least one person from every thousand will be sick, and given the hit rate, we know that this will account for one of the positive results. Of the remaining 999 persons from every thousand (who are *not* sick), 5 per cent will return false positives.

Therefore, of a thousand people tested, the *rate* of positive results will be the one-in-a-thousand actual sick persons plus the 5 per cent of the 999-in-a-thousand *well* people who falsely show up as positive on the test. In the end, all of this yields us the following computation:

$$[1/1{,}000] \times [1] / [(1/1{,}000) + (999/1{,}000 \times 0.05)]$$
$$= 0.01962708537$$
$$\approx 0.02$$

But an even quicker way to answer the question is to reflect on the fact that, for every one sick person who tests positive (out of every thousand), there will be a further 50-ish false positives (5 per cent of 999). Therefore, around one in 50 (or so) patients with positive results will in fact be sick; hence, the probability of close to 0.02.

In the context of the conundrum, a person with a positive result in the test will actually have just a 2 per cent chance of having the disease. However, when this problem was presented to a group of 60 physicians, residents, and fourth-year trainees at Harvard Medical School, half of these medically educated respondents estimated the risk to be 95 per cent. This was close to being the *exact opposite* of the correct answer. The average of all the answers given was 56 per cent, which is still 28 times bigger than it should be. Only 11 of the respondents, around one in six of them, came up with the right answer. In short, if you had canvassed the views of these 60 physicians, neither the majority response nor the average of all responses would have been *even close* to being correct. If the consequence of a positive diagnosis were to recommend some kind of limb amputation, virtually all the patients with positive test results would have lost a limb unnecessarily.

Such drastic consequences are not confined to fictitious scenarios in research studies. In public health contexts, nearly all screening tests have false-positive rates. This is because they are designed to be sensitive enough to pick up (rather than ignore) disease markers in all cases, including the physiologically unusual ones. In other words, the tests tend to err on the side of caution. The dilemmas arise when the costs of over-diagnosis appear to outweigh its benefits. For example, with regard to population-based breast cancer screening, several analyses have shown that false-positive test results lead many women to undergo unnecessary surgeries, with far fewer women benefiting from a true diagnosis of cancer. In the most recent analysis (Gøtzsche & Jørgensen, 2013), the investigators traced the outcomes of over half a million women who had undergone breast cancer screening across several countries, and came to the conclusion that:

> ...for every 2000 women invited for screening throughout 10 years, one will avoid dying of breast cancer and 10 healthy women, who would not have been diagnosed if there had not been screening, will be treated

unnecessarily. Furthermore, more than 200 women will experience important psychological distress including anxiety and uncertainty for years because of false positive findings. (p. 2)

This finding – that for every woman whose life is lengthened, ten will undergo treatments such as chemotherapy and mastectomy *unnecessarily* – is very robust and has been reproduced in many different studies. Similar findings have been observed with regard to other forms of screening, such as for prostate cancer and depression. Such findings help us realize that screening resources need to be continuously developed and refined. Nonetheless, there often appears to be resistance to accepting their implications, both among the public at large *and* among sections of the medical profession (Ehrenreich, 2009).

Some of this resistance reflects a moral view: that saving even one person from death will justify the damaging, but non-fatal, consequences of over-diagnosing several others. However, some objections relate not to differences in perceptions of the moral balance, but to disbelief that things could actually be this way. In one UK national newspaper, a specialist declared that the conclusion formed by these statistical findings 'just doesn't stack up', and that this divergence between prediction and data presented a 'constant annoyance' to those who wish to promote screening uptake (Laughland, 2012). The fact that the statistical findings are so often replicated is seldom taken as evidence of a consistent underlying reality that might reveal the distortions in a single practitioner's idiosyncratic career experience (in fact, by creating an annoyance that is described as 'constant', it would appear to be part of the problem). Instead, these observers persist in attaching greater weight to personal impressions garnered from immediate encounters, and lesser weight to the medically documented outcomes of millions of patients they have never met. This bias toward personal perspectives and away from objectivity highlights the need for public health information campaigns to be especially careful in explaining screening systems by describing their disadvantages as well as their advantages (Campbell, 2013). Even with the benefit of experience and training, human judgement seems significantly challenged by the task of predicting, and explaining, probability.

The challenge of flukes

Perhaps the core underlying problem with relative likelihoods and pure probabilities is that the human psyche seems ill-equipped to deal with the notion of randomness. 'Random' is the word we use to describe things that appear to have occurred unpredictably or unsystematically, or by pure chance. Millions of random events occur in our lives every day. The vast majority of these, such as the gusts of wind in the air around us, or the precise sequence of street noises that rumbles along in the background of our immediate environments, are totally unremarkable. Only a tiny fraction of random events will catch and maintain our attention. Of course, the fact that something interesting has just happened does not mean that something meaningful *made* it happen. Yet, when faced with an occurrence that seems contrived but is not, our natural tendency is to identify some kind of meaning within it, even if only fleetingly, despite the fact we know the event was indeed truly random.

Take, for example, those times when somebody phones you at the precise moment that you were getting ready to call them. The fact you were thinking about them had no influence on their choice to call you, and yet it is hard to avoid being struck by the coincidence that has occurred. Of course it is possible that you were thinking about the other person for a particular reason (such a recently made arrangement to contact them 'sometime soon'), and that it was this same reason which prompted them to call you at that moment. But sometimes an almost forgotten acquaintance will get in touch for no reason other than they just happened to decide to do so on the day you happened to recall that they existed. All that has occurred is that two thematically comparable events have coincided. Yet if the timing is correct, such coincidences can lead to vivid feelings of surprise, if not even awe.

If *more* than two comparable elements coincide, then this awe can become dazzling. Consider the example of the couple from Birmingham who gained international media notoriety as the world's unluckiest tourists: they somehow chose to holiday in New York, London, and Mumbai on the precise dates that each city would be targeted by a major terrorist attack (Crompton, 2013). Or think of the Irish nurse-stewardess who was serving on the luxury ship RMS *Olympic* in 1911 when it sank near the Isle of Wight, only to sign up for a second job a year later on the similarly doomed RMS *Titanic*. After surviving both disasters, she later accepted a third commission on board the hospital ship *Britannic*, with it too sinking when it struck a mine in the Aegean (luckily, our heroine emerged unscathed on each occasion; Jessop, 1997). Or consider the Croatian man who between 1962 and 1995 survived a train crash, a plane crash, a bus crash, and a car crash, before winning $1 million in his country's national lottery with the first ticket he ever bought (Hough, 2010). Many of these events are noteworthy in their own right, but it is the fact that they occur in such thematically coherent sequences that makes them truly memorable. In general, people understand that coincidences can just be flukes. But when coincidences begin to stack up – in other words, when we start to see coincidences *of* coincidences – people's suspicions begin to rise. We hear rationalizations about how the protagonists must be inherently lucky or unlucky people, or perhaps even blessed (or cursed) by supernatural influences. Or maybe the coincidences did not really happen, and these characters just made up those stories about their exploits. Or maybe there is more to these stories than meets the eye: maybe the heroes were exerting unseen influences on their own good (or bad) fortunes all along. In short, whatever happened, it could hardly have been a simple coincidence.

In fact, sometimes you hear people complain that they just 'don't believe in' coincidences. Perhaps people who hold such a view feel they are being sensibly cautious. However, an unwillingness to believe in coincidences can have catastrophic consequences. In 2003 and 2004, Lucia de Berk, an innocent Dutch nurse, was declared guilty of seven murders and three attempted murders primarily because the prosecuting authorities were unwilling to believe in coincidences. De Berk worked as a paediatric nurse in several hospitals in The Hague. After a baby died in her care, one of the hospitals noticed she had previously been on duty when other children died or required resuscitation. Having studied the various records, they concluded that the likelihood of de Berk's shifts coinciding with that many deaths and near-deaths was so remote as to be impossible to result from chance alone. A statistician computed that the probability of such events occurring by chance as 1 in 342 million (Derksen & Meijsing, 2009). After the finger of suspicion had been pointed, a number of de Berk's former colleagues came forward to cast aspersions on her character. They said she was unfriendly, and drew attention to her eccentric personality and her interest in tarot cards. De Berk was sentenced to life imprisonment even though there was no forensic evidence or eye-witness account to corroborate the statistical testimony. Following the conviction, a number of scientists and statisticians began to question the reasonableness of the prosecution's case. As coincidences do happen, and as human observers find it difficult to intuit probabilities, basing a murder conviction primarily on one estimation did seem quite parlous. Academic statisticians began to comment publicly that they could not understand the prosecution's computations, and ultimately it was established that the calculations were conducted incorrectly. The true chances of Lucia de Berk working those particular shifts were in fact closer to 1 in 25 than they were to 1 in 342 million (Gill, Groeneboom, & de Jong, 2010). A prominent physicist pointed out that not only were the required probabilities computed incorrectly, they were also interpreted badly: while the prosecution asserted that natural deaths were rare, they failed to appreciate that murder was even rarer, and instead falsely assumed it to be the more likely eventuality (Buchanan, 2007). Eventually, at the initiative of a professor of mathematics, a petition to re-open the case was launched, and in 2010 de Berk was acquitted of all charges. The appeal court found that the prosecution had relied on tunnel vision and statistical misunderstandings, and that de Berk had provided very effective care and saved many lives throughout her career. Despite the fact that no material evidence was ever presented – and the fact that nothing statistically outrageous had actually occurred – the initial unwillingness of the authorities to 'believe in' a series of grim coincidences led an innocent nurse to spend seven years incarcerated as a convicted child murderer.

A similar fate had befallen British solicitor Sally Clark, who in 1999 was convicted of the murder of her own two children. Clark had had two babies who died similarly in early infancy, two years apart. An eminent paediatrician produced statistical evidence in court, asserting that the chances of two infants from one family dying suddenly in such circumstances were beyond 1 in 73 million. As with Lucia de Berk, academics and scientists emerged to argue that this statistical evidence was flawed (Hill, 2004), and eventually Clark was acquitted of all charges. The eminent paediatrician had been regarded as an expert on the probabilities of sudden infant death; now the court found him to have been so wedded to his (flawed) statistical conclusions that he felt compelled to disregard relevant medical information in order to preserve the coherence of his position. For this reason, he was investigated by the General Medical Council for misconduct, and for a time was struck off the medical register. A number of his previous cases were re-opened, and two other women who had been jailed for murdering their own babies were acquitted and walked free. But all this was of little comfort for Sally Clark. After spending three years in prison being treated as a child killer instead of as a bereaved mother, she developed a series of life problems. She died of alcohol poisoning in 2007, aged just 42.

Most of us find it difficult to reason through coincidences in everyday circumstances. Not only this, but such a task can go disastrously wrong even for established experts, especially when they are tempted to provide summary opinions that draw on their experience instead of their training. Quite often, the true complexities require deliberate and systematic mathematical analyses based on statistically nuanced principles. Juggling probabilities is challenging because the randomness that underlies coincidence is existentially profound. After all, it is hard to fathom how random events can be truly unpredictable, without giving credence to the spurious notion that inanimate objects can take action on their own. Even coin tosses have not always been seen as authentically random: historically, the practice of tossing coins was used not to model randomness, but as a way of divining the true intentions of the gods who controlled the destiny of the universe (Arbuthnot, 1710).

The systematic nature of error

A common assumption regarding the way human beings reason through evidence is that they absorb information accurately, evaluate it accurately, and use it to draw accurate conclusions. However, it would be truer to say that people absorb information crudely, evaluate it inconsistently, and use it to draw conclusions that are often – but not always – *roughly* correct.

Sometimes the way our brains return premature conclusions after taking cognitive shortcuts through the evidence can be revealed in a striking manner. An example is when we look at clouds or tree-bark or some other quasi-random pattern and feel, compellingly, that we can 'see' an image represented within its contours. A perceptual effect known as *pareidolia* is said to occur when we attach unwarranted significance to meaningless visual or acoustic stimuli. Significantly, paraeidolia often involves 'seeing' a human face in some non-face pattern or 'hearing' a human voice in some non-vocal sound. Many cases involve objects in our natural environment. Cliff edges or landscapes, shaped by centuries of gradual erosion, can strike us as if they resembled artistically produced carvings. Examples include the Pedra da Gavea rock near Rio de Janeiro in Brazil, the Sphinx rock in the Bucegi Mountains in Romania, or the (now collapsed) Old Man of the Mountain cliff face in New Hampshire. While such landmarks have been identified (and occasionally worshipped) for centuries, new ones continue to be discovered. One series of valleys near Alberta in Canada was generally ignored throughout its history until 2006, when users of internet-based maps noticed that aerial views looked eerily similar to a human head wearing Native American headwear. Following media publicity, the formation, which comprised a type of dry terrain known as 'badlands', was re-named the Badlands Guardian by the local county council.

Technologically produced sounds have often been described as containing human voices where none really exist. A famous category of this auditory form of pareidolia is the claim that certain recordings, when played in reverse, contain secret voice-based messages. While musicians since the 1960s have used a process known as *backmasking* to incorporate reversed voice recordings as sound effects in their music, this was largely in order to capitalize on the unusual noises that were created rather than as a means of transmitting secret messages to listeners. More intriguing are the cases where artists have gone to no such effort, but have had their music accused of containing backmasked messages nonetheless. In 1982, a prominent American religious radio host claimed that the fifth verse of Led Zeppelin's 'Stairway to Heaven' contained an extensive message valorizing Satanic worship, only audible when the track is played in reverse. Similarly, the chorus of Queen's 'Another One Bites the Dust' is said to sound, when played backwards, like a paean to cannabis. It is worth restating that these instances are quite different to true backmasking, in that neither Led Zeppelin nor Queen made any attempt to embed reverse-recorded messages into their music. The fact that the reversed form of these tracks phonetically resembled coherent human speech was just a coincidence.

But just as a human face in a cloud merely *resembles* a human face, the noises in these music tracks merely *resemble* human voices. Anyone listening to them would realize that they were not actual voices (as anyone looking at a cloud would realize that it does not contain a real face). We seem primed to recognize faces and voices from images and sounds that are by no means complete facsimile likenesses. So what is it that our brain does in these situations? Psychologists typically argue that such visual and acoustic priming has arisen because, throughout evolutionary history, any inclination to be fast at detecting faces or voices has conferred survival advantage on those of our ancestors lucky enough to possess it. It is these ancestors who have succeeded best in passing their genes on to subsequent generations. Over time, the trait of being quick to detect faces or voices has been retained in the gene pool, and the majority of us alive today possess it to some extent. Indeed, the trait has proved so useful that the most hypersensitive forms of it have been best retained, such that we now even see faces and voices that are not really there. We over-compensate: we see faces in face-like objects, and hear voices in voice-like sounds. This is certainly more useful than having *under*-sensitive face and voice detection abilities, and so has, in evolutionary terms, given our species the most lucrative kind of survival advantage. In short, the dynamics of survival advantage have led our species to evolve a perceptual shortcut, where we identify certain shapes and sounds and quickly tag them as 'faces' and 'voices', without spending the thinking time necessary to work out what it is exactly we are seeing or hearing. This is more often useful than it is useless: after all, most of the face-like shapes we see *are* faces, and most of the voice-like noises we hear *are* voices. The shortcut sometimes leads us to make false classifications, but its advantages in helping us make correct classifications far outweigh the disadvantages of this error rate.

We mentioned another of these evolutionary shortcuts in Chapter 3, when we referred to the human tendency to attribute greater significance or complexity to familiar subjects than to unfamiliar ones. This is known as the *availability* or *recognition heuristic* (a *heuristic* being a common-sense shortcut used for making decisions or solving problems, sometimes idiomatically referred to as a 'rule of thumb'). The availability heuristic is highly pertinent to our earlier discussion of the assessment of risks in air travel. For example, you probably believe that more people are killed in plane crashes than are drowned in their own homes. In reality, however, far more people drown in their baths at home every year than die in plane crashes. We form our aviophobic conclusion because we can more easily bring to mind memories (or fictional depictions) of plane crashes. This is largely because of the extensive media coverage that plane accidents receive. In contrast, we find it much harder to remember any specific case of a domestic drowning.

Similarly, when research participants are asked to draw accurate maps of their own and surrounding countries, most end up making their own country disproportionately large (Lorenzi-Cioldi et al., 2011). It is as though we assume our own country must be bigger because we know in so much more detail all the regions, cities, and features it contains. The availability heuristic gives our species the tendency to assume that easily remembered information must simply be the most relevant, regardless of the wider unseen context. This reflects the fact that, throughout evolutionary history, easily remembered information (such as where to find food, how to avoid danger, who is friendly, and who is hostile) often just *was* the most relevant, and so acting on this assumption could be expected to pay off more than it would cost. When tested against the strict requirements of accuracy, especially those that have emerged as our world becomes more technologically driven and reliant on literally correct information, the error-rate associated with this strategy can nowadays catch us out. For example, an easily remembered case of an apparent health risk can govern the behaviour of millions of people long after it has, less memorably, been shown to be baseless by scientists. We will consider such a case in Chapter 10.

Psychologists who study the way people reason through information have developed an understanding of several different types of cognitive heuristic. Much of this work has been associated with the aforementioned research duo, Amos Tversky and Daniel Kahneman. A prominent example is the *representativeness* heuristic (Tversky & Kahneman, 1974). This is when people assume that events will take a certain course because they appear similar to other well-known events, or that people will behave in a particular way because they resemble a well known representative stereotype. In some senses, the decision-making processes used by participants in Tversy and Kahneman's Linda-the-bank-teller study were based on the representativeness heuristic: Linda was perceived to be a *feminist* bank teller because she conformed to popular stereotypes describing feminists *and* bank tellers. This affected people's judgements more than the statistical reality that it was simply more probable Linda was merely a bank teller.

Another heuristic is encapsulated by a particular model of subjective probability known as *support theory*. This refers to the way we attach more significance to detailed information than we do to simple information. Consider the following two questions (Tversky & Koehler, 1994):

What is the probability that you will die next year?

What is the probability that you will die on your next summer holiday from a disease, a sudden heart attack, an earthquake, terrorist activity, a civil war, a car accident, a plane crash, or from any other cause?

When people are asked these questions, they tend to rate the probability of death as higher when the longer question is put. This is despite the fact that the longer question relates only to death during your summer holiday, which is a mere subset of the timeframe covered by the shorter question (the phrase 'any other cause' in the latter means that the actual cause of death is irrelevant in any comparison between the two). It appears as though the additional detail in the longer version elevates the *feeling* of probability in the listener's mind. With this heuristic, people infer relevance based on the time and attention given to the point by the person asking the question. Also, the attention-grabbing details encourage the listener to think more about the subject, and so to imagine it as a realistic prospect.

A third heuristic is called *anchoring*. This is when we base a judgement artificially close to a level that has already been articulated (the 'anchor'), such that the prior mention of that level comes to contaminate our judgement. For example, if participants are asked 'Is the percentage of African countries which are members of the United Nations larger or smaller than 65 per cent?' they usually estimate the actual percentage to be somewhere close to 65 per cent. But if they are asked 'Is the percentage of African countries which are members of the United Nations larger or smaller than 10 per cent?' their subsequent estimates move closer to 10 per cent. However the question is varied, the estimates duly vary in a way that correlates with the question. This is the case even if the percentage used in the question is randomly produced using a spinning carnival wheel in full view of the participant who provides the estimate (Tversky & Kahneman, 1974). The anchoring heuristic also affects judgment when the initial anchor is clearly implausible. When participants were asked to estimate the year of Albert Einstein's first visit to the United States (actually 1921), those who were presented with the (obviously false) anchor of 1215 provided earlier estimates than those who were presented with the (equally obviously false) anchor of 1992 (Strack & Mussweiler, 1997).

Anchoring biases can be subtle but highly relevant to everyday life. In one study, a sample of legal experts was asked to recommend a sentence for a fictitious criminal who had been found guilty of shoplifting. The experimenters instructed them to roll a dice after reading the case study but before formulating a sentence. Unknown to the participants, the dice was rigged so that it always resulted in either a '3' or a '6'. Even though the participants believed that the dice throws were random, and thus logically irrelevant to any sentencing decision, and even though they were told to base sentences on the normal legal considerations and nothing else, they nonetheless issued heavier average sentences after rolling the *higher* number on the dice. When the dice showed a '6', they recommended an average of 8 months imprisonment; after rolling a '3', they sentenced the fictitious criminal to 5 months in prison (Englich, Mussweiler, & Strack, 2006). The legal experts who participated in the study were all early-career specialists (and so had relatively up-to-date training on sentencing practice) and, in fact, they had all worked as real court judges in the world outside the laboratory. One wonders whether any of their previous defendants were made aware of the results of the experiment.

Another study placed signs next to the tomato soup section in a supermarket. When customers were told there was a limit on how many cans of soup they could purchase, most bought the number of cans that was mentioned in the sign. Customers who were told they could buy no more than four cans tended to purchase four cans; customers who were told they were limited to twelve tended to purchase twelve. Customers who were told there was no limit at all tended to buy two or fewer cans of soup, even though the information they had to hand would have justified many more purchases. In short, the number referred to in the sign seemed to influence the purchasing decision of customers; the higher the number, the more that sales were made (Wansink, Kent, & Hoch, 1998). That '10 items or less' queue in the supermarket might look as though it is there to expedite customer throughput, but it might also have the effect of encouraging small-scale customers to purchase more items, making it a profitable initiative for the supermarket's owners.

All these heuristics share the same underlying evolutionary dynamic: their benefits have exceeded their costs in terms of survival value, and so they have persisted through the generations and become prevalent within our gene pool. They also have in common the fact that, while useful as shortcuts much of the time (in that they often lead us to broadly accurate decisions), some of the time they lead us to draw erroneous conclusions. And sometimes they generalize poorly into contexts, such as courtrooms or supermarkets, which evolution did not really have in mind. The error these heuristics produce is systematically present: far from being reliably accurate thinkers, it is *normal* for human beings to recurrently make certain types of mistakes.

The constructive nature of evidence

At a cognitive level, it is clear that human beings find it difficult to balance likelihoods, to judge pure probabilities, and to reason without systematic error. In short, human beings find it difficult to figure out the reasons for something, especially when it is abstract, unseen, or hypothetical. However, even when they are presented with all the information that is relevant, they can still encounter difficulty in explaining what they have encountered. When asked to describe something they have personally witnessed, human beings are notoriously poor at recording, storing, and retrieving the necessary information. The most famous empirical studies of this were conducted by American psychologist Elizabeth Loftus. She asked participants to watch a film depicting a car accident, after which she presented them with a number of questions. As this was an experiment, different groups of participants were asked different questions. When participants were asked 'About how fast were the cars going when they hit into each other?' the average answer given was 34.0 miles per hour. When participants were asked 'About how fast were the cars going when they smashed into each other?' the average answer given was 40.5 miles per hour. It is important to bear in mind that all participants had seen the same film and so had identical experiences; nonetheless, due to the procedures used by the experimenters, they relayed systematically different *accounts* of those experiences. Given that all other aspects of the procedure were unchanged, the variance in speed estimates could only have resulted from the recasting of the question to incorporate the word 'smashed' instead of 'hit'. In short, while participants recorded and stored their memories in broadly the same way, they retrieved them differently.

What's more, when participants were brought back to the laboratory a week later for further questioning, the distortion of memory caused by this retrieval context was found to have lasting effects. Twice as many participants in the 'smashed' group than in the 'hit' group reported seeing broken glass in the film, despite the fact that the film depicted no broken glass at all (Loftus & Palmer, 1974). So, not only did participants *retrieve* memories differently in the first place, doing so interfered with the substance of their memories by the time further retrieval was required a week later. By asking a particular question, the experimenters had shaped the content of memory in a permanent way, even making participants recall elements that did not exist. Subsequent experiments using this type of procedure have shown that participants can be made to 'remember' other non-existent elements in such films, including road signs and traffic lights. All of this shows why significant caution is needed when considering eye-witness accounts – especially in court proceedings that deal with driving offences – while simultaneously highlighting the insidious damage that can be caused when witnesses are asked leading questions.

Most people intuitively feel that their memory works in much the same way as a video camera: as events take place, our point-of-view experiences are somehow recorded contemporaneously in our brains. When we wish to recall the event later, all we need to do is locate that recording and, metaphorically, replay it in our mind's eye. However, a large body of research over many decades suggests that human memory does not work like a video recording at all. Instead, as we experience events, we select some of the more pertinent details of it and store those in our brain, along with a rough temporal sequence of how the event unfolded, and some kind of broader meta-code to help us understand the overall context. When we wish to recall the event, we bring to mind all these elements simultaneously, and then combine them in such a way as to reconstruct the event in our minds. Most of the non-essential details in our new 'reconstructed' account are not specifically stored for that event, but are instead filled in intuitively based on our general knowledge of how things in the world tend to be. For memories of traffic accidents, some of our recollection will be of the event in question, the rest (such as road markings and dimensions, the location of nearby objects, the colours of the various vehicles involved, and so on) will be based on our prior expectations of the overall context (such as what we expect roads to look like). If these expectations are influenced, after the fact, by the phrasing of a particular question or by the details of another witness's account, then our own recall is likely to be influenced too. If somebody leads us to believe that the cars 'smashed' into each other, then our recollections will evolve to take account of that detail: we will remember that the cars were going quite fast, and that it led to a lot of broken glass. We might even remember that the motorists appeared to be driving recklessly just before the incident.

Much of our cognition takes this form. For example, when we step back and look at the environment around us, we see that it is made up of very many fine details. To truly absorb every single one of them would require a particularly high-resolution photographic memory or, failing that, an enormous amount of time in which to meticulously focus on every macro and microscopic feature that is there. In reality, our brain does not work this way. Instead, when visual input enters our pupils (in the form of light) and stimulates the optic nerves that connect our eyes to our inner brain, only some of the information contained therein will end up in our conscious awareness. As with memory, our brain would rather save time by identifying only the pertinent elements of what we have seen, and creating the rest of our total experience using background knowledge. This approach generally serves us well: we go through life successfully navigating our environs without noticeably failing to see the things that matter.

However, on occasion, our brain's tendency to rely on shortcuts can be revealed. For example, sometimes we see a person (perhaps from a distance) and falsely believe them to be somebody they are not. Only when we do a double take – that reflex action where we take a sudden second look – do we realize that this person is in fact somebody else. At first, our brain registers only partial details of the person we were looking at (maybe their overall body shape, hairstyle, or choice of clothing). Assuming it to be somebody familiar, our brain then fills in the blanks to create a visual impression of that person such that we feel we actually 'see' them. It is only when we take a second more focused look, where we absorb the rest of their physical details, is it revealed they are a different person altogether. Nonetheless, that first glance truly *feels* like we are seeing our acquaintance – after all, our brain has indeed derived that very conclusion, and is pushing it through to our conscious awareness. Even after our double take, we feel a little confused at having been so convinced in our error. But sometimes we don't get that second glance and are left with the overwhelming sensation that we have just seen someone who was actually not there at all. It is undoubtedly the case that this reconstructive nature of cognition accounts for at least some of those confident claims that a celebrity has been sighted in the local high street just the other day, even though this person's death had been reported by the media many years previously.

Two other aspects of the way we make sense of experience are worth mentioning here. Both are biases of cognition, although they are not really heuristics. Instead they relate more to the constructive nature of cognition. The first is *hindsight bias*. This is where we form a falsely confident conclusion that events which have occurred in the past were actually predictable to begin with. Colloquially, psychologists have referred to this as the *I-knew-it-all-along* effect (Roese & Vohs, 2012). It relates to the constructive nature of cognition because it affects how we selectively recall events; we are often prone to more readily focus on those details that bolster our assumed prescience. We are prone to hindsight bias when, after our favourite team loses a match, we claim that we knew this would happen and so are not now disappointed, even though we cheered them on committedly as the match itself was taking place. (Of course, had they emerged victorious, we would have instead claimed that we just *knew* they had it in them to overcome their opponents.) We are also guilty of hindsight bias when we drive down a street that leads us to a traffic jam caused by lights having been struck by lightning, and then say to ourselves, 'I *knew* I should have gone the other way' (as if we genuinely could have predicted lightning). At a collective level, society at large can adopt hindsight biases too, often in order to make sense of past historical events. One example is when we give myopic attention to a single cause as being the precursor to an entire episode of history, such as the popular belief that a single assassination led to the onset of the First World War. This catastrophic war is likely to have had multiple causes: the build-up of tension among European powers with competing imperialist tendencies; technological advances in weaponry and the emergence of a commercial arms industry; the growth of nationalism as a political self-concept; and so on. As such, the reductionism involved in attributing the war to a single assassination is at the very least misleading, and may impede our collective understanding in ways that make us less likely to avoid similar wars in the future. Hindsight bias may give us a satisfying sense of certainty about past events, but it may do so in a manner that leads to overconfidence and poor vigilance.

The second type of bias is called *confirmation bias*. This occurs when we attach particular significance to information that corroborates our beliefs and pay relatively less attention to information that contradicts them. In addition, it occurs when we arbitrarily interpret ambiguous evidence as being not ambiguous at all, but actually supportive of *our* point of view. Sometimes people consciously skew the facts in order to support a case. However, in the psychological sense of the term, confirmation bias occurs spontaneously and without our deliberately setting out to mislead (Nickerson, 1988). When somebody lists off all the positive reasons in favour of a proposition, without even considering to think of any negative ones, they are guilty of confirmation bias (anyone who has ever sat on a committee is likely to be aware of how common such one-sided discussions are in everyday corporate governance). Similarly, when smokers focus attention on people who have lived to a long age despite having smoked all their lives, but fail to consider the much larger number of smokers who have died prematurely (and so, ironically, are easier to forget), this too is an example of confirmation bias.

Confirmation bias underlies many stereotypes, a problem as relevant to psychological research as it is to wider society. Seeing as psychologists are just as prone to stereotyping as anybody else is, it follows that confirmation bias can affect the way psychological research is interpreted. For example, research on sex differences usually yields results that are truly ambiguous. However, despite this vagueness, it has been observed that sex difference findings are typically interpreted as supporting hypotheses that reflect some or other cultural stereotype (Fine, 2010). American social psychologist

Carol Tavris (1993) has presented a number of examples. One concerns the study of self-esteem. In research that compares women and men's scores on psychometric tests of self-esteem, women usually show lower scores than men. This sex difference is typically interpreted as revealing women to have a *low* sense of self-worth, with men having less difficulty in valuing themselves highly. In other words, women lack a positive attribute – pride in one's value as a person – that men more readily exhibit. Tavris points out that this interpretation is arbitrarily unflattering to women. The finding could just as easily be interpreted as revealing men to have *excessive* feelings of self-worth, such that they are the more conceited and arrogant gender. If so, then it could be said that men possess a *negative* attribute – a proneness to egotism and smugness – that women (thankfully) don't exhibit. The standard interpretation that women lack something positive is not intrinsically supported by the sex difference in test scores. As such, the pattern of interpretation that appears in the scientific literature more likely reflects a stereotyped cultural assumption about how men and women differ psychologically. This is confirmation bias.

Similarly, research documenting sex differences in task satisfaction tends to be interpreted as suggesting that women don't value their efforts as much as men do; Tavris points out that it could just be that men *overvalue their efforts*. Sex differences in anticipated performance tend to be interpreted as evidence that women lack confidence; Tavris observes that it could equally imply that men *are not as realistic as women in assessing their abilities*. Sex differences in attitudes to social networks tend to be interpreted as evidence that women find it difficult to develop autonomous identities; Tavris suggests that it could just be that men *have difficulty forming proper relationships with other people*. In other words, all such findings could as easily be interpreted in ways that flatter women and denigrate men, instead of the standard approach of flattering men and denigrating women. As there is no inherent reason to interpret such findings in a particular direction, the tendency for psychologists to *choose* such a direction more likely reflects stereotyped reasoning than empirical merit in the data. We will return to this subject in Chapter 8.

Human cognition is far from perfect. It is characterized by several heuristically driven biases and distortions as well as a limited comprehension of randomness and probability. That said, it would be incorrect to suggest that human beings are unable to ever make proper decisions through their own judgement and reasoning. Indeed, while laboratory studies often reveal apparent limitations on cognition, the fact that they are conducted in laboratories (or in other artificial contexts) is itself pertinent: such studies highlight how human reasoning is particularly challenged when tested in ways that are divorced from everyday life. Many psychologists have suggested that the shortfalls of heuristic reasoning can be somewhat attenuated when laboratory tests are redesigned to be less artificial and abstract. When psychologists re-conducted some early cognition experiments by replacing standard laboratory stimuli (cards containing arbitrary text) with more familiar props (envelopes containing stamps and addresses), the errors caused by reasoning heuristics were indeed greatly reduced (Johnson-Laird, Legrenzi, & Sonino Legrenzi, 1972). This is consistent with the view that limitations on cognition reflect the way in which our naturally evolved brain handles our technologically modern environments. When we are given the chance to concentrate, our reasoning abilities tend consistently to be more effective (Kahneman, 2012).

The problem we face is that we usually fail to realize when concentration is required, and instead make quick reasoning decisions without regard to the various pitfalls in our midst. More often, we allow social and communication cues from others to disarm our scepticism, and end up succumbing to the various errors caused by cognitive sloppiness. It is to these social and communication influences that we now turn.

The social dimension of error

Cognitive processes such as computation, memory, perception, and reasoning provide many explanations for erroneous thinking. However, they are unlikely to fully account for why errors happen. For one thing, people can avoid cognitive error if they are appropriately motivated to do so, and as long as they are somewhat aware of how errors emerge. So while lapses are likely, they are not inevitable. In daily life, there often appears to be a broader subtext to the occurrence of errors, one that is intertwined with incentives and punishments. Errors don't occur in a vacuum; they typically have consequences. Many mistakes confer benefits on some people while hurting others (or at least depriving them of the aforementioned benefits). Other mistakes influence the social homoeostasis: when an error occurs, the cohesion of a group can be threatened or – if cohesion is dependent on error – can be reinforced. Other mistakes are considered innocuous because their individual effects are trivial even though, when allowed to accumulate, they create adverse outcomes for society at large. And other mistakes systematically make people happy, where logical reasoning would make them sad.

Day-to-day social interaction warps the very transmission of knowledge. By the time we hear about something, we can be sure that some kind of interpretation has been applied. Elizabeth Loftus's research on eye-witness testimony showed directly how a responder's understanding of events can be affected by a questioner's implied assumptions. This reflects a broader process of distortion that characterizes person-to-person communication, one that is unsurprising when you consider the challenges involved in truly relaying a full human experience using words alone. In its own way, the verbal transmission of an experience by one person to another represents the same type of constructive process as occurs when a single human brain attempts to make sense of incomplete perceptions and memories. In short, when we tell somebody what happened, we report a mere subset of the total detail, leaving it to our listener to fill in the blanks in order to make sense of the account. When he or she retells the story to someone else, they do the same. With each retelling, the level of approximation accumulates. Insofar as this approximation undermines a full appreciation of what the original event involved, a degree of error becomes gradually and permanently embedded into the explanation that ultimately circulates.

This error component is seldom truly random. When choosing what details to incorporate in the summary of your experience, you are likely to choose ones that help convey a particular narrative that you have in mind. You may emphasize, or even exaggerate, certain details (a tactic that sociolinguists refer to as 'sharpening'), while choosing to downplay, or even ignore, others (a tactic known as 'levelling'). Your intention is not to mislead; on the contrary, your intention is to ensure that your audience appreciates what you believe to be the true gist of the story. The problem is that your listeners, who start off knowing nothing at all, are unable to tell which details are being sharpened, which are being levelled, and which remain unfiltered. This inherent asymmetry of insight explains why the integrity of information deteriorates with its transmission: recipients cannot be expected to know the exact nature of how senders have nuanced their accounts.

In one study, participants were asked to retell a story about a student who had failed a mathematics test. In the written story used in the experiment (an account of some 125 words), the student was described as reacting to the event in a number of ways, including crying, talking to a friend, taking a long bath, displaying aggression by kicking something, retiring to the kitchen to do some baking, and playing video games. Each participant was asked to retell this story to someone who had not heard it, and that person was then asked to tell it to someone else. The researchers found that in this process of double-retelling, some details of the story became sharpened while others became levelled. What was particularly informative was the precise pattern of sharpening and levelling. This was determined by whether or not the original story was written in terms of a *male* student or a *female* student. When participants were told that 'John' had reacted in these ways, they chose to retell the story by emphasizing John's aggression and video-game playing, while de-emphasizing (or omitting) his crying and baking. However, when participants were told that 'Sylvia' had reacted in this manner, the retelling featured more emphasis on her crying, baking, talking to friends, and behaviours reminiscent of female gender stereotypes. Moreover, just as in Loftus's eye-witness experiments, participants concocted additional details to add to the original story in ways that confirmed these (gender-laden) assumptions. Participants talking about 'John' added that he drank beer, played basketball, and broke things, even though none of these details appeared in the original story. Likewise, participants talking about 'Sylvia' added that she hugged her teddy bear and worried about her weight (Ganske & Hebl, 2001). Such non-random sharpening, levelling, and embellishing is a normal feature of human communication, and high-lights how social factors affect the quality of what circulates as evidence for a given claim.

Another type of social influence on the quality of evidence relates to our assumption that, in general, other people know what they are talking about. When we hear somebody explain something to us, we have a tendency to assume that they wouldn't be speaking with such apparent confidence were they not pretty sure about what it was they were saying. Unless their utterances are profane, we take it for granted that they have some basis in fact and we are unlikely to ask the speaker to provide evidence to support their statements. But sometimes speakers say things in a much more tentative way than their audience appreciates. Sometimes a speaker will present a statement as a kind of hypothesis, in order to see how it is received by others. When listeners show signs of passivity, the speaker can infer this as endorsement. Meanwhile, the listeners themselves are forming the view that there must be *some* merit to the statement, simply because the original speaker was willing to make it. From such tentative exchanges can committed positions emerge.

On other occasions a speaker will make a statement that they *know-ingly do not believe*, because he or she wishes to ingratiate themselves with a particular audience. The members of the audience themselves might not believe the statement either, but hearing it uttered before them, they come to assume that *everyone else* must actually do so. Now viewing the statement as widely believed, they assume it must have some merit; or at the very least, they conclude that they should, for appearances' sake, concur with it publicly. Ultimately, it can arise that a group of people end up endorsing a collective view that no individual member actually holds, simply because each member falsely believes that everyone else in the group supports it. This almost surreal set of circumstances has become known as *pluralistic ignorance* (Miller & McFarland, 1987), and has been found to be very common. For example, studies have suggested that pluralistic ignorance underlies many expectations regarding people's sex lives (Lambert, Kahn, & Apple, 2003). In such research, young adults typically report a perception that their peers are more satisfied with sex than these other people's own ratings actually suggest. Moreover, such young adults also tend to believe that they themselves are different, in that they are uniquely dissatisfied when compared to their peers. In such cases, pluralistic ignorance creates a universally held false impression of what other people are thinking, such that the resulting consensus is inconsistent with the individual views that it supposedly represents. In fact, the consensus reflects the *opposite* of what people are really thinking. The impact this has on people's anxieties about sex and sexuality is surely stark: assuming everyone else is more satisfied than you is bad enough, but when everyone you speak to agrees that the majority are indeed satisfied – even though they are not – the pattern seems particularly regrettable.

However, more often than not, the influence of social context drives us to develop overly optimistic views of our own predicaments and attributes, rather than ones that cause us misery. The tendency for most of us to see ourselves as rather good at things has been widely touted. Several studies have suggested that humans hold statistically irrational assumptions regarding their own finer qualities. For example, it is relatively straightforward to demonstrate that the majority of people see themselves as 'above average' in various domains (the implausibility arising from the fact that it is not arithmetically possible for a majority to be above the average). A number of years ago, a study of one million American high school students showed that not only did 70 per cent believe themselves to be 'above average' in leadership ability, but a further 28 per cent believed themselves to be 'average'. This meant that only 2 per cent believed themselves to be *below* average on this trait (Gilovich, 1991). In the same study, the one million students were asked to rate themselves for their ability to get along with others. Fewer than 5,000 students rated themselves as either 'average' or 'below average', meaning that, when rounded to the nearest whole-number percentage, it appeared that *all* the students – 100 per cent of them – rated themselves as 'above average' on this quality. Humans have repeatedly been shown to exhibit this sense of *illusory superiority* across a diverse range of domains. People reckon themselves to be better drivers (Svenson, 1981), to have better memories (Schmidt, Berg, & Deelman, 1999), to be less likely to require medication for mental health reasons (Reid, 2011), and even to have a better sense of smell (Gilbert, 2008) than their average peer.

People's belief in the merit of their own standing often appears impervious to contrary evidence that ought to be obvious to them. Several studies show how people can persist in believing themselves to be highly proficient at a task even though they are actually awful at it. This 'miscalibration of the incompetent' has become known as the *Dunning-Kruger effect*, after the two psychologists who first described it (Kruger & Dunning, 1999). Further studies show how we can wrongly believe that others feel the same as us even when they diametrically disagree with our position, a circumstance referred to as the *false-consensus effect* (Gilovich, 1990). Still other studies show how we adeptly avoid negative feedback by being selective in our socializing: in essence, we dodge criticism by spending less time with critics, and instead spend more time with people who reinforce our views (Hart et al., 2009). For example, cigarette smokers spend more time with other smokers, and so are less likely to encounter convincing reasons to give up smoking (Hwang, 2010). Smokers also over-estimate the number of smokers in the population (Cunningham & Selby, 2007), while non-smokers underestimate it, showing us how selective exposure to contrary opinion can become intertwined in its own form of pluralistic ignorance.

Our pattern of self-serving thought extends to our attitudes toward inanimate objects and random events, and sometimes the two in combination. People are willing to gamble more money on the roll of a dice if they are allowed to touch the dice themselves compared to when a croupier rolls it for them (Fleming & Darley, 1986). And typically, we are more likely to take credit for events when they turn out well, but to blame external forces beyond our control when things go badly (Gilovich, 1991). Students who do well in tests tend to attribute their success to their own efforts and study methods, and they tend to believe the test to be a valid measure of their abilities; at the same time, their teachers tend to feel good for having taught them well. Students who do poorly in tests are more likely to describe the outcome as having at least partly resulted from unfortunate circumstances (such as the test itself being unusually difficult), and to consider the test as being arbitrary and unfair. These students' teachers tend to express regret that the students didn't work harder on their preparation.

When looked at starkly, these self-serving biases can make us appear almost delusional. After all, failing to recognize one's own true abilities, and believing situations to be other than they actually are, does not *sound* like a promising basis from which to tackle life's challenges. For centuries, having a poor grasp on reality (along with unhappiness, reduced productivity, and a difficulty relating to other people) has been considered a major signifier of mental illness. A person who is described as having 'lost touch with reality' is regarded a person to be pitied. Meanwhile, having accurate perceptions of your life, your relationships, and the circumstances in which you find yourself have all been seen as being integral to effective and healthful cognition. Many major psychotherapies seek to treat mental illness by correcting cognitive distortions. However, the fact that self-serving biases are so prevalent should lead us to infer that such modes of thought must have advantages as well as disadvantages. In fact, it could be the advantages that are the more salient.

In an early influential review, American psychologists Shelley E. Taylor and Jonathon D. Brown (1988) posited a theory that, in fact, distorted cognitive reasoning was something that *promoted*, rather than compromised, mental well-being. They argued that a bias towards optimistic thinking had the effect of perpetuating happiness, in that it makes you happy even when you have no particular reason to be especially so. In terms of the welfare of humanity, such happiness can exert social benefits by stimulating us to help other people. And seeing ourselves as being talented encourages us to attempt things that we might otherwise avoid, thereby promoting innovation, entrepreneurship, and creativity. All in all, then, a human species that lacks such a tendency toward optimistic self-appraisal might end up dwindling towards extinction due to apathy. In corroboration of their theory, Taylor and Brown cited a number of studies showing that people who are low in self-esteem or who are moderately depressed consistently display a *better* ability to interpret self-relevant information in an unbiased way. In the succeeding years, there has now emerged a large body of work confirming the existence of a pattern of thinking that can be called 'depressive realism' (e.g., Fu, Koutstaal, Poon, & Cleare, 2012). In other words, the research shows that people with depression *lack* the tendency to see themselves as being unrealistically in control, as having statistically unlikely good fortune, or as being implausibly 'better than average'. It is mental *ill*-health that is characterized by the ability to grasp reality with accuracy; positive mental health requires a systematic distortion in the direction of self-praise (Mezulis, Abramson, Hyde, & Hankin, 2004). Not only do our cognitive systems expose us to the risk of error by relying so much on constructive reasoning and heuristics, but the social contexts in which we live actually *require* certain types of error in order for us to thrive.

The perils of mass media

Part of our innate tendency to see ourselves as better than we actually are involves a belief that we are duly equipped with the ability to avoid the pitfalls of erroneous human reasoning. In fact, even when distortions such as pluralistic ignorance and self-serving bias are explained to us, we generally see ourselves as less susceptible to them than we consider other people to be. You might say that we believe ourselves to be 'better than average' at withstanding bias. However, a body of research confirms that such views are just as unwarranted as the notion that we are all better-than-average drivers (Pronin, Gilovich, & Ross, 2004). For example, when researchers test the extent to which people feel manipulated by mass media advertising, they usually find the prevailing view to be 'Others are influenced, but not me'. However, experimental work shows that this self-serving perception is far from valid (DeLorme, Huh, & Reid, 2007). Indeed, our tendency to think of ourselves as being able to withstand such influences is subject to a pattern of depressive realism: depression is associated with a reduced tendency to believe oneself to be uniquely immune to the effects of direct marketing (Taylor, Bell, & Kravitz, 2011). Such findings also help us to realize another facet of how people reason through evidence. This is because much of the information we receive is transmitted to us through the modern mass media, and is thus subject to the market forces and logistical quirks that characterize that industry.

When social psychologists talk about reasoning, they are usually referring to the direct contact that occurs between individuals and groups, a process fraught with competing ambitions, contrasting appearances, and the need for ongoing impression management. However, most of the information we grapple with has travelled quite a long way, not just by word of mouth through a network of personal acquaintances. For example, when people worry about the risks of air travel, but fail to worry about the risk of a building collapse (even though more people are endangered by civil engineering catastrophes), they are far more likely to be influenced by material they have read online or seen on television than to have spoken directly to someone who has been involved in an air accident. Similarly, when large numbers of people develop a view that sedentary and solitary pastimes such as video gaming are unhealthy for children, but fail to worry about pastimes such as reading (which is also sedentary and solitary), it is more likely to reflect a mass media zeitgeist than something that has truly occurred to them and their friends alone.

Many audiences are at least somewhat conscious of the fact that you can't always trust information you see on television or read on the internet. Nonetheless, when attempting to form views about what is happening in the world, or about what it is that humankind knows about a particular subject, it is hard to imagine audiences realistically doing *without* these sources. According to the US Bureau of Labor Statistics, the average American adult spends around 3 hours per day watching television, with the average teenager spending over 2 hours per day doing so. It is estimated that across school-going years, the average American will spend 30 per cent more time watching television than they do sitting in a classroom. Furthermore, the average teenage American spends nearly an hour per day 'using a computer for leisure', compared with just 0.11 hours per day either 'relaxing' or 'thinking'. Between the ages of 15 and 35, Americans spend more time using computers for leisure than they do relaxing, thinking, and reading *combined* (US Bureau of Labor Statistics, 2014). In the UK, data from the Broadcasters' Audience Research Board (2014) suggest that the average Briton watches 4 hours of television per day. Other UK statistics suggesting that a further 80 minutes per day – one in every 12 waking minutes – is spent online (Chalabi, 2013).

As such, it is worthwhile briefly considering that when the mass media directs information to its audiences, it does so within several constraints. The banalities of media production cycles often place artificial pressures on journalists and other information providers. A constant effort is made to ensure that material is produced quickly, is attention-grabbing, and is more attractive than that available from competitors. Airtime on broadcast media and space in news outlets (including news websites) are always in short supply, and even colleagues within the same media organization will end up competing with one another to have their efforts given whatever prominence can be afforded (Russell, 2010). These pressures encourage the reliance on pre-prepared materials, such as press releases, which are often actively promoted by lobby groups with ulterior commercial or political motives (Davies, 2009). It is also very much part of media culture to look for so-called 'hard' news, which focuses on the core impact and visual appearance of an issue rather than on associated analysis or explanation. While it is often noted that the cacophony of competing voices on the internet, epitomized by blogs and social network sites, is essentially unregulated when compared to print or broadcast media (Russell, 2010), it is also true that the modern professional journalist is expected to be competent in so many diverse subjects as to regularly blur the distinctions that might traditionally have been made between expert and novice reporting. In the end, consumers of information transmitted through the media have little authentic guidance in choosing what is more likely to be true.

When transmitting information, the media also fall back on a number of common shortcuts. We might consider these to be akin to heuristics. For example, when attempting to refine an issue down to a visualizable story that maintains an audience's attention, the media will often employ metaphorical forms of explanation. A good example of a media metaphor is when the term 'hole in the ozone layer' is used to describe the depletion of spring and summer ozone concentrations that is annually recorded over the Antarctic (Ungar, 2000). Another good example is when changes in certain economic indices are said to reflect the 'green shoots' of an economic recovery (Porto & Romano, 2013). Such metaphors may enhance the degree to which an audience can think about a particular news report or article, but the very nature of metaphors is that they require the alteration of technical details. Altering details to make a subject more accessible creates a risk that audiences will come away with a flawed, rather than enhanced, understanding. The suggestion that there is a 'hole' in the ozone layer is misleading when considered against the nature and function of ozone in the atmosphere. Likewise, the idea that certain events constitute 'green shoots' of recovery implies that such a recovery will inexorably grow in the very near future, when in reality such growth is not assured.

A second media heuristic, which is related to media metaphors, involves the use of pre-existing frames of reference to contextualize new events or information. These narrative frames encourage readers to think about certain topics in certain ways, such as when advances in some branches of science (like nanotechnology or genetics) are repeatedly discussed with regard to their social, rather than technological, implications (Stephens, 2005). Another example of framing is when increases in immigration are discussed as though they indicate an automatic threat to the economy, when more typically immigration is both a sign of, and contributor to, economic prosperity. Once again, while such media heuristics make it easier to categorize and transmit information, they also require a degree of information distortion. Whether or not the distortions are trivial is impossible for consumers to determine.

One further media heuristic relates less to the content of information, and more to the selection of what does and does not get transmitted. As audiences, we are often unaware of the extent to which the mass media filters the flow of information throughout society. We may feel that certain events or ideas become popular simply because of their importance or relevance, and that the media reflects this significance by dutifully reporting them to us as they arise. However, in reality, all forms of media choose what materials to relay and what to ignore. They also decide when to report something quickly, and when to do so slowly. Through this type of practical and aesthetic screening process, some events or ideas get to be reported as news (or even as fact) while some do not. Often, the choices made reflect the aforementioned tendency to resort to media frames. For example, media reporting of terrorist attacks increased greatly during the decade following the 11 September 2001 attacks on New York, even though the incidence of terrorism in Western countries was much lower than during the preceding decade (Russell, 2010). On a perhaps more banal level, empirical studies have also shown that media coverage of climate change science varies in line with local weather conditions. When air temperatures rise, the number of newspaper stories concerning global warming also rises, even though global warming is not itself responsible for day-by-day fluctuations in temperature (Shanahan & Good, 2000). Such findings highlight the way one-off events (like terrorist attacks), or trivial ones (like the weather), can serve as triggers for the transmission of thematically associated information. Ultimately, the array of issues and ideas discussed in the media at a given time does not always reflect their relative importance or, indeed, their factual soundness.

Evidentiary reasoning and psychology

Psychology offers an insightful basis from which to consider the relationship between science and pseudoscience. Research in psychology highlights the sheer difficulty that lies in the task of distinguishing that which is evidence-based from that which is not. Contrary to centuries of historical assumption, man (or, rather, humankind) is far from the rational being that gives *Homo sapiens* its Latin name. Human reasoning is a skillset that evolved to be adaptive in highly particular environments, and so is beset with limitations that reflect the degree to which those environments have changed across historical time. Our reasoning faculties were honed in the same biological context in which buttercups evolved to be yellow, and in which peacocks evolved to have huge tails. These faculties are not pre-designed according to some logically refined template, but instead reflect accumulated mutations from much more primitive beginnings. While evolutionary processes have helped preserve the types of human reasoning that best perpetuate the species, these reasoning styles do not cover all eventualities with equal efficiency. In particular, when applied to contexts that exemplify our modern habitats (and so which were absent in our evolutionary environments) – features such as technology, abstract resources, symbolic systems, and even language itself – the limits on our reasoning methods become easily exposed. We have difficulty understanding randomness and probability, we perceive and remember things by constructive processes, and we take instinctive shortcuts when trying to make sense of evidence. In short, we muddle through. And yet, when we need to, we are capable of switching to a much more deliberative style of reasoning that allows us to solve highly complex problems.

As if these cognitive complexities were not enough, we are also subject to the limitations that come with being part of a social species. Some of the information we are told is shaped and readjusted by those who tell it to us, sometimes because the speaker wishes to look good in our eyes, sometimes because the speaker just doesn't know any better. We cannot really know whether we can truly rely on what we hear, and yet we attach unwarranted credibility to other people's utterances simply because they have made them. In doing so we offer false reassurance that we endorse what it is they have said. Through successive iterations of such unfortunate choreography, we can end up in a group – or even a society – that collectively purports to share a view that none of its individual members actually agree with. Layered upon all of this is our in-built bias to see things in self-flattering ways. We hold unrealistically optimistic views of ourselves, our place in society relative to others, and what other people think about us. And we rely on doing so in order to maintain our mental health: the tendency to develop accurate self-appraisals is actually characteristic of depression. It is as though we are hard-wired to think illogically for social reasons as well as cognitive ones.

And we grapple with these social and cognitive influences while trying to handle information that swirls through the vast quagmire that is our contemporary mass media. Human knowledge is compressed and modified and refined and filtered in countless unknowable ways before we even encounter it. While we live in an information age, we are largely unaware of the extent to which the huge infrastructure that supports information flow is itself involved in shaping and distorting that which flows in our direction.

In considering these matters, psychology is relevant for at least two reasons. Firstly, the subject area of psychology includes the study of cognition, social influence, and communication. Psychologists conduct research that helps shed light on how all these phenomena operate, and so should help us to understand the various limitations that result. Equally, by identifying limitations, psychological research should duly help to isolate the ways in which these shortfalls can be avoided or, at least, incorporated into our strategies for interpretation. By knowing what constitutes flawed reasoning, we can learn what constitutes effective reasoning too. Psychology, therefore, is the subject area that best assists us in making sense of ambiguous evidence.

The second relevance of psychology relates to the very fact that psychology, as a science, is itself a major producer of ambiguous evidence. If ever there was a science whose subject matter required careful and considered interpretation, then psychology is certainly it. Moreover, much of the subject matter of psychology is itself intertwined with our self-image, and so is very difficult for us to assess with real objectivity. As a result, the risk that our handling of evidence in psychology will be hampered by our proneness to social influences and our tendency toward self-aggrandizement is surely very high. This relates as much to the conclusions we draw from data, as it does to our assessment of the methods we use in our research. Similarly, the subject matter of psychology is of great interest to the mass media. While psychologists often pride themselves on the place of psychology within academia, it has to be noted that the majority of ordinary citizens acquire information about human thoughts, feelings, and behaviour from sources that are far from academic. The extent to which people garner such knowledge from mass media impacts directly on the extent to which they take psychology, and the advice of psychologists, seriously.

But most of all, the psychology of evidentiary reasoning helps to explain why so many non-scientific ideas persist in mainstream society despite the scientific revolution. Very often, when somebody wishes to defend a pseudoscientific idea or practice, they fall back on a simple claim. According to their plea, 'If so many people believe it, then it must be true'. However, as we have seen, there are significant grounds to be sceptical toward this line of defence. In reality, there are very many reasons why lots of people (and even a majority) may end up believing a claim that is quite far from true. Indeed, many ideas circulate within society and percolate through time despite being very obviously false. A good number of these false ideas concern the ways human beings experience the world, how they interact with each other, and how their well-being can be achieved. As such, they are of significant interest to psychologists. Some emanate from mainstream psychology itself, and are fostered by erroneous academics, self-deluding scientists, and poorly informed practitioners. However, many such ideas come from the margins of formal psychology, and are promoted by contributors who, in the main, are not psychologists at all. In seeking to consider cases of psychology-related pseudoscience in detail, we will turn to this category first.

Part II

Psychology and Pseudoscience in Practice

LEABHARLANN
CO. CILL DARA

Chapter 5

Examples from the Fringes: From Healing the Mind to Reading the Body

Introduction: Bad ideas do not necessarily fail to prosper

The psychology of evidentiary reasoning helps us to understand why ideas that don't work can become just as popular as ones that do. It also shows us how two contradictory ideas can become *equally* popular. Indeed, it is commonly the case that two contradictory ideas come to be believed by the one person. This should not be surprising. After all, if the popularity of a belief is determined by how it *feels* to the believer, rather than by whether it is sound, then contradiction is not going to be much of a barrier. Right across our culture, ideas that work and ones that don't are frequent bedfellows; science and pseudoscience live cheek by jowl. Even though physicists have a useful paradigm for understanding thermodynamics, thousands of people subscribe to the entirely contradictory possibility of perpetual motion. Despite the fact that astronomers have shown how, through parallax effects, celestial bodies move in paths that are different to what might be assumed from simply looking at the night sky from earth, millions still attach merit to claims that zodiac signs – which are entirely skewed by their earthly perspective – can be used to help predict future events affecting individuals born at particular moments. And while there is a scientific consensus regarding the biological basis of thought as an output of physiological brain activity, and that human brains decay after death, it is entirely usual for people to believe that we can communicate with deceased persons using séances, necromancy, prayer, or spiritual channelling.

One irony is that people who champion unscientific claims about the world, as well as those who advocate a view that science itself is nothing special, usually live lives utterly dependent on the technological fruits of scientific enterprise. Critics who warn that science is unreliable rely daily on their computers, televisions, and smartphones, not to mention their cars, their refrigerators, and their access to clean water. The fact that they are surrounded by signs that methodical empiricism regularly leads to very complex, subtle, and useful ideas is not immediately appreciated. Yet bad ideas can take hold very quickly. Simply put, bad ideas do not necessarily fail to prosper.

Talk of prosperous ideas inevitably brings us to psychology. Psychology is undeniably a target of mass fascination. The subject is one of the most popular at universities around the world, as well as at secondary (or high) schools in those countries where it is taught. In a typical year, over 60,000 full-time students take psychology at UK universities, with a further 30,000 taking the subject part-time. This is around three times the number of students taking biology, and around five times the number taking physics (Higher Education Statistics Agency, 2014). Many psychologists consider such uptake statistics to indicate psychology's robust health as an academic subject. However, as it is psychologists themselves whose research demonstrates that the popularity of an idea is a poor index of its logical value, perhaps they should approach psychology's own prosperity with a particular degree of caution.

If popularity is something of a warning flag, then it will be no surprise to learn that psychological subject matter attracts a high level of pseudoscientific attention. Many of the most frequently discussed pseudosciences concern themselves with the vagaries of human thoughts, feelings, and behaviour, as well as with mental health and social relationships. Several of them – such as the aforementioned astrology, as well as the many aspects of crystal therapy that were described in Chapter 2 – are very far removed from the content that makes up the 'ordinary' field of psychology as it is studied in universities. However, this is not always obvious to people outside those ivory towers. It might even be argued that many formally trained psychologists are quite receptive to pseudoscientific ideas, as though the boundary between science and pseudoscience were incidental rather than fundamental. Therefore, in this chapter we will look at some of the most important examples of psychology-related pseudoscience that exist at the fringes of the field.

We will consider four specific cases. Firstly we will examine the nature and extent of pseudoscientific therapies as they are recommended for people's psychological problems. From this we will consider the related area of how the human mind is frequently romanticized in a pseudoscientific manner, as if consciousness were some quantum phenomenon lying beyond the realm of ordinary nature. We will then look at ways in which the mind has been depicted as being capable of extraordinary feats, such as telepathy and psychokinesis. And finally, we will evaluate common claims that people's traits, abilities, and moral characters are correlated with the shape and appearance of their physical bodies, a worldview that encompasses a belief in the existence of fundamental race differences in psychological attributes. While these areas are each pseudoscientific, all have generated claims that have been believed, however tentatively, by very many people for very many years. For this reason, it is not only important that psychologists be familiar with them, but also that society at large appreciates their implications.

Complementary and alternative therapies for psychological problems

Intellectual brilliance in one domain does not guarantee competence in all. Take for example Steve Jobs, the entrepreneur who brought us the iPhone. Jobs was a renowned visionary credited with enhancing the lives of millions through his insight and invention. He was a pioneer of personal computing, and had a particular talent for identifying ways in which emerging technologies could be exploited as mass consumer products. In various leadership roles, he was responsible for significant enhancements of laptop computing, laser printing, mobile telecommunications, online music platforms, and even cinematic animation. In 2003, his extremely busy and varied career was interrupted when he received a diagnosis of pancreatic cancer. His form of the disease, an islet cell neuroendocrine tumour, would have offered excellent survival prospects had it been addressed using ordinary medical treatment straight away. However, after his tumour was discovered, Jobs at first chose to eschew mainstream scientific medicine and instead turned to a series of esoteric approaches that he discovered on the internet. His treatments included various forms of dietary restriction, bouts of colonic hydrotherapy, some acupuncture, visits to a psychic, herbal remedies, and many other unorthodox services. It is perhaps ironic that one of the world's greatest ever technologists turned to a genre of medicine that was so resoundingly bereft of any scientific or technological basis. His health declined rapidly over the succeeding months, and he eventually came to regret his decision to avoid mainstream medicine (Isaacson, 2011). After a poignant change of heart, he eventually underwent ordinary medical surgery, but only after his cancer had unnecessarily progressed. When he ultimately died from his disease in 2011, aged just 56, many cancer experts offered the view that his initial decision to choose alternative therapies had effectively cost him his life (Offit, 2013). Throughout his career, Steve Jobs had a reputation for being a demanding perfectionist. But when it came to his health, he seemed willing to rely on therapies that were far from perfect in empirical terms. In fact, the therapies he chose were each scientifically baseless.

When it comes to therapeutics, the terms *complementary* and *alternative* refer to those therapies 'not presently considered to be part of conventional medicine' (National Center for Complementary and Alternative Medicine, 2012). Therapies that fall outside the conventional mainstream do so for two reasons. Either they lack scientific plausibility (i.e., they are purported to work on the basis of a mode of action that is incompatible with what is scientifically known about nature) or they lack empirical evidence of efficacy (i.e., there are no rigorous studies to show that the treatments actually lead to a cure). Often both conditions are met. While such treatments are frequently directed at physical illnesses, they are also regularly recommended for mental health needs and psychological challenges, ranging from depression, anxiety, and pain to developmental disabilities and autism. In fact, the consumption of such therapies by persons with psychological needs is so common that some psychologists have argued for their inclusion as part of ordinary clinical psychology, and for psychology trainees to be schooled in their provision (Bassman & Uellendahl, 2003). However, a not insignificant barrier to such a proposal is the very fact that these treatments are fundamentally pseudoscientific, whereas psychology (in principle) is not.

Take, for example, the suggestion that sensory integration therapy (SIT) be considered suitable for use with persons with autism whose behaviour has become unmanageably challenging. SIT is based on a theory that autistic symptoms result from neurological abnormalities that interfere with the way sensory information is processed in the brain. Interventions are intended to retrain the brain to process sensory inputs in ways that are, essentially, less abnormal. This is achieved by assisting the person with autism to participate in a series of physical activities, such as jumping, swinging, and rocking. SIT also involves tactile treatments, like body massage, which are aimed at regularizing sensations. Some SIT interventions require the recipient to wear a specially designed vest with weights stitched into its seams, so that the person's body can feel constant sensory input in the form of weight pressure. Other interventions require the person to ride on a scooter board, so that they can practice the task of processing sensory input relating to physical balance. These approaches have become very common among certain professions, such as occupational therapists (National Board for Certification in Occupational Therapy, 2004). Psychologists working with autism will certainly encounter them in clinical contexts. However, there are two significant drawbacks with SIT. Firstly, the idea that there is a link between sensory integration and behavioural problems is entirely speculative. It has no basis in the biological sciences relating to sensation or to behaviour. The associated claim that SIT-type interventions are successful in 'reprogramming' the brain is equally without any empirical foundation. There is, in effect, no scientific reason to expect SIT to work in the manner intended. Of course this could simply be because scientists do not yet understand what is going on in a person's brain during SIT treatments. This brings us to the second drawback: there is no robust evidence to suggest that SIT interventions are actually effective in reducing challenging behaviours. While a plethora of studies have been conducted, the most methodologically rigorous ones appear to show no difference between administering an SIT intervention and simply leaving the child alone (Lang et al., 2012). The reputation of SIT may have grown in recent years, but it appears to be largely a subversive kind of reputation. SIT is used despite, rather than because of, the scientific consensus.

As an alternative example, consider Tai Chi. Tai Chi is a popular martial art that originated in China, and involves slow movement, meditation, and deep breathing. It is often recommended as a kind of relaxation therapy for stress, anxiety, depression, mood disturbance, and even low self-esteem (Wang et al., 2010). It is generally well accepted that the movements and discipline involved in Tai Chi exert positive effects on a person's mobility and co-ordination. It is also accepted that Tai Chi helps people to relax. There are other benefits too. The personal application required to master the complexities involved in being a proficient Tai Chi practitioner will likely help a person to attain positive feelings of achievement and success. Overall, Tai Chi is a pleasant hobby in which many people enjoy participating. However, you could pay similar compliments to any pleasant hobby that requires learning, practice, co-ordination, and self-discipline. Activities such as cycling, swimming, and playing the piano also aid mobility and co-ordination, and also help people feel a sense of competence when they do it well. So the question here is not so much whether Tai Chi helps people feel pleasure, but whether it does so in an especially therapeutic way that other forms of relaxation are unable to achieve. Unfortunately for Tai Chi fans, there is little or no evidence that it has anything unique to offer in terms of relaxation benefits. Studies that have compared Tai Chi to other forms of relaxation have shown that it is no more effective at reducing stress than embarking on a brisk walk (Jin, 1992). (With training in Tai Chi often taking several weeks and many classes under the supervision of a professional tutor, cynics might observe that it would be much easier, and cheaper, just to learn how to walk.)

A related question is whether Tai Chi succeeds in ameliorating the specific psychological problems for which it is often recommended, such as depression and anxiety. Again, there is little or no rigorous empirical evidence to suggest that this is the case, with the relevant studies often lacking even rudimentary placebo conditions (Hughes, 2008b). And perhaps the most vexed question of all is whether Tai Chi exerts its effects in the way that it has historically been purported to be doing. Tai Chi is not in and of itself a physical mobility or relaxation technique. Rather, Tai Chi arises from the Chinese philosophy of *taiji*. This is based on an assumption that our well-being is affected by the way we physically channel an invisible energy known as 'qi', said to circulate both within and without human bodies. Essentially, the graceful poses required by proficient Tai Chi are intended to align the practitioner's body so that it most efficiently intercepts this energy as it moves in alignment with the earth's polarity. Tai Chi is but one of a number of related therapeutic practices that are based on this notion of channelling, and, if necessary, unblocking, the vital energy 'qi'. Suffice to say, however, this energy is not in a form that mainstream science can either detect or explain. It is neither electricity nor magnetism, nor is it heat, light, radiation, or gravitation. As far as mainstream physics or biology is concerned, there is no such thing as 'qi'. In this regard, not only is Tai Chi inert as a treatment modality, it is also pseudoscientific at a conceptual level. And yet it is tremendously popular, not least among psychologists.

As a further example, consider chiropractic. In industrialized societies, many people visit chiropractors in search of a manipulation-based treatment for low back pain. This is not altogether surprising when you consider that chiropractors are specifically trained to focus on the shape and alignment of a patient's vertebrae, and to attempt to correct faulty alignments (known in chiropractic vocabulary as 'subluxations') by vigorously pressing down on the patient's back. What is less well known is that chiropractic is a form of holistic medicine intended to be of benefit to all aspects of human health and well-being, and not just for the treatment of sore backs. In chiropractic theory, different areas of the spine are responsible for maintaining health in different systems of the body; as such, virtually any health problem can be traced back to a subluxation, however slight, in the relevant vertebra. For example, according to chiropractors, the vertebrae numbered T5 to T10 (the six lowest in the spine's thoracic region) include those responsible for the liver, pancreas, and kidneys. As such, liver disease can be treated chiropractically, by using physical manipulation to achieve optimal alignment of these particular vertebrae. Likewise, other specific vertebrae are said to be implicated in psychological disorders such as depression and anxiety and so are used as the target of chiropractic therapy for these conditions. However, once again, such claims are biologically implausible. There is no empirical corroboration for the purported connection between the spinal column and conditions such as anxiety, depression, or for that matter liver disease. Furthermore, research on chiropractic treatment provides no convincing evidence of therapeutic efficacy for psychological conditions (Ernst & Canter, 2006).

A similar situation applies to two conceptually related techniques, craniosacral therapy and reflexology. Craniosacral therapy is akin to a specialized form of chiropractic that focuses only on the hard joints linking the cranium to the spine. Reflexology, as is popularly known, focuses on the sole of the foot. Both approaches endorse the assumption that the entirety of human health can be addressed by manipulating localized areas of the body, and both are commonly recommended for psychological challenges. Craniosacral therapy is often used to address particularly difficult problems, such as dyslexia, intellectual disability, and even criminality; whereas reflexology is typically recommended as a method of stress-reduction. As with chiropractic, both approaches are based on biologically unsupported models of health, and both lack a body of research that demonstrates their therapeutic efficacy (Ernst, Pittler, Stevinson, & White, 2006).

Each of the preceding examples – SIT, Tai Chi, and chiropractic-style manipulation – involves the idea that psychological outcomes can be achieved by simply changing the way the human body moves or behaves. Other complementary and alternative therapies are more invasive. Acupuncture is based on much the same notions as Tai Chi: thin metallic needles are used to pierce the skin at specified points of the body in order to redirect the flow of 'qi' so that health can be restored. Acupuncture is of particular interest to many psychologists because of its reputed efficacy in addressing problems of addiction, such as smoking. However, perhaps unsurprisingly given its conceptual premises, the empirical research on the usefulness of acupuncture for such conditions is disappointing to say the least (White, Rampes, Liu, Stead, & Campbell, 2011). The data are no more encouraging when it comes to acupuncture treatments for pain (Ernst, Lee, & Choi, 2011) or for depression (Wu, Yeung, Schnyer, Wang, & Mischoulon, 2012).

Another set of invasive therapies requires patients to consume substances orally. For example, several herbal remedies have been recommended for psychological problems. While naturally occurring substances drawn from plants such as St John's wort are known to contain chemicals with psychopharmacological effects, their specific effects on serious psychological conditions such as depression are unlikely to exceed those of placebos (Rapaport et al., 2011). Indeed, because these plants grow organically, the calibration of their chemical ingredients is uncontrolled, and so their ultimate psychological effects (and side-effects) are difficult to predict. The assertion that they constitute useful treatments for highly nuanced psychological symptoms greatly exceeds that which can be known in terms of their pharmacochemistry. Homoeopathic remedies, as discussed in Chapter 2, also involve the ingestion of specially prepared substances. However, unlike herbal remedies, homoeopathy is based on an assumption that therapeutic substances work better the more they are diluted. Indeed, it is nowadays reasonably well known that homoeopathic solutions are diluted far beyond what can be physically tolerated by any chemical. As a result, homoeopathic pills can be confidently said to contain no active ingredient at all. Should homoeopathic remedies for anxiety, shock, and the like ever actually work, it is almost certainly going to be due to a placebo effect (Shang et al., 2005).

These are just a small selection of complementary and alternative therapies relevant to psychological health. The exact number of such treatments is difficult to determine. In the UK, the Complementary and Natural Healthcare Council (2014) registers fifteen different therapies. However, as a voluntary register, its coverage is limited to those practitioners who wish to become involved, and it would appear that a large majority would prefer to remain unregistered. According to some sources, there are in fact many hundreds of complementary and alternative therapies, ranging from acupuncture and aromatherapy all the way to Zero Balancing and zàng-fǔ theory. In the US, the National Center for Complementary and Integrative Health (2015) is not a registration body, but a research one. As such, its remit is to scrutinize all forms of complementary and alternative healthcare rather than just a subset. In describing its activities, the NCCIH delineates two main categories of therapy, namely, 'natural products' (which includes herbal remedies) and 'mind-and-body practices' (which includes acupuncture, chiropractic, SIT, and Tai Chi). It also identifies a third group of 'other approaches' (in which homoeopathy is listed).

While these therapies are truly multitudinous and varied, they all have in common the two main defining features of complementary and alternative practices described above: they lack scientific plausibility; and empirical evidence of their efficacy remains forthcoming.

Therapies whose therapeutic efficacy is difficult to either explain or demonstrate might not be expected to be commercially viable, never mind globally popular. And yet, as noted in Chapter 2, the business of complementary and alternative healing has become a multibillion pound industry with a worldwide reach. As such, its relevance to psychology extends beyond its possible use to treat psychological problems. While some such therapies will elicit placebo effects (in which inert treatments initiate psychobiological processes that cause actual changes in pathology), in most cases consumers will falsely attribute outcomes to their chosen therapy, addled by the probability problems, cognitive heuristics, social communication biases, and mass media effects described in Chapter 4. In other words, these therapies offer an object lesson in how pseudoscience itself actually works.

If our symptoms appear to improve shortly after we undergo a therapy, our discomfort with coincidences might encourage us to infer that the therapy *caused* that change, however biologically implausible such an inference might be. In reality, there are many reasons why symptoms might improve after we receive a particular treatment. For one thing, we typically look for help when we feel at our worst, without fully appreciating that most of our symptoms (including pain and depression) are to some extent cyclical. This means that we often seek out treatments precisely at the moment just prior to a natural improvement in our well-being. Therefore, when we experience such an improvement, we should not assume it was caused by the treatment; there is a statistically favourable chance it will have resulted from a natural symptom fluctuation. The next problem is that cognitive shortcuts can then coax us into attaching unwarranted credit to unusual therapies. For example, if we undergo acupuncture and take a prescribed anti-depressant drug, the availability heuristic might lead us to attribute any improvements in our mood to the more vividly memorable of the two treatments. Our tendency to view ourselves in a positive light might reinforce this conclusion: we are likely to feel that actions based on their own decisions (such as choosing an alternative therapy) are more effective than actions based on the advice of other people (such as medical professionals). Afterwards, these biased convictions will affect the way we tell other people about our experiences. And our testimony will transmit well through social and media networks due to the fact it fits several popularly accepted narratives (such as the idea that there is always hope, even when mainstream medicine tells you otherwise). In summary, even if a popular consensus emerges in support of a particular scientifically controversial therapy, there is very little reason to assume that the therapy must be right and that science must be wrong.

Occasionally defenders of complementary and alternative therapies point out that all the above applies to *mainstream* treatments too. Even if a given therapeutic approach, such as cognitive behaviour therapy, has a positive empirical record in research trials, it does not prevent a particular CBT patient from drawing false conclusions as to the causes of their symptom improvements. This is of course true. It is also true that many mainstream activities in everyday clinical psychology are compromised by practicalities and fall short of the standardization expected of genuine rigour. In sum, much of what goes on in 'mainstream' psychological therapy is pursued without a compelling base of evidence to support it. However, this does not mean that complementary and alternative practices are just the same as mainstream ones. There is in fact a critical difference: primarily, mainstream therapies are *not* scientifically implausible. Mainstream psychological therapies do not invoke modes of action that scientists cannot explain, but rather are derived from a conviction that human thoughts, feelings, and behaviours are amenable to empirical scrutiny and, in time, will prove to be eminently explicable. There is no reliance on mystery life forces or on models of well-being that are divorced from mainstream conceptualizations of the human mind and body. Complementary therapies, on the other hand, are dependent on a therapist's ability to ignore the fact that their models of well-being are idiosyncratic and uncorroborated. The therapist must also ignore the fact that efficacy studies show their techniques to be inert. The shortfall between theory and implementation in clinical psychotherapeutics results from the practicalities of replicating scientific study protocols in real-life contexts; the shortfall in complementary therapeutics result mostly from a fundamental rejection of science itself.

This latter point makes the problem of complementary therapies in psychology one of ethics as much as one of empiricism. Whether intentionally or not, proponents of such therapies often end up advocating a form of epistemological double standard (Hughes, 2015). They are quite happy to cite research studies where they exist to support a treatment; but where no such evidence is available, a given therapy can be classified as 'complementary and alternative' and simply used anyway. Seldom do practitioners point out (or realize) the central inconsistency of such an approach. For one thing, as discussed in Chapter 3, virtually all codes of ethics governing psychological practice call for scientific rigour in the psychologist's work. It is hard to see how rigour can be reconciled with a take-it-or-leave-it attitude to basic epistemology. Furthermore, these same codes of ethics require psychologists to obtain informed consent from clients before proceeding with an intervention. As such, when discussing complementary and alternative therapy options, psychologists should be sure to point out that such treatments are based on assumptions that contradict our scientific understanding of nature itself, and, in any event, have no data to support their use. Psychologists do not have the luxury of simply ignoring the fact that a suggested therapy is pseudoscientific.

Consciousness on a quantum pedestal

As described in Chapter 3, it is sometimes asserted that the human mind is qualitatively unlike any other entity studied by science. Some see the mind as sacred, profound, and inherently mysterious, others see it as irredeemably complex. Many view it as unfettered by ordinary notions of determinism: the mind is said to possess free will, in the sense that it can decide for itself what to think next. This, it is argued, is why human behaviour is ultimately unpredictable (and thus why psychology itself is ultimately unscientific). However, such arguments are essentially cosmetic. Putting the human mind beyond the reach of natural science is equivalent to arguing that it occupies a realm that is, literally, *super*natural. It does not explain why human minds are different from other things; it simply declares the subject to be out of scientific bounds. While this is consistent with our dispositional tendency for self-aggrandisement, it might also be construed as a form of special pleading. After all, other animals have minds; are they supernatural too? And if the human mind is truly autonomous, and thereby unpredictable, does this not logically imply that its behaviour must be random? Human minds that conform to *non*-random – that is to say, systematic – patterns of thought are hardly truly 'free': sufficient knowledge of such patterns would allow us to predict a person's thoughts in advance. Indeed, such knowledge would even allow us to *control* a person's thoughts, if we could control the causal factors that precipitate them. So for free will to exist in the human mind, it requires a separation between systematically knowable causal factors and subsequent action. In other words, it requires randomness. And as we concluded in Chapter 3, proponents of free will rarely accept that human behaviour is random, which is akin to saying that free will isn't actually 'free' at all. If a person's will is connected to context, causes, and conscience, then it is 'bound', not 'free'. In the end, it seems more likely that free will is an illusion rather than a reality: we simply *feel* as though we have free will when in fact we (probably) don't.

While it might be difficult to sustain a claim that the human mind is autonomous, it is much easier to agree that its dimensions are complex. While most areas of psychology have approached this complexity as an inherent and inevitable challenge, some have sought ways around having to deal with it. Most notably, the behaviourist school has long argued that 'private mental events' should be seen as lying outside the scope of psychology. While these psychologists are often accused of talking as though the conscious human mind doesn't even exist – British psychologist Cyril Burt (1962) once derided behaviourism for having 'lost all consciousness' – it is probably more true to say that behaviourists just don't consider notions like *mind* and *consciousness* to be particularly useful. (In fact, contemporary behaviourists often make mental events their subject matter, considering them to be discrete behaviours in their own right.)

That said, some psychologists *do* cast doubt on the very existence of the human mind, at least in the popular sense of the term. For example, the American cognitive scientist Daniel C. Dennett (1991) proposes a theory that there is no such thing as a person's unitary consciousness, and that the consciousness we *think* we have is the end outcome of several different processes of cognition in the brain. His theory of consciousness is famous for its view that there is no distinction to be made between the various sensory inputs we handle at a cognitive level and the integrated stream of subjective experience that we commonly refer to as consciousness. In other words, the whole is not any greater than the sum of the parts. Dennett argues that our thinking occurs not as some kind of monologue-based serial sequence, but in multiple parallel semi-independent strands drawing simultaneously on various different parts of the brain. In this theory, our feeling of having a single integrated 'mind' is just an illusory by-product of complex information processing, as is our sense of personal subjectivity.

One source of evidence in support of this proposition comes from experimental studies of patients who have undergone a surgical procedure known as corpus callosotomy. This operation is usually intended to minimize the impact of epileptic seizures. It involves severing the neural fibres that connect the two hemispheres of the patient's brain. A side effect of the surgery is that it prevents the two brain hemispheres from communicating with each other, but in most situations this is not debilitating. Since the 1960s, studies of split-brain patients have yielded a number of fascinating findings about the way humans construct unified experiences from fragmented sensory input. For example, in one landmark experiment (Gazzaniga, 1967), split-brain patients were shown a strip of flashing lights that were visible in both their left and right visual fields. In this set-up, the way visual perception is routed through the brain would mean that lights presented to a person's left will be perceived by neurons in their right brain hemisphere, while lights presented to their right will be perceived by neurons in their left brain hemisphere. When the split-brain patients were asked to point out the lights with their fingers, they pointed at all the lights that were presented (i.e., those on the left *and* those on the right), revealing that both hemispheres of the brain had been successful in detecting them. However, when the patients were asked to verbally report what they had seen, they could only report seeing lights coming from their right-hand side. They did not refer to lights coming from their left, and as far as they were able to report verbally, they simply *had not* seen them. Unlike finger-pointing, which involves both hemispheres of the brain, verbal reporting involves just the left hemisphere. Therefore, the patients' verbal reports referred only to what was seen by that hemisphere. Essentially, the right hemisphere was unable to communicate with the relevant speech area to report that it had seen lights as well – however, it *was* able to communicate with the fingers in order to get them to point. As far as the patients were concerned, they had maintained a single unitary experience of what had taken place. They were oblivious to the fact that they had essentially undergone two separate experiences, and had reported on them separately to the experimenters. It is often said that the surgery had left these patients with *two* brains rather than one. In the present context, we could say it left them with two *minds* as well.

Such experiments highlight the fact that consciousness is heavily dependent on biological underpinnings. They also show us that far from being central to our identity and existence, our minds might simply be something of a coincidental after-effect, a way of constructing a sense of personal reality from a plethora of biologically necessary, but essentially dispensable, perceptual processes. From an evolutionary standpoint, this ability to maintain a single account of our personal activities undoubtedly assists us in managing our lives. However, it does not necessarily reflect a single conscious mind in which all our thoughts and experiences come together, an internal venue where our psychological lives are lived. Rather than constituting a kind of independent mental version of our self that resides inside our brain, what we believe to be our conscious mind might simply be a continuously updated progress-report summarizing – very approximately – our latest cognitions.

This conclusion is far removed from the traditional view of human minds as holding a precious place in the realm of human existence, if not indeed the universe itself. Far from relegating human consciousness to a kind of mundane side-effect of simply being alive, many commentators have promoted the idea that the human mind is virtually all-powerful in its capacities. One common strand of this approach has been to allege that human minds operate according to the principles of quantum mechanics. For example, the bestselling Indian-American writer Deepak Chopra has attained a worldwide following for his holistic model of health and well-being, where he uses quantum physics as a paradigm within which to discuss the human mind. Two of his most famous books are *Quantum Healing: Exploring the Frontiers of Mind/Body Medicine* (1989) and *Ageless Body, Timeless Mind: The Quantum Alternative to Growing Old* (1993). According to Chopra, the quantum nature of human consciousness can be harnessed in ways that enable people to prevent biological ageing. It can also be used to re-programme their behaviours so that they can accumulate vast wealth. His theories make much use of the quantum notion of *superposition*, which describes how a physical entity (for example, an electron) can exist in all its possible states at once. When applied to the human mind, it is possible (at least metaphorically) to argue that human consciousness can achieve any potential outcome, given the right conditions.

Chopra is not the only proponent of what has become known as *quantum consciousness*. Several other authors have argued that the human mind behaves not deterministically, but in ways that reflect its unique complexity and nature. Such allusions to quantum phenomena refer to how, in the early 20th century, physicists discovered that objects at the nanoscopic scale appeared to behave differently to those at larger scales, in a way that the principles of traditional Newtonian physics were unable to explain. This need to accommodate new findings about nanoscopically sized entities (such as photons and electrons) led to the development of quantum mechanics. Essentially the point being made is that human minds might fall into a similar category. Their unique forms of activity might not conform to that of other phenomena in nature, and so may require a new *type* of science (or, at least, a new type of mechanics) to explain. In its broadest form, such a position treats the quantum nature of consciousness as a metaphor: the mind is said to occupy the same epistemological context as a nanoscopic particle. However, few quantum consciousness proponents limit themselves to metaphorical speech. More often, their arguments espouse a view that quantum approaches are needed to explain how human brains, minds, and consciousness really work because the actual physical events that take place inside the brain, and drive psychological outcomes, do so in a manner that is *literally* subject to quantum principles.

The views of British theoretical physicist Roger Penrose and colleagues exemplify this approach. They base their arguments on the premise that human consciousness operates in ways that traditional scientific models cannot account for, but which can be explained using the 'new' physics represented by quantum mechanics. For example, they argue that human consciousness is the result of quantum gravity effects that occur inside the microtubules of brain cells, where particular types of electrons become subject to an effect known as quantum entanglement (Penrose, 1997). Quantum entanglement occurs when seemingly independent particles are found to affect each other in unexpected ways, such as when the act of measuring the location of one appears to determine the movement of the other (Albert Einstein famously nicknamed the effect *spukhafte Fernwirkungen* or 'spooky action at a distance'; Mackay, 1991). In broad terms, if quantum entanglement occurs in brain cells, then this would allow the brain to sustain processes that would not be expected to conform to traditional cause-and-effect sequences. Penrose and his collaborators have offered a number of theoretical papers that attempt to mathematically model the activity of brain cells in order to demonstrate the feasibility of quantum effects. However, their work has received much criticism from across the scientific and broader academic community. Some cognitive scientists have suggested that the basic assumptions upon which Penrose's ideas are formed are themselves based on misunderstandings of how cognition occurs (Minsky, 1991), while philosophers of science have questioned Penrose's interpretation of the computational basis to logic (Putnam, 1995) and quantum physicists have been critical of the way Penrose's theories deal with the speed of neural activity (Tegmark, 2000).

Nonetheless, the notion of quantum consciousness has become widely popular as a framework for the development of self-help manuals and other therapeutic services. One of the most famous self-help manuals in recent years, Rhonda Byrne's (2006) *The Secret*, makes extensive reference to quantum physics in order to explain how the human mind is possessed of unlimited potential for success. This book spent 150 weeks at the top of the *New York Times* bestseller list and is estimated to have generated nearly £200 million in sales around the world. Byrne's two subsequent volumes, *The Power* (2010) and *The Magic* (2012) have also been runaway bestsellers, and a spin-off movie has made over £35 million in DVD sales. Despite the fact that Byrne's linking of quantum physics to personal life success has been widely parodied and criticized, it is apparent that there remains an enthusiastic popular audience for her ideas.

Invoking a basis in quantum mechanics allows for some dramatic claims to be made about the human mind. It also facilitates a related argument about the merits of science and scientists. According to many proponents of quantum ideas about the mind, the emergence of quantum mechanics not only marked a development of a new explanatory paradigm in physics, it also marked the obsolescence of all paradigms that preceded it. It is suggested that psychologists who persist in applying standard scientific methodologies to their subject matter are pursuing a doomed enterprise. Quantum physics, it is said, has demonstrated that the universe is not deterministic after all, and so psychology is adhering to out-of-date standards when it seeks to establish cause-and-effect relationships in its exploration of human behaviour (Davies, 2004). In fact, such critics consider quaint psychology's usual view that linear systems of cause-and-effect take place in forward-moving time, because 'mainstream science has evolved considerably and is no longer based on classical physics' (Roll & Williams, 2010, p. 16).

Within this narrative, attention is often drawn to the 'uncertainty principle', an idea put forward by German physicist Werner Heisenberg (1927). Broadly speaking, the uncertainty principle refers to the limit on precision faced by physicists who wish to measure different physical properties of a particle at the same time. It arose from the observation that it was impossible to measure the position of a subatomic particle without losing track of its momentum, and vice versa. Therefore, the more certain we are of one measure, the less certain we are of the other. But while the uncertainty principle describes a serious challenge facing quantum physicists, its extrapolation to suggest that the *profession* of physics is hereby itself afflicted by a malaise of 'uncertainty' – wherein its traditional methods have lost their ability to provide reliable information – is specious (Gross & Levitt, 1994). The principle was formulated to help describe electrons and photons, and its implications relate to such nanoscopic physical matters. To take a description of subatomic particles and apply it to the everyday academic attitudes of professional scientists amounts to rhetoric, and draws the connotation of the term 'uncertainty' far away from its original context. It is no different to alleging that the development of a new theory of gravity (taking 'gravity' to be a force that attracts objects to the centre of the earth) somehow makes physicists think more earnestly about their jobs (by simultaneously referring to 'gravity' in its *other* sense of importance or seriousness). Using a principle called 'uncertainty' in this way to cast doubt on the soundness of the mainstream empirical tradition that spawned it reveals, at best, a case of scientific illiteracy, or, at worst, a deliberate strategy of obfuscation (Stenger, 1992).

It is worth remembering that, despite its widespread promulgation, the suggestion that human consciousness is a quantum phenomenon enjoys no direct empirical support. The closest that proponents (such as Penrose) have come has been to argue for its *plausibility* based on some tenuous mathematical modelling. And, in the main, these arguments have been faced with strong rebuttals and with arguments for its *im*plausibility. The idea that human consciousness is somehow special, and that special methods are required to study it, more likely reflects our sentimental bias towards illusory superiority: we are instinctively seduced by the suggestion that our minds are special in ways that place them above ordinary natural things. In any event, just because we find consciousness difficult to understand does not necessarily mean that quantum physics must be the answer. Such reasoning is little more than a New Age manifestation of Bertrand Russell's argument from ignorance (and anyhow, references to quantum dimensions more often appear contrived to add, rather than remove, confusion). In summary, claims as to the special place of human consciousness in nature are tenuous. They rely on non-falsifiable assertions, in that they disregard the assumption of determinism. They respond to scientific ignorance not by calling for an intensification of research effort, but by advocating the abandonment of empirical certainty. In the same manner, they cast aspersions on parsimony by championing maximum complexity. And quite often, they use obscure jargon and misappropriated scientific references to defend assertions about why we human beings are in fact *special*. These claims of exceptionalism, notwithstanding their popular appeal, are squarely pseudoscientific.

The normalizing of paranormality

One of the ways in which the human mind has been said to be particularly special is its alleged capacity to interact with the ordinary world through extraordinary channels. While the ordinary depiction is of a mind that receives and sends information using the recognized physical senses, the extraordinary depiction allows it to do so while bypassing those physical senses in some (as yet unexplained) manner. Such so-called extrasensory perception (or ESP) can involve the ability to know another person's thoughts (telepathy), the ability to detect distant objects without sensing them (clairvoyance), or even the ability to perceive events before they actually happen (precognition). More broadly, such perception can include the ability to detect and interact with deceased persons by communicating with a purported spirit world (mediumship). A further extension of the notion includes the idea that some information arrives into the mind as a result of the person having lived a previous biological life, or even many such lives (reincarnation). As well as receiving sensory input that bypasses ordinary sensation, the mind is also said to be capable of exerting physical influence upon the world without true physical interaction. For example, the mind is said to be able to move physical objects without touching them (psycho-kinesis), or to aid a person in elevating their bodies in defiance of gravity (levitation). Other people are said to be able to cure illnesses by manipulat-ing the organs inside a sick person's body without touching them (psychic surgery), or by communicating thoughts from an entirely remote location (distance healing). All these purported abilities of the mind to achieve sensory or physical functions without requiring direct sensory or physi-cal contact have collectively been referred to as 'psi phenomena', 'psychic' abilities, or 'paranormal' phenomena, and their study has been referred to as 'parapsychology'.

As mentioned in Chapter 3, public interest in psychic possibilities was intense in the early days of academic psychology, which in its modern form emerged as the nineteenth century was drawing to a close. At that time, many people wanted to see science used to establish an empirical basis for mysticism. This reflected a broader appreciation that there may be merit in non-religious explanations for human existence. New technologies like the telegraph and the X-ray hinted at the existence of otherwise invisible reali-ties, and made communication from a distance seem technically feasible (Coon, 1992). In Europe, but especially in North America, generational divisions arose wherein young people saw modernity as fashionable while older people put greater store in superstition. Some historians point to the particular impact of the American Civil War (1861–1865). This highly destructive conflict (which killed more Americans than all other wars, before or since, combined) helped spawn a new cohort of citizens who yearned to look to the future and to dispense with the norms of the past. Because of the extreme death toll, virtually all American families mourned the premature loss of multiple relatives and friends. This in turn led to the growth of a vibrant commercial séance circuit across the United States, from which the burgeoning of opportunities for professional psychics saw the rise of a number of celebrity mediums. It was against this heady back-drop that the first psychology departments opened in the American univer-sity system.

Many of those departments involved themselves, at least peripherally, in the study of the paranormal. William James (1842–1910), Harvard profes-sor and author of one of the first formal psychology textbooks, maintained a long-held interest in spiritualism, and discussed with his philosopher col-league Thomas Davison the possibility of using scientific evidence of an afterlife as the basis for establishing a new secular religion (Coon, 1992).

James wrote prolifically on many areas of psychology. Within this was included a large strand of work concerning such matters as telepathy, clairvoyance, mediumship, and psychokinesis (cf., James, 1986). Not all of James's contemporaries were comfortable with his obvious enthusiasm for such subject matter. Given his worldwide fame as a psychologist, they felt his approach threatened to undermine the standing of what was still a fledgling discipline in academia (e.g., Jastrow, 1889). Over time, psychologists' attention switched from the paranormal dimensions of psi phenomena to their assessment using rigorous scientific methods. Gradually, the scientific principle of parsimony influenced psychology to dispense with theories that invoked unproven forces as explanations for psi abilities.

The first half of the 20th century saw much research into various forms of psi phenomena. Specialized laboratories and research centres were established, including at Stanford University (1911) and Duke University (1930), and later at the University of Edinburgh (1962). Many of the earliest studies seemed to corroborate claims that psi abilities did indeed exist. However, the methods used were not always foolproof. Typical studies involved straightforward observation, with little attempt to control for either threats to validity or the possibility of fraud. When studying mediumship, for example, researchers would simply attend séances and take notes, and so were poorly equipped to detect the chicanery that was truly widespread at the time. Gradually, methods developed to include a variety of experimental controls, and to adopt a more sceptical approach when studying people who claimed to have psychic abilities (O'Keeffe & Wiseman, 2005).

One example related to the problem of 'sensory leakage'. The realization that alleged mediums or telepaths might be able to make accurate inferences from a bystander's appearance or body language, or indeed from the speed with which they responded to questions, led researchers to focus on the need to prevent psychics from receiving verbal or nonverbal cues. They developed standard protocols for such evaluations, which included the requirement that mediums or telepaths be placed alone in sound-attenuated cubicles during tests.

A second example concerns the way we interpret claims made by others about us. When attending public performances of purportedly psychic persons, audiences can sometimes attach unwarranted specific meanings to statements that are in fact ambiguous and broad-sweeping. Many general descriptive statements – such as 'You are often critical of yourself' or 'You pride yourself as an independent thinker' – can strike us as personally relevant, especially when we trust the source and are expecting their insights to be accurate. This tendency to find personal relevance in generality is often called the 'Barnum effect', after Phineas T. Barnum (1810–1891), the famous 19th-century American showman who made his name promoting hoaxes. The Barnum effect is frequently cited as the reason why magazine horoscopes are successful: they can so often seem intriguing. For example, at the time of writing, a horoscope published by a national newspaper advises me not to become too worried about pressure from others to be serious, and to allow my inner playfulness to serve as a counterbalance. While I may find such advice sensible and, in its own way, relevant to my life right now, I realize that it is so nonspecific as to be applicable to just about anyone at just about any time. But my realization is informed by prior awareness of the problem. Unless readers have grounds to be suspicious, they may be open to the suggestion that a particular horoscope was written with them specifically in mind. Perhaps the most famous demonstration of this point was offered by American psychologist Bertram R. Forer. He presented his students with individual horoscope readings and asked them to rate the readings for accuracy. Only after the students had provided an average accuracy rating of 4.26 out of 5 were they informed that the readings were not individualized at all: in fact, the students had all been given the *same* generically worded description (Forer, 1949). The fact that magazines still carry horoscopes today, over six decades after Forer's original finding was published, helps to illustrate the power of the effect.

While the Barnum effect relates to statements that are wholly generic, even relatively specific claims (such as 'You have a scar on your left knee' or 'Someone in your family is called "Jack"') will turn out to be true of large numbers of people, simply on statistical grounds (O'Keeffe & Wiseman, 2005). Overall, it is relatively easy for a person offering psychic readings to appear convincing much of the time. A person sufficiently gifted at exploiting these trends can convince many an audience that they have true telepathic powers or can really speak to the spirit world. Accordingly, scientific scrutiny of such claims requires more than simple observation or hearsay evidence. For example, in empirical tests, researchers often present the purportedly psychic readings alongside a selection of decoy readings, and ask the person about whom the readings are being offered to rate each one for accuracy. If the psychic person is truly psychic, then his or her readings should attract greater accuracy ratings over time. Suffice to say, this tends not to happen.

The early studies of psi phenomena often took place under the formal auspices of university psychology departments. However, research on such subjects has evolved over time to become a niche interest at the fringes of psychology, one with very little interaction with other areas. Today, whether sympathetic toward psi abilities or not, parapsychologists are no longer considered to be working in the academic mainstream. Most parapsychology research is sequestered into self-contained conference circuits and narrow portfolios of obscure academic journals. Insofar as scientific peer review is common, it typically involves parapsychologists reviewing one another's output. This seems to be the case at both ends of the belief spectrum: believers interact with other believers, review their work, champion their conclusions, and only rarely publish papers in orthodox psychology journals; sceptics follow a largely corresponding pattern.

Overall, the research into psi phenomena has shown resoundingly that no such phenomena really exist. Studies of telepathy suggest that people are not really able to read one another's thoughts (Hines, 2003). Studies of clairvoyance show that people cannot detect objects or events removed in space or time (Milton & Wiseman, 1999). Studies of precognition have found no evidence of an ability to see into the future (Wagenmakers, Wetzels, Borsboom, & van der Maas, 2011). There is no empirical evidence of a spirit world or of reincarnation (O'Keeffe & Wiseman, 2005). Research into psychokinesis fails to support claims that people can move objects by the power of their minds (Bösch, Steinkamp, & Boller, 2006). And psychic surgeons and distance healers are unable to cure anybody (Masters, Spielmans, & Goodson, 2006).

Of course, while the preponderance of rigorous research points consistently to such conclusions, there remain many advocates willing to argue that these investigations are flawed or biased, or who choose to point to that minority of studies that have produced contrary or inconclusive findings. (Sometimes the conventions of scientific practice appear ill-equipped to balance the record: as described in Chapter 3, a prominent psychology journal once declined to publish a study that failed to corroborate a previous precognition finding on the grounds that it 'doesn't publish replications'.) Other proponents fall back on the claim that psi phenomena reflect aspects of the human mind that lie beyond the scope of scientific scrutiny. Perhaps predictably, many cite quantum physics as something of an all-powerful explanatory trope (e.g., Roll & Williams, 2010). However, claims that psi abilities exist are pseudoscientific in several respects: key terms are ambiguous and loosely described, thereby undermining falsifiability; ignorance is customarily treated as evidence of possibility; parsimony is discarded in favour of hypercomplexity (such as with the citing of quantum mechanics); peer review is incestuous; and anecdotal evidence is highly valued (Alcock, 2010).

This prevarication might help explain why popular belief in psi phenomena remains strong in wider society, at least insofar as can be judged from the popularity of media psychics, telephone clairvoyants, and magazine horoscopes. Occasionally, the belief in psi phenomena can rise to influence the most serious contexts. People claiming to be psychics have been recruited by detectives to aid the search for missing children and aircraft (Hickman, 2011). It has been reported that military researchers – in major industrialized nations – have attempted to harness the powers of clairvoyants in order to gather intelligence (Ronson, 2004). And horoscope readings have been used to inform decision-making by a number of prominent figures, including football managers (Gilmour, 2008), businesspeople (Lennard, 2012), and Presidents of the United States (Wheen, 2004). While certainly illustrating the extent to which psi phenomena are accepted by all sorts of people, it is notable that these real-world attempts to exploit their benefits have uniformly failed. In other words, their track record in practical contexts matches that seen in the laboratory.

Physiognomy – The good, bad, and ugly

Sometimes we do not need telepathy to tell from a distance what other people are thinking. We can just look at them. People's faces, and even their postures, are very responsive to mood, and human observers are highly adept at reading nonverbal cues. If our friends are worried or preoccupied, we can often tell so from looking at their faces. We can even tell if they are trying to *hide* the fact that they are worried or preoccupied. When they are happy, we can form a reasonable assessment of precisely how happy they are – whether they are thrilled, satisfied, or merely content – from looking at their eyes. Beyond the realm of emotion, we are pretty good at detecting whether somebody is genuinely concentrating on us when we are talking to them, or whether they are thinking about something else. On a more basic biological level, we can judge by looking at people whether they are tired, and sometimes whether they are hungry. The habit of staring at other people and studying their appearances is deeply ingrained in our behaviour: as babies we are particularly distracted by faces from when we are just 30 minutes old (Pinker, 1997). It is little surprise that by adulthood, we feel extremely confident we can form all sorts of useful opinions about people's moods, concentration levels, and physiological well-being, by simply scrutinizing their faces and bodies.

It is sometimes argued that we can infer much more than just their mood, concentration levels, and physiological well-being. A plethora of websites claim that we can also make judgements about their personality type, intelligence, and even their moral character. According to these sources, it is a systematic matter to tell whether people are conscientious, rebellious, or good at listening simply by classifying the shapes of their noses, ears, or chins. The cover of one relevant book tells us that we can conclude a person 'has an active mind' if he or she has a high forehead, and a 'cautious' nature if he or she has small nostrils (Webster, 2012). Smartphone owners can download several different apps and have their own faces analysed. According to one such app, my own nose shape reveals me to have a high level of energy, my forehead indicates that I am good at spotting danger, my eyebrows suggest that I am an innovative thinker, and my chin indicates that I am eager to indulge in many different pursuits (Innoplexus, 2013). While some such readings are offered merely for interest, and maybe even for entertainment, some are presented in the context of profound implications. For example, one popular self-help manual suggests that hirsutism indicates a tendency to blame others, arthritic fingers represent a desire to punish people, and round shoulders tell us that a person feels helpless – the advice then offered is that these physical characteristics (or illnesses) can be attenuated by altering the associated thought patterns (Hay, 1984).

The idea that human physical appearance can reveal psychological character is thousands of years old. The formal practice of reading faces dates back to the palaeo-Babylonian period in Mesopotamia, and ancient Greek philosophers such as Aristotle frequently wrote about the links between countenance and character (Jenkinson, 1997). In scientific contexts, the biopsychological approach of Descartes helped to link brain activity to behaviour, and from there it was an incremental generalization to conclude that there may be scientific merit to claims that physical characteristics might truly reflect psychological ones. In the 18th century, Swiss pastor Johann Kaspar Lavater (1741–1801) developed a detailed theory of physiognomy, which he felt represented a more scientific approach than the popular superstitions that had been handed down anecdotally up to that point. Lavater structured human nature around three major aspects: the intellectual (which was marked by different features of the head); the moral (epitomized by the chest, heart, and facial expressions); and the instinctive or animalistic (as embodied by the stomach and the 'organs of generation'). One of his most influential principles was that the psychological features of human character were just as hereditary as physical traits, such as eye colour, height, or build (Collins, 1999). While aiming to be scientific, Laveter was nonetheless deferential to his pastoral background. He ended up focusing his system on the religious significance of human nature and the physical features that indicated people's morality. His extensive written works were published in four volumes in the late 1700s (Lavater, 1840 [orig. 1775–1778]), and went on to be extremely widely read, after his death, well into the 19th century.

Spurred on by such thinking, the late 18th-century German physician, Franz Josef Gall (1758–1828) developed a number of similar, but nonetheless contrasting, theories. Based on recollections of his fellow students at medical school, he concluded that protuberant eyes were a sign of intelligence. His examination of a female patient with a highly active sex drive led him to conclude that having thick flesh at the base of the skull and upper neck was associated with amorousness. He noticed that many of the local boys he had hired to run errands had bulges in the skin above their ears; as these boys were from socioeconomically underprivileged backgrounds, he considered the bulges to be markers of their (undisputed) potential for criminality. Unlike Lavater, Gall sought to corroborate his judgements using empirical observation, and so went about systematically examining patients in order to link such features with psychological traits. Gall focused his attention almost exclusively on the surface of the human head, guided by an assumption that the underlying brain was the organ of primary interest. On this basis he claimed to have identified dozens of different parts of the head that could be used to determine the size of the brain areas relevant to several distinct psychological characteristics. Gall's was as much a theory of brain activity as it was of diagnosis: by attributing different roles to different parts of the brain, Gall was one of the first theorists to posit the idea of modularization of brain function. Perhaps for this reason, his followers began to refer to the practice as 'phrenology', after the Greek term for 'study of the mind' (Gall himself had emphasized the methodological aspects in preferring the term 'cranioscopy'). Under this title, the theory attracted public fascination, and even today ceramic busts depicting the layout of the various different phrenological brain areas are widely available (albeit primarily as ornaments).

However, Gall's phrenology provides us with a very good example of confirmation bias. Even though Gall felt he had gathered plentiful evidence in support of his theory, anatomists were later to establish very clearly that phrenology could not possibly be correct. This is because, contrary to what Gall believed at the time, the shape of the human skull is unrelated to the shape of the brain that underlies it. As such, conclusions about the sizes of different brain areas cannot be drawn from examining the outer contours of a person's head. Despite this, Gall and his followers had become convinced that their theory had merit, mainly because they succeeded in finding so many cases that supported their beliefs. They found many thieves whose head shapes conformed with what they had predicted for persons who were high in 'acquisitiveness', and many belligerent people whose heads suggested a high level of 'combatitiveness'. They found noble intellectuals who had bulges in that part of the head associated with 'ideality', overweight people with bumps indicating 'alimentiveness', and charming people whose head indentations suggested 'suavity'. The phrenologists' biggest problem was that they concentrated almost exclusively on documenting confirmatory cases. They overlooked (or ignored) people who had various bumps and contours but who did *not* exhibit the behaviours associated with them. In addition, their definitions lacked specificity: what exactly was meant by terms such as 'ideality', 'suavity', 'acquisitiveness' and so on, was not always clear. By succumbing to confirmation bias and by relying on vague terminology (today more readily recognized as signs of a pseudoscience), the phrenologists became convinced of an elaborate and seductive theory that simply wasn't true.

While psychologists often focus on the historical relevance of phrenology, it is wrong to suggest that Gall's theories led people to focus exclusively on the brain as the centre of behaviour. Various theories linking other parts of the body to behavioural and intellectual characteristics continued to exert influence on popular beliefs. The prominent Italian criminologist Cesare Lombroso (1835–1909) put forward a well-regarded theory of the nuances of criminality, which he believed were discernible from various physical defects of a criminal's body. For example, he argued that burglars had different shaped noses to, say, murderers (Gould, 1996). His underlying reasoning was that criminal behaviour constituted an abnormality of the human condition. According to Lombroso, it reflected a degeneration of human development such that a criminal's body would exhibit primitive physical features that were reminiscent of the human species in earlier stages of its evolution.

Lombroso's theory contained a further unique claim, influenced no doubt by the then very recent emergence of Darwin's theory of evolution.

Specifically, Lombroso felt that criminalistic traits were transmitted from parents to offspring and on through the generations, but that they were often deeply embedded in an individual and may not, in fact, become manifest within a person's lifetime. As such, a criminal may have parents who were perfectly law-abiding, and may even have fine and upstanding grandparents too. However, given the hereditary nature of criminality, Lombroso believed it was certain that further scrutiny of a criminal's family tree would eventually, and inevitably, lead to the identification of a criminal ancestor. This idea, of behavioural tendencies lying dormant within a family line, was most unlike anything referred to in phrenology. By positing such a mechanism, Lombroso was effectively arguing that criminals were born and not made.

In the late 19th and early 20th centuries, Lombroso was far from alone in arguing that human bodies varied in shape in accordance to psychological attributes, or that these patterns reflected an evolutionary basis for undesirable behaviour. The French theorist, Arthur de Gobineau (1915) wrote how in certain people, 'the animal character that appears in the shape of the pelvis' provided conclusive evidence of their low intellectual abilities. In the 1920s, a number of American universities began to teach courses on 'anthropometrics', covering methods for measuring the qualities of people's hair, lip sizes, and skin colour. All of this was explicitly intertwined with the belief that such measurements would facilitate relevant conclusions about people's behavioural patterns and tendencies (Guthrie, 2004). Of course, these beliefs reflected another important influence of the times, one that is inextricably linked with just about all physiognomic approaches. By linking physical appearance to psychological character, these theories provided a clear rationalization for claims that different ethnic groups could be expected to behave, think, and function differently, together with a basis for viewing different races as being of different levels of worth. Allusions to heredity also offered a justification to believe that the different ethnicities stemmed from genetically distinct biological origins, hinting none too subtly that different ethnicities were akin to different species. Essentially, the physiognomic approach created a scientific template for racism.

In his essay mentioning people's pelvis shapes, de Gobineau – the developer of the theory of the Aryan master race – was explicitly advocating a racist worldview. His belief was that European Caucasians were superior in intellect to all non-Caucasians. Such beliefs would have been very common to anyone raised as a nobleman in late 18th-century Europe; as such, his argument that the pelvises of non-Caucasian persons revealed a biological basis for their then downtrodden status amounted to a scientifically framed rationalization for what was an already widely held doctrine. It deployed the pseudoscience of physiognomy to retrospectively justify centuries of aristocratic privilege. De Gobineau's views were hardly unique among scholars of his time. In a movement now referred to as 'scientific racism', the prevailing discourse of research and theory took as its assumption a range of fundamental distinctions between races, most of which conflated physical appearance with psychological nature, and all of which were framed to explain – and defend – the social hierarchy of the day.

Historically, many scientists worked from the taxonomy of racial distinctions first offered by the 18th-century Swedish botanist, Carl von Linnaeus (1707–1778). As part of an exercise of taxonomizing all species in nature, Linnaeus classified humans into four groups: homo Europaeus, homo Americanus, homo Asiaticus, and homo Afer (Guthrie, 2004). Without explicitly arranging these into a hierarchy, Linnaeus nonetheless attached flattering characteristics to homo Europaeus (describing them as muscular and led by their opinions) while attaching unflattering ones to homo Afer (whom he depicted as indulgent, cunning, negligent, and led by fickleness). Notably, skin-colour was a core defining characteristic for each grouping, which had the effect of associating physical traits with psychological ones. Linnaeus's student, the German naturalist Johan Blumenbach (1752–1840) extended the system to include a fifth category (representing the Malayan race), to expand the range of relevant physical characteristics to include head shape and size, and, crucially, to develop an organizing conceptual framework that implied a hierarchy. Specifically, Blumenbach organized the classification in the order of closeness to the ideal of humanity as created by the Christian god (Tyson, Jones, & Elcock, 2011). Using a somewhat circular argument, this ideal was presumed to be matched most closely by homo Europaeus.

Before long, prominent anthropologists were measuring head circumferences and skull capacities in order to discern the intelligences of different races. A pioneer in this field was the American naturalist Samuel George Morton (1799–1851), who accumulated a collection of over a thousand human skulls, which he kept in his house. He methodically examined their dimensions in order to test the theory that different races were characterized by different brain sizes, with the most worthy races possessing the largest brains. By pouring mustard seeds into the cavity of each upturned skull, and then decanting the seeds into a standardized container, he was able to determine the volume of the brain the skull once contained. He published reams of data which, he claimed, confirmed his view that Caucasians had systematically larger brains than other races. Indeed, his findings were that 'English' skulls housed the largest brains, followed by 'Anglo-American' and 'German' ones; with by far the smallest brains belonging to 'Peruvians', 'Hottentots' and (indigenous) 'Australians' (Gould, 1996). Wherever his suppositions were contradicted, his theoretical position simply shifted to suit. For example, when he discovered that the ancient Egyptians had relatively large skull capacities, he declared that they must have been Caucasians rather than Africans, thereby preserving his original position that Caucasian peoples were the most intelligent. Ultimately, however, Morton's data were less than reliable. When, in the 1970s, the American palaeontologist Stephen Jay Gould replicated Morton's methods to re-measure the original skulls in Morton's collection, he found that the originally published findings were strewn with error and thus wholly misleading (Gould, 1978). Nonetheless, the idea that Morton helped to popularize – namely, that there exist empirically demonstrable race differences in intelligence, which correlate with the physical morphology of the human body – has had a lasting influence on popular belief, and on a number of prominent psychological researchers.

Often heralded as a pioneer of psychology, the English psychologist Francis Galton (1822–1911) held firm beliefs about the comparative worth of different races. He developed statistical procedures for understanding correlation, regression, measurement error, and the normal curve, introduced the modern field of psychometrics, established the area of differential psychology and, thus, the study of personality, and provided numerous insights regarding the way nature and nurture differently affect human psychological development. However, he also wrote extensively about how the 'childish, stupid, and simpleton-like' behaviour of African people made him 'ashamed of [his] own species' (Galton, 1962 [1869], p. 395). He reminisced in his diaries about physically cruel punishments that he meted out to African servants during his many tropical expeditions (Richards, 1997). He theorized that African people were doomed to imminent extinction as a result of their shortcomings (Galton, 1883). And he established the field of eugenics, the social movement aimed at enhancing the human race through, among other things, selective breeding. Although it might be harsh to implicate Galton directly in the many atrocities that were eventually carried out under the purview of eugenic reasoning (which included policies of enforced sterilization, racial segregation, and, ultimately, genocide), it is hard to dispute that Galton's original views on race were, in their own right, deeply unsavoury.

Some commentators have suggested that psychology retains a physiognomic imprint of these historical attitudes. It is true that many psychometric tests of intelligence, despite having gone through several modern revisions, were originally designed at a time when belief in the merits of physiognomy was strong. Early psychometricians such as Alfred Binet (1857–1911) and Theodore Simon (1872–1961), whose IQ test continues to be widely used in its latest edition today, strongly advocated the use of observation of physical characteristics as part of any assessment of intellectual ability. In particular, they advised psychologists to identify abnormally shaped ears, hairlines, teeth, and eyes as probable indicators of low intelligence (Collins, 1999). However, by and large, the core claim that psychological traits can be inferred from physical ones – including those pertaining to ethnicity – has been thoroughly discounted, time and again, by the accumulated empirical evidence.

Nonetheless, the idea that bodies, and in particular faces, reveal something about inner character or ability continues to enjoy support in wider culture. Many studies show that people retain a firm belief in the stereotype that physical beauty is associated with generosity, even though there is no empirical evidence to suggest that this is the case (Lemay Jr, Clark, & Greenberg, 2010). There is also a very common, but unfounded, belief that adults with 'babyish' faces are likely to exhibit childlike behaviour

(Zebrowitz, Montepare, & Lee, 1993). Such stereotypes inform the judgements of children (Rennels & Langlois, 2014) as well as those of adults (Zebrowitz & Franklin, 2014), and are witnessed in different cultures around the world (Zebrowitz et al., 2012). The idea that people with wide faces are less trustworthy is also popularly held, and is the subject of recurring research. Indeed, in a somewhat ironic turn, there is even a body of research examining whether face shape can be used as an indicator of a person's tendency toward racial prejudice (e.g., Hehman, Leitner, Deegan, & Gaertner, 2013). The attribution of psychological traits and moral worth to people with different body-types and faces remains a common trope in the arts and popular culture more generally (Wegenstein & Ruck, 2011).

Physiognomic claims are pseudoscientific for several reasons. Historically they have been perpetuated by confirmation bias, offering few consistent or specific predictions that might be falsifiable through empirical scrutiny. Physiognomic claims circulate mainly through popular channels, such as websites and mass media, with very little by way of peer-reviewed scientific investigation to support them. The idea that different shaped nostrils or ears or chins – or, for that matter, different skin colour – might reflect underlying moral tendencies or personality traits is biologically implausible, in that no (mainstream) scientific theory exists that might account for it. In other words, it relates to no existing body of knowledge in either anatomy or psychology. In this way, it lacks parsimony: its explanation relies on as yet unheard of assumptions about how nature works. The extension of physiognomic reasoning to suggest that physical traits might in fact be alterable by the adjustment of thinking styles, or to suggest that there exist broad-sweeping and deep-seated racial differences in cognition, behaviour, or aptitude is unwarranted by any scientific investigation. In fact these extensions highlight the way speculation that is unfettered by scientific rigour can open a door to very worrying social belief-systems and political movements (correspondingly, they highlight the merit of scientific reasoning is in its capacity to assess such claims, and to reveal them to be unwarranted where this is the case). They also remind us of the continuous thread of time that connects the modern world with some very dark periods of human history. Ostensibly trivial activities like face reading and phrenology might seem harmless when availed of in the context of entertainment. However, it is worth bearing in mind the common conceptual frameworks, and ancestry, they share with eugenics and scientific racism.

Conclusion

Complementary and alternative therapies that defy scientific explanation, claims that consciousness is a quantum or otherwise extraordinary entity, theories that psi phenomena exist, and appeals to physiognomy are all bad ideas. However, it is certainly true to say that none of them has failed to prosper. All these domains reflect common beliefs that have been endorsed by millions of people for hundreds of years, and which continue to enjoy widespread support despite being unable to attract scientific corroboration. And it is no surprise that scientific support has been lacking. These topics are mired in non-falsifiable assertions, vague terminology, inconsistent definitions, deference to guru figures and anecdote, and unjustified hyper-complexity, and avail of a worldview where an absence of empirical evidence is seen as no impediment. In short, they are pseudoscientific.

It is a matter of definition that pseudosciences are activities that purport to actually be *real* sciences. To some extent, while the case studies in this chapter are true to this definition, the fact that their scientific standing is open to debate is pretty conspicuous. These areas are easy to recognize as controversial. Their outputs are presented and published separately from those of mainstream science. Their activities take place away from mainstream laboratories or conferences. These fields are regularly the subject of high-profile criticisms. To the extent that the case studies in this chapter are drawn so clearly from the fringes of scientific thinking, you could say that they represent somewhat easy targets for a critique of pseudoscience. It is not difficult to raise questions about subjects that are so obviously questionable; a more difficult task, both technically and socially, would be to raise corresponding questions about subjects that do not ordinarily attract such negative attention.

It is certainly important for psychologists, and society at large, to be familiar with these obviously controversial case studies. However, it is also important to consider activities from mainstream psychology that might be accused of exhibiting pseudoscientific features (such as circular reasoning, non-falsifiability, confirmation bias, vague definitions, lack of parsimony, and all the rest). Unfortunately, this is not because psychology is immune to criticism in this regard. In fact, it might be argued that the core of psychology is as much a magnet for pseudoscientific thinking as are its fringes. The coming chapters seek to consider a number of examples of how this is the case.

Chapter 6

Examples from the Mainstream: Biological Reductionism as Worldview

Introduction: We're all animals after all

We have seen that scientific psychology is descended from many ancestors: philosophy is one, religion is another, and ordinary human curiosity about why people are the way they are is a third. Dictionary definitions of *psychology* variously characterize the field as offering a scientific study of the 'mind' or 'behaviour', and even the etymology of the word echoes a focus on the ephemeral: taken literally, ψυχή-λογία means 'laws of the soul'. From these perspectives, psychology seems primarily concerned with the outwardly displayed, culturally experienced aspects of the human condition; those which 'make us human' in the first place. This narrative recognizes the view of most (religiously informed) cultural taxonomies that regard humanity to be unique in the world, if not indeed the universe, and to be both distinct from and dominant over other species in nature. We are humans after all, not animals.

However, notwithstanding this, it is generally well regarded – even by the most socially conservative commentators – that human beings are nonetheless a *type* of animal. While we engage in activities that are uniquely human, our bodies resemble those of non-humans in several respects. Anatomically and biochemically, our organs and physiologies are simply variants of those seen in other species: nothing about human structural biology is authentically unique to humans. Moreover, the role of the animalistic aspects of the human species in influencing its behaviour is well signposted. Drinking alcohol, taking drugs, suffering a brain injury, growing out of childhood – experiences that change us biologically change us psychologically too. Accordingly, the view that biological dimensions are intertwined with the human condition has been recognized for centuries, making biology another ancestor of modern scientific psychology.

From Descartes onwards, the study of the biological basis of human behaviour also helped to focus attention on the role of scientific reasoning in psychology, and to blend psychology into the wider scientific revolution.

130

It is worth remembering too that the earliest formal sciences had related to pure physics, with even chemistry being slow to be recognized as a bedfellow. The latecomer biological sciences, seen at first to be so immersed in uncontrollable nature as to lie permanently outside the reach of laboratories, needed to elbow their way in somewhat. As such, biology not only stood as a role model for scientific research practice, but also as a model of how to set out one's stall as a science. Thus the extent to which psychology was seen as an offshoot of biology largely served to enhance its reputation in modern scholarship.

Of course, with such a history comes a number of pitfalls, the idea that *biological* is synonymous with *scientific* being one of the most salient. Just because an aspect of psychology is being linked with something biological (for example, genes, hormones, or the brain) does not guarantee that sense is being made. Indeed, insofar as audiences presume the opposite to be true, the laying on of bio-jargon can create a comfort zone that actually discourages, rather than encourages, scientific rigour. Assuming *biological* to be synonymous with *scientific* removes the need for scientists to concentrate on the epistemological aspects of what they are doing. And as we have seen, unregulated reasoning all too often leads to sloppy logic. When it occurs in science, the risk of pseudoscientific practice is heightened.

We have already discussed the way some pseudoscientists think the human mind lies beyond the realm of ordinary nature. In this chapter, we will consider the way some mainstream scientists think the human mind is so bound up in ordinary nature that it is effectively untouched by other influences. In this regard we will begin with the most biological of distinctions: that between males and females.

The battle of the battle of the sexes

In the spring of early 2013, several international newspapers carried excited reports on some new research into the psychology of parenting. The study, published in the prestigious science journal *Nature Communications* (Gustafsson, Levréro, Reby, & Mathevon, 2013), involved 29 babies, aged between two and five months, who had been monitored across three consecutive days. The researchers made a number of audio recordings of the children, from which they isolated six 8-second clips of continuous crying by each child. They then invited each of the babies' parents into a laboratory to listen to a series of 30 of the recordings. Six of the clips were of the parent's own baby, with the other 24 recordings taken from similarly-aged infants. The parents' task was fairly straightforward: all they had to do was pick out the six particular recordings that featured their own baby's cries. Overall, the study revealed them to be quite good at identifying their babies, with parents correctly identifying an average of 5.4 of the six target recordings. Even the least accurate mothers and fathers identified four of the six. Nonetheless, some parents were better than others. Further analyses showed that success was determined by the number of hours parents spent with their children each day: parents who spent more time with their babies were better able to distinguish their cries from those of other infants.

Now this was hardly the world's most extensive research study, nor did it produce the most remarkable of findings. After all, it might ordinarily be expected that the more time you spend listening to something, the more familiar with it you will become. But, albeit no doubt assisted by the journal's press office, this modestly scoped study succeeded in attracting international headlines. The results were invariably regarded as having revealed a surprising outcome, perhaps even a shocking one. So much so, in fact, that sub-editors at the *Daily Mail* felt the need to resort to capital letters in order to properly highlight the basis for their amazement, resulting in the following headline (Innes, 2013):

Mothers and fathers are EQUALLY good at recognising their baby's cry

Arguably, the phrasing used in the actual journal article exhibited only marginally less astonishment. As it appeared in the original, the paper was titled 'Fathers are just as good as mothers at recognizing the cries of their baby'. It seemed that both newspaper and scientific journal felt it important to emphasize the simple fact that two groups of similar human beings, having undergone a similar protocol, went on to exhibit similar abilities (with neither publication seeking to afford headline prominence to the rather more tangible finding relating to parental time spent with children). The fact that one group of around 30 people had been found to be pretty much the same as another group of around 30 people had – apparently – shocked everyone.

Clearly, the narrative frame used to report this story was informed by an underlying assumption about what the study *should* have revealed. Simply put, the study *should* have revealed a difference in how mothers and fathers responded to their children's cries. You possibly don't need me to tell you the direction of this expectation, but for completeness it might be useful to state it outright: it was assumed that mothers would be better than fathers at recognizing the sounds of their offspring. After all, women benefit from a maternal instinct, do they not? And women are more sociable, have greater empathy, are more natural communicators, and are generally more helpful, are they not?

The idea that women are psychologically different from men, and seek different things from life, has certainly been a long-running theme in mainstream psychology (Shields, 1975). At the turn of the 20th century, German neurologist Paul Julius Möbius explained that women's mental incapacity assisted the survival of the human race, as their low stimulation thresholds ensured they were more than gratified by the humdrum chore of looking after children. Indeed, he warned that 'If woman was not mentally and physically weak ... she would be extremely dangerous' (Möbius, 1901). In the 1920s, prominent British psychologist William McDougall noted that women's 'physical weakness and delicacy' created a 'resemblance to a child', all of which amounted to Nature's way of ensuring that nearby men would be motivated to 'protect and shield and help her in every way' (McDougall, 1923). Somewhat later, in the 1960s, prominent American child psychologist Bruno Bettelheim famously wrote that 'as much as women want to be good scientists or engineers, they want first and foremost to be womanly companions of men, and to be mothers', while fellow psychoanalyst Erik Erikson argued that women's bodily designs conferred upon them 'a biological, psychological, and ethical commitment to take care of human infancy' (Weisstein, 1971).

Such quotations are regularly cited in order to illustrate the historic male bias of mainstream psychology, with the intention of drawing a contrast with the feminist enlightenment we enjoy today. However, it can be argued that little has really changed except the frames of reference within which such views are expressed. Psychologists at large seem to generally accept the proposition that there exists an extensive range of psychological differences between men and women, across a breadth of cognitive, social, personality, and mental health domains (Hyde, 2014). This acceptance stretches to dozens of gender-specific assumptions that underpin our everyday discourse. For example, women (and girls) are regularly said to possess superior verbal skills compared to men (and boys). Girls are said to reach intellectual maturity faster than boys, and young women regularly outscore young men in high school examinations. Women are said to remain more verbally adept into adult life. According to Louann Brizendine, a professor of neuropsychiatry at the University of California at San Francisco, women use three times more words per day than men (Brizendine, 2007). Women are also said to be less likely to be sexually promiscuous or to even think about sex, less likely to hide their emotions, less domineering, and less aggressive than men (Brizendine, 2010). Men are said to be so prone to aggression that they even transmit this tendency to their pets. In another recent behavioural research study to make international news headlines, researchers in the Czech Republic found that dogs were four times more likely to try to bite other dogs when both animals were being walked by male handlers, compared to dogs being walked by women (Řezáč, Viziová, Dobešová, Havlíček, & Pospíšilová, 2011). According to widespread media coverage, the study was particularly extensive, having involved nearly 2,000 dogs and their owners ('Dogs walked by men are more aggressive', 2011).

Women are also characterized as coping differently with mental stress when compared to men. Classically, psychological stress research (especially that focusing on the impact of stress on health) has followed the model of Walter Cannon's *fight-or-flight* response (Cannon, 1932). This elegant theory was among the first to describe how human beings might react physically to stressors, with autonomic responses that assist us in fighting or fleeing: increases in blood pressure, for instance, help to stimulate blood flow and transmit oxygen to our musculature, to better equip us for throwing a few punches or making a quick getaway. The problem is that, while such effects would have been very useful in helping our primitive ancestors react to the sudden appearance of a sabre-toothed tiger, they are less useful in our modern language-driven and technological world. Today's stressors are more likely to involve tight deadlines, money problems, and traffic jams, than wild animals that we can fight or run away from. As a result, the fight-or-flight response carries more costs than benefits for modern humans: our blood pressure gets elevated with no clear advantage accrued in return. This basic model has led to nearly a century of productive research on psychosomatic disease processes, and has helped clinicians identify several medically relevant psychosocial risk factors for illness (not least of which is mental stress itself). However, recently some theorists have declared it to be insufficient. They question whether the model truly captures the entire human experience. Specifically, they argue that while men may well go around fighting and/or fleeing, women actually behave quite differently. Women, it is said, respond to stress by engaging in nurturing activities that promote children's safety, and by looking to create and exploit social support networks in their vicinity. Echoing the terminology of Cannon's model of (male) stress responsivity, this newer theory of the female reactions has become known as the *tend-and-befriend* model of stress (Taylor & Master, 2011).

However, while tending-and-befriending certainly *sounds* nicer than fighting-or-fleeing, not all modern views about gender differences present women in a positive light. Young women outperform boys in high school, but it is widely claimed that young boys have greater capacities in mathematics and in aspects of geometric reasoning necessary for careers like engineering and architecture (such as three-dimensional mental rotation). It is also argued that men possess personal qualities that make them more suited to leadership roles in the corporate world. Simon Baron-Cohen, professor of developmental psychopathology at Cambridge University, asserts that men's brains are better equipped for analysing, exploring, and constructing complex systems. This makes the average man more inclined and better able to understand systems, to predict their outcomes, and if necessary to invent new ones (Baron-Cohen, 2004). (Baron-Cohen's theory carefully states that it is possible for women to have 'male' brains too, but it is clearly far more likely that men will possess them.) And those assertions that women make better parents, while superficially positive, are not necessarily flattering. One of the effects of insisting that women are more adept at child-rearing is to convey the intrinsic additional assumption that it would be better for children if their mothers stayed at home: society itself prospers when women keep out of the public square.

Psychologists' assumptions about sex differences permeate professional bodies as well as empirical science. In 2006, the British Psychological Society's Standing Committee for the Promotion of Equal Opportunities nominated a working group to examine the issue of sexism in academia. Their brief report appeared in the society's house journal *The Psychologist* (Riley, Frith, Archer, & Veseley, 2006). Among other factors of concern, the authors noted the masculinist nature of many academic workplaces, the impact of motherhood on women's availability to produce the outputs rewarded by university promotions systems (such as journal articles), and the underrepresentation of women among university managers and major research councils. They also offered the following observation:

> With regard to women in psychology, discourses around what is important: peer-reviewed journals, positivist epistemology, and quantitative methods work ... to reduce women's participation in psychology. (p. 96)

Embedded within this view would appear to be the assumption that women are uniquely disadvantaged by expectations that they participate in activities involving 'quantitative methods', a clear allusion to an expected sex difference in mathematical ability. For this reason, the authors go on to commend the BPS for having recently established a special subsection devoted to *qualitative* methods, offering the view that such a subsection would have important implications for the promotion of opportunities for women.

While modern psychologists appear perfectly willing to allude to such sex differences in formal contexts, they seem less inclined to attribute them to the mental delicacy of females, their longing to be womanly companions of men, or their ethical commitment to the care of human infants. More likely they will articulate such views within one of three theoretical frameworks: evolutionary psychology; cognitive social learning theory; or sociocultural theory (Hyde, 2014). Evolutionary approaches to sex differences focus on the idea that natural selection has favoured different behaviours and attributes for men and women. The divergence arose from historically distinct roles in the human environment of evolutionary adaptation, and the need for men and women to approach sexual selection somewhat differently. By virtue of natural selection, these differences have been preserved and accentuated along gendered lines within the human gene pool (Buss, 2013). Cognitive social learning theory argues that an individual's behaviours and attributes emerge as a function of the rewards and punishments encountered in one's environment. Assuming gender norms and roles to be endemic in society, boys and girls are said to be presented with different psychological approaches from very early in their lives, with gender-specific behaviours being extensively and consistently reinforced (Bussey & Bandura, 1999). While such a view helps to explain why men and women might reasonably be expected to develop distinct patterns of thoughts and behaviour in contemporary society, it still begs the question as to how these powerful gender norms actually originated. Sociocultural theory seeks to address this by referring back to biology, arguing that gender norms in psychological attributes emerged from the way society historically divided roles and workloads as a result of physical differences between men and women (Eagly & Wood, 1999). Because men are bigger, stronger, and less physically involved in child-bearing, they were more likely to engage in warfare, gather assets, and avail of other means of accumulating power and status. Through the aeons, as male roles became more dominant, female roles became more subordinate, with the overall result that each successive generation of men and women systematically becomes socially and culturally reinforced to develop particular gendered styles of behaving and thinking.

However, notwithstanding the sheer volume of discourse concerning psychological gender differences, there exists a significant problem with these assertions. With regard to the most common claims of differences between men and women – those relating to mathematical ability, verbal skills, temperament, emotionality, aggression in all contexts, promiscuity in sexual behaviour, leadership effectiveness, moral reasoning, conscientious-ness, self-esteem, and the extent to which they are motivated by tangible rewards – the problem is this: according to the data, *all* such claims are unfounded. With regard to other claims – such as the assertion that women are more helpful than men – the data appear to show the opposite to be actually the case. Only in very few instances are commonly cited sex dif-ferences corroborated by empirical data, but even in these cases – such as male superiority in mental rotation or abstract reasoning – the data show the true differences to be extremely small, and the stereotype to be greatly exaggerated. These insights do not come from isolated and hard-to-repli-cate individual studies. Male and female differences have been studied for more than a century: the evidence refuting them is drawn from entire bod-ies of research, where datasets have been assembled, pooled, and analysed in their totality.

Perhaps the best examples of this total-research approach are the meta-analytic reviews published by American psychologist Janet Shibley Hyde (2005; 2014). Having scrutinized the cumulative research evidence on psy-chological sex differences, Hyde has concluded that assumptions of gender differences should give way to what she refers to as a 'gender similarities hypothesis'. In essence, this hypothesis says that psychologists should begin with a working assumption that they will more frequently *not* encounter gender differences in their research. What is striking about Hyde's reviews is that they cover research that was originally conceived to test what were assumed to be *likely* sources of gender differences. Generally speaking, studies in the gender differences literature do not randomly select their tar-get variables, investigating whether men and women differ in attributes about which there is no prior suspicion or theory. Instead they aim for variables that are considered to be likely prospects, such as mathematical performance. And yet it is precisely this literature that Hyde shows is producing a big fat null result.

Take, for instance, mathematical performance specifically. Hyde describes an early meta-analysis of data drawn from over three million persons, in which the computed gender difference in mathematical performance equated to a total effect size of $d = 0.05$ (Hyde, Fennema, & Lamon, 1990). This tiny statistical effect is virtually trivial. The effect size statistic 'd' represents the difference in two scores divided by the standard deviation of the data as a whole, and a d of 0.05 equates to a shared variance of just 0.06 per cent. This means that in this dataset, 99.94 per cent of the variability in maths scores resulted from factors *other than* the participants' genders (for example, individual differences in ability or education). But that sample was of 'only' three million people. A more up-to-date meta-analysis, covering over seven million boys and girls, found differences that were similarly minute: in this sample, ds ranged from 0.02 to 0.06 depending on participants' ages (Hyde, Lindberg, Linn, Ellis, & Williams, 2008). The non-existence of a sex difference is seen in adults too, and in high-level mathematics (as opposed to junior school maths). The relevant meta-analysis (Lindberg, Hyde, Petersen, & Linn, 2010), focusing on how young men and women in the final year of secondary school performed in several different domains of complex mathematics, yielded a sex difference never greater than $d = 0.07$. This equates to a shared variance of 0.12 per cent, meaning that 99.88 per cent of the variability in these students' maths performance was *unrelated* to gender. In short, on the basis of studies of literally millions of people, it can be stated that any claim that men and women differ in mathematical ability is simply wrong.

Across both reviews, Hyde details similar findings for dozens of other psychological traits, abilities, and behaviours, all of which are normally depicted as differing markedly across the genders. The overall conclusion seems empirically compelling: the gender similarities hypothesis, which states that men and women are similar in most (although not all) psychological variables, is essentially correct. There are some exceptions, of course, but they are small in number – and it is simply unclear whether any of them reflect embedded genetic differences as opposed to differences in environments and contexts. For example, men typically outperform women in mental rotation tests, with one meta-analysis reporting an effect size of $d = 0.57$ (equating to a shared variance of 7 per cent between gender and mental rotation ability; Maeda & Yoon, 2013). However, we also know that mental rotation improves with repetition; therefore, the gender difference seen in the data may reflect different participation levels in extracurricular activities (such as video games or sports) that involve the practicing of mental rotation for fun (Hyde, 2014). Another example is the empirical gender difference in self-reported sexual behaviour (men are more likely to report a greater number of sexual partners). However, whether self-reported behaviour equates to *actual* behaviour is difficult to establish. It could just be that there exists a gender difference in people's willingness to boast about such things. A third example relates to emotional stability. Large-scale personality studies often report sex differences in neuroticism and agreeableness (Costa, Terracciano, & McCrae, 2001). However, Hyde (2014) points out that while most of these findings arise from data gathered from US adults, the sex differences are not consistent across countries and are virtually non-existent in some. The very fact that these sex differences vary cross-culturally suggests that they are the *result* of cultural contexts, rather than of intrinsic natural differences between male and female human beings.

It turns out that nearly all commonly claimed sex differences are either unfounded or else based on very flimsy foundations. Brizendine's assertion that women use more words than men appears to be unrelated to any published study, and contradicted by several (Liberman, 2006). That Czech study on dog walkers, while ostensibly including nearly 2,000 dogs in total, actually included only 28 dogs relevant to the hypothesis-test concerning aggression while being walked by a man. Moreover, it failed to control for the possibility that, because of their relative strength, men might be more likely to be given the task of walking the most aggressive dogs (Hughes, 2011). The tend-and-befriend model of stress has been critiqued by several authors who question its over-reliance on the idea that men and women had distinct roles in evolutionary history, and on observational findings that modern men are not inclined towards tending or befriending (Geary & Flinn, 2002). Moreover, its inherent claim that women resort to helpful behaviour during stress appears to conflict with at least some of the research evidence. Meta-analyses suggest that not only are men more likely to engage in helping behaviour than women, but that this divergence is accentuated when a situation involves danger (Eagly & Crowley, 1986). Baron-Cohen's idea that male brains are more equipped for thinking about systems has been widely criticized, not least for its failure to specify what is meant by a 'system' or to provide a consistent means of measuring the pattern of thought he refers to as 'systemizing' (Fine, 2010). In addition, one of its key underlying empirical studies, which appeared to demonstrate sex differences in the preferences of day-old babies (Connellan, Baron-Cohen, Wheelwright, Batki, & Ahluwalia, 2000), was methodologically flawed in allowing the experimenter to know in advance the gender of the babies being tested (Nash & Grossi, 2007). Subsequent studies that avoided this confound detected no sex differences in babies' preferences (Leeb & Rejskind, 2004). And in light of the aforementioned non-differences in men and women's maths performance, why female academics should have to worry about quantitative methods in psychology appears very unclear.

Beyond whatever is said about the *existence* of commonly touted gender differences, there is even the empirical question of whether such differences can be properly tracked in the population. Statistical studies have shown that most commonly cited sex differences refer to variables that cannot reasonably be compared along gender lines: they are 'dimensional' rather than 'taxonic' (dimensional variables are ones that range continuously from low to high; taxonic variables are ones that involve non-arbitrary categories). Because they refer to characteristics that occupy ranges instead of categories, these variables are distinct from biological gender itself, which is primarily a two-category construct. This means that such characteristics do not easily lend themselves to separation into binary groupings. As such, when conceptualizing sex differences, it is seldom statistically justified to declare that men are 'like this' while women are 'like that' (Carothers & Reis, 2013). In the end, people vary in all sorts of ways *as individuals*. Some men are highly sexually active while others are celibate: is it therefore truly meaningful to ask whether 'men' differ from 'women' in this behaviour? But even this conundrum, intended to encourage a more nuanced approach, has itself been construed as a target for binary sex categorization. Many theorists have argued that it is *psychological variability* in which men and women differ, with men exhibiting much more variability than women. However, this view too has been tempered by scrutiny of large datasets (Hyde, 2014).

It should be pointed out that the flimsiness of commonly cited sex differences is not a recent discovery. While the media might report a recent study on male parental sensitivity as 'news', that finding really isn't new at all. The fact that fathers and mothers respond to children with similar levels of parental sensitivity and with equivalent physiological profiles has been demonstrated repeatedly for the best part of two decades (Geary & Flinn, 2002). The observation that the time a parent spends with their children is a more important predictor of their sensitivity than their gender has been observed for at least 25 years (Risman, 1987). Even the specific point that men and women react equivalently to crying babies has been well established, some decade and a half prior to this more recent study being reported as 'news' (Storey, Walsh, Quinton, & Wynne-Edwards, 2000).

Likewise, the broader gender similarities hypothesis has a very long history. Prior to Janet Shibley Hyde, essentially the same hypothesis was proposed nearly a century ago by American academic, Chauncy N. Allen (1927). And before him, American psychologist and gender researcher Helen Thompson Woolley had written extensively about the empirically unfounded nature of alleged sex differences:

> The general discussions of the psychology of sex, whether by psychologists or by sociologists show such a wide diversity of points of view that one feels that the truest thing to be said at present is that scientific evidence plays very little part in producing convictions. (Woolley, 1914, p. 372)

She also foreshadowed the resistance of purported sex differences to contradiction by research findings, reserving particular exasperation for her fellow psychologists:

> There is perhaps no field aspiring to be scientific where flagrant personal bias, logic martyred in the cause of supporting a prejudice, unfounded assertions, and even sentimental rot and drivel, have run riot to such an extent as here. (Woolley, 1910, p. 340)

It appears that little may have changed in the century that has passed since Woolley offered these observations.

The fact that common social prejudices about sex differences appear so resistant to falsification raises the question of whether psychologists who propound such claims are, in fact, favouring hypotheses that are *non*-falsifiable. If so, then they are hovering at the borderline between science and pseudoscience. It has repeatedly been shown that most commonly cited sex differences have been undermined by empirical data. Even the fact that most target variables are dimensional rather than taxonic suggests that hypotheses about between-group differences are, at best, poorly conceived. But all this hardly seems to matter. The study of all-embracing gender differences is alive and well, and continues to expand. Its momentum is at least partly sustained by the re-issuing of old findings as if they were new ones, and the projection of surprise (hardly a scientifically quantifiable judgement) at data that show men and women are, in many respects, not that different after all.

Much of the field of sex differences research appears also to be weighed down by confirmation bias. Weak studies and flimsy evidence escape criticism because the findings produced are so consistent with the prevailing social view. A study reporting sexual excess among teenage boys is presented as yet further confirmation of their relative promiscuity, even though the data are drawn entirely from self-reports. A study suggesting that male dog-owners make their dogs aggressive is welcomed into the narrative frame, despite the fact that its sample is tiny and its statistical analyses underpowered. It can also be noted that psychologists often appear comfortable citing personal experiences to support their views that boys and girls are genetically different in clichéd ways (Fine, 2010). When psychologists lower their sceptical guard because a finding is consistent with their prior beliefs, they are eschewing the notion of parsimony: overall theoretical positions are bolstered by the inclusion of untested assumptions. And when they resort to anecdote, they are essentially attaching value to hearsay evidence. These approaches too amount to flirtations with pseudoscience.

There is little doubt that men and women differ in many ways, some of which are psychological. However, the extent of these psychological differences has been exaggerated for centuries. By rights, the interpretation of empirical research findings should lead, rather than follow, public opinion; but, as we will see in Chapter 8, psychology's attempts at objectivity often fall foul of a presumptive masculinist bias. In virtually all regions of the globe, women face social, economic, and cultural discrimination, if not outright violence and subjugation. The peddling of artefactual sex differences in behaviour styles, thought patterns, emotionality, and cognitive abilities (including mathematical competence) does little to help matters. When presented with a veneer of scientific approval, it runs the risk of providing a justification for those who wish for discrimination to continue. Non-falsifiable, quasi-anecdotal claims to the effect that women make better parents or are intrinsically demure creatures or think about children instead of fighting-or-fleeing, serve to reinforce the public view that women have a 'place' in our societies, and that that place is not the same place that men have. The costs of a pseudoscientific approach to this question are likely to be stark.

The invisible hand of evolution

In seeking to rationalize presumed gender differences, scientists have invoked many different forms of biological reductionism. As described in Chapter 2, reductionism is the epistemological view that complex systems can be adequately considered in terms of the sum of their constituent parts. *Biological* reductionism occurs when complex psychological phenomena in biological organisms are presumed to result from simple biological causes. The main problem with biological reductionism is that it fails to take account of the situational influences, social and cultural contexts, experiential factors, personal motivations, aesthetic choices, intellectual creativity, and other things that affect people's behaviour (Hughes, 2012). If lots of young women choose to study nursing at university, it could be because of the gender roles with which they have been inculcated by family and peers, it could be because of a fear that choosing different career paths would expose them to hostility or discrimination in the workplace, it could be because of a calculation regarding future salaries and working conditions, or it could be because of a perception that nursing is an interesting career choice, even a fashionable one, at a given moment in time. Indeed it could be because of several of these factors, or many others not mentioned. In contrast, a position informed by biological reductionism would focus on biological causes: perhaps it is because women's brains are hardwired for empathy.

Already we have seen that historical psychologists sought to justify female exclusion on the grounds of delicacy. More specifically, many made explicit reference to a feebleness of the brain (with several adopting the craniometric position that women's brains were smaller than those of men). However, other forms of biological reductionism were also common. In the early 20th century, prominent psychologists worried that the toll of menstruation was impeding girls' progress through mixed gender education; not only that, they warned that educating girls created demands for blood flow to the brain that could, in due course, render a young woman unable to menstruate at all (Shields, 1975). Other figures more or less resorted to physiognomy, likening the verticality of the female face to that of 'primitive' races (Russett, 1989).

Over the past century, scientific psychology has of course revealed much about the biological basis to complex human behaviour. And anyone who has reflected on the way drugs and brain injuries can affect behaviour and cognition will quickly appreciate the fundamental intertwining of biological factors with a person's psychological functioning. The problem of biological reductionism becomes acute when it is implied that biological factors simply override all other considerations, when the depiction of biological dimensions of behaviour are oversimplified or exaggerated, when mundane biological occurrences are attributed with mystical causative influence, or when unwarranted leaps of reasoning are made from the biochemical and anatomical to the macro-level and sociocultural. In short, biological reductionism occurs when psychologists overstate what they know about the biological nature of psychology beyond that which can be justified by the science.

Ways in which psychological concepts can be subjected to biological reductionism are plentiful. At the broadest level, biological reductionism frequently occurs when evolutionary frameworks are used to explain psychological phenomena. That is not to say that natural selection has nothing to do with behaviour; clearly it does. Given that humans are biological organisms, and given the intertwining of biological and psychological life, it is inevitable that many aspects of the human experience will be shaped by natural selection across time. The ability of young babies and children to acquire spoken language without specifically being trained to do so represents an elaborate cognitive skill that cannot reasonably be accounted for by contingent reinforcement in their immediate environments. The fact that language arises in much the same way (and with many of the same malapropisms) for all typically developing babies around the world, regardless of the cultural settings into which they are born, strongly implies it to be an innate skill, which – given our understanding of genetic inheritance in biology – is hard to explain except by invoking the notion of genetic inheritance, and thus natural selection, in psychology too.

However, determining precisely what aspects of thoughts, feelings, and behaviour are so shaped is far from straightforward. Clearly, the environment has some impact on language development: a baby born in China will (most likely) acquire Mandarin, while a baby born in Greece will acquire Greek. While composition and comprehension skills might emerge through the natural course of things, what the baby hears and experiences in his or her surroundings is also integral. And, culturally, languages themselves evolve over time, in the sense that they change in content and delivery through selective processes: new phrases and words which prove to be useful (like 'selfie') are retained; while those that prove to be less so (like 'gongoozle') will be forgotten. In the end, human evolution and cultural evolution are intertwined, with adaptations in one being shaped by adaptations in the other.

A second problem is that evolutionary progress is extremely slow, at least in terms of the human frame of reference for time. The first primates diverged from other mammals around 85 million years ago; within this group, the so-called 'great apes' (or *hominids*) emerged around 20 million years ago. A subset of these developed into creatures that could walk on two legs – but that change, comprising infinitesimally minor increments across each successive generation, took around 15 million years. Within this group, some populations gradually developed into the first *homo* species, who figured out how to use stone tools, but that took a further 3 million years. Our own branch of this family, which developed the attributes of *homo sapiens*, only arrived about 400,000 years ago. It took another 300,000 years – the first three quarters of our history – for us to learn how to talk. We were then talking to each other for 95,000 years before we figured out that it might be useful to develop a way to write things down; a practice that we have now been engaged in for just the past 50 centuries (although some authorities reckon even that to be an overestimate). To depict all this using a classroom example: if the entire history of evolution was compressed into a single calendar year, then *homo sapiens* would not arrive until 11:30 pm on 31 December, while human speech would only emerge at 11:58 pm. Writing would not begin until six seconds before midnight. This slow pace of change along such an elongated timeframe has three important implications: (a) the vast bulk of what has evolved into the modern human genotype reflects selective pressures that far pre-date the modern environments in which we now live; (b) even tiny differences between organisms could end up having big influences on the way the species evolves given enough time; and (c) because these differences are so tiny, we wouldn't actually realize they were there within our own lifetimes.

Evolutionary perspectives help to explain much about our perceptual and cognitive processes, including, for example, the types of reasoning errors and heuristics that we considered in Chapter 4. They can also help us to understand communication activities, including language, emotion, and even facial expressions. Aspects of visceral aggression, sexual behaviour, and how stress affects our health (as per the fight-or-flight response) are also elucidated by evolutionary approaches. For example, it is useful to bear in mind the implications of the idea that exaggerated stress responses are unlikely to be adaptive in the long term. Decades of research has confirmed that sustained elevations in blood pressure due to stress are indeed damaging to physical health (Hughes, 2013) and nowadays there is widespread appreciation of the fact that events that elevate your blood pressure will not be good for you. However, an evolutionary framework would also suggest that an *under*active stress response should be maladaptive, in that for a trait to be amenable to natural selection, it must vary in the gene pool in both suboptimal and super-optimal directions. Only recently have researchers investigated the consequences of underactive blood pressure responses to stress; and have indeed discovered that they, too, are associated with ill-health (Phillips, Ginty, & Hughes, 2013). In all such cases, the evolutionary underpinnings will be intertwined with the relevant cultural context. But when approached with this in mind, they are invariably informative.

As mentioned, biological reductionism arises when it is argued that biological factors override all other considerations. One example relates to the role of major histocompatibility complex (MHC) molecules in human sexual interaction. MHC molecules are involved in helping immune cells communicate with other cells in the human body and, in particular, determining which cells can communicate with each other. For example, MHC cells are among those that determine whether a transplanted organ will be successfully accepted by a new host body. All of this takes place at the microscopic level within a person's bloodstream, unreachable by conscious awareness. However, recently, there has been speculation that people can in fact make subtle determinations about the nature of MHC cells. In particular, it is argued that they can detect the MHC cells *of other people* in various ways. And one way they can do this is by kissing them. According to Robin Dunbar, a professor of psychology at Oxford University, human saliva contains over a thousand proteins that provide information about a partner's physical compatibility and hardiness:

> Kissing is probably all about testing the health and genetic make-up of prospective mates ... So a good kiss, as much as a good sniff, not only allows you to check out how closely someone is related to you (related people will have smells in common) but also who isn't related to you and so might make a good mate ... kissing – and I don't mean a coy peck on the lips – provides a great deal of other information about a prospective partner .... (Dunbar, 2012)

Dunbar's allusion to 'a good sniff' relates to another way we can determine the MHC status of a sexual partner – by smell. True enough, a number of studies have found that people rate their partners as sexually more interesting when they share fewer MHC alleles with them, consistent with the view that such genetic diversity in parental MHC should lead to healthier offspring. Meanwhile, romantic couples whose genes lack MHC diversity are reported to be more likely to be unfaithful (Garver-Apgar, Gangestad, Thornhill, Miller, & Olp, 2006). However, few if any of these studies have tested whether MHC compatibility is determinable by kissing or smelling; and so there may be other reasons why reported sexual attraction comes to correlate with such variables. In today's world, developing a romantic relationship with another person (what biologically reductive people call 'mate selection') – not to mention the weighty decision to embark on infidelity – is likely to be heavily reliant on a host of social, cultural, and opportunistic factors. The point here is that there is almost certainly more to kissing and canoodling than the performance of biological cross-checks for genetic compatibility.

Let's recall that elongated evolutionary timeframe. Because of the excruciatingly slow pace of natural selection, even a tiny difference in our physiology or psychology will end up making a big difference to our evolutionary trajectory over time. So even if a tiny number of people could detect a tiny difference in the smell or taste of potential lovers, in a way that ended up causing even just a tiny benefit to the health of their offspring, then it is likely that such a difference would be favoured by natural selection. But because all these factors would be so tiny, we wouldn't be able to notice them. Their influence would be swamped by torrents of more immediate cultural and social inputs; their progress into our genotype would take thousands, if not millions, of years. So while some genes that affect immunocompatibility also affect the way we smell, it doesn't mean that we would ever be consciously aware of it, or personally swayed by it. Secondly, even if we *could* navigate our sexual relationships by smelling our way along, such a skill would be of little use. The underlying evolutionary processes will have, by their very nature, preserved features that were advantageous to our ancestors rather than to us. And because the vast majority of our personal ancestors lived in a world before language or civilization, eating uncooked meat and encountering prehistoric bacteria, what was important for their immunocompatibility might be totally irrelevant to ours. On the other hand, our present day romantic considerations – encompassing, as they do, physical appearance, mirroring of attitudes and interests, and mere propinquity – are very salient indeed. Compared to the imperceptibly minuscule effects of MHC, they are surely dominant.

Evolutionary narratives can be seductive without necessarily being robust. The main problem is a lack of appreciation for the complexity of what evolutionary biology involves. For example, the common view that evolution often takes care of nature by conferring it with backchannels and safeguards – such as the idea that evolution provides childless adults in order to ensure that orphans can be looked after – is almost certainly always wrong (Myers, 2014). Evolution is not an agent that has objectives for humanity, nor does it care about orphans. Nor does it manage the overall size of the population of species. Nor does it organize a competition in which only the fittest survive. Evolution is simply a process in which species thrive, or fail to do so, for different physical reasons. When it comes to psychological phenomena, evolutionary explanations become deficient if they are divorced from the nexus of interrelated social and cultural factors within which the human experience is enmeshed. Claims that particular human thoughts, feelings, or behaviours are 'the result of' evolution, even distally, are almost certainly incomplete.

Organisms as machines

Another major type of biological reductionism relates to the undue linking of psychological phenomena with physiological or anatomical features of the nervous system. This often involves the attribution of causative effect to hormones or other bodily biochemicals, or, perhaps more likely nowadays, to the structure or function of the brain. One example of the former is the hormone oxytocin. Oxytocin is a brain neuromodulator, which means that it diffuses throughout the nervous system and moderates the effects of multiple complexes of neurons. Its name comes from the Greek term for 'quick childbirth' (ὀξύς τοκετός), reflecting its clinical use for stimulating contractions during labour. The American biochemist Vincent de Vigneaud was awarded the 1955 Nobel Prize in Chemistry for isolating and identifying its structure, and since that time synthetic oxytocin has become widely used in obstetrics. Later researchers linked oxytocin to aspects of social interaction in various animal species: blocking the hormone made some animals more promiscuous and others neglect their offspring (Yong, 2012). From such beginnings the view gradually emerged that oxytocin was a chemical that caused animals to be nice to each other.

Interest in oxytocin in humans really exploded after a group of researchers administered it to college students during a laboratory-based economics experiment, finding that these students then trusted a partner with more money compared to students who had received a placebo (Kosfeld, Heinrichs, Zak, Fischbacher, & Fehr, 2005). One of the authors of that study, Paul Zak, a professor at both Claremont Graduate University and Loma Linda University in California, went on to develop an extensive programme of related research, and to publish a best-selling book, *The Moral Molecule*, in which he summarized the findings (Zak, 2012). In it, Zak describes his work as follows:

> My research had demonstrated that this chemical messenger both in the brain and in the blood is, in fact, the key to moral behavior. Not just in our intimate relationships, but also in our business dealings, in politics, and in society at large. (p. 11)

He goes on to explain that oxytocin is what 'makes us moral' (p. 274), and to prescribe a regime of eight hugs a day in order to ensure that sufficient amounts of the stuff is released into our blood flow to make us happy; not only that, but 'the world will be a better place because you'll be causing others' brains to release oxytocin' (p. 273). Most broadly, supporters of this approach suggest that people who live in countries associated with higher levels of oxytocin are more likely to report greater levels of trust in their communities (Zak & Fakhar, 2006), and to enjoy the benefits of greater economic growth and GDP (Zak, 2008).

However, the idea that one molecule could be responsible for one aspect of human nature (and such a profound one at that) is questionable. Once again, we run into the problem of biological reductionism: a single, and arguably trivial, biological dimension is presumed to override all the situational, cultural, environmental, and social features that might affect behaviour. For example, many of the studies into differences between countries rely on proxy measures, such as national rates of breastfeeding, to estimate oxytocin levels in populations. Surges of oxytocin due to breastfeeding are very short-lasting, and it is difficult to see how they would create community-level trust of a kind that would boost the national economy. Observed associations between breastfeeding and per capita income (e.g., Zak & Fakhar, 2006) could be due to any number of non-biochemical factors, not least the possibility that people with higher income might be more likely to choose to breastfeed in the first place. In any event, as there are no available data on whether such upward shifts in oxytocin can even be detected at the population level, the use of proxy measures makes these conclusions highly speculative indeed. And with regard to the laboratory studies that have linked administered oxytocin with prosocial behaviour, it is worth noting that several researchers have questioned their naively simplistic design (Yong, 2012), the statistical proficiency of their reported data analyses (Conlisk, 2011), and their interpretations of findings (Bartz et al., 2010). More recently, studies have shown that people administered with oxytocin can behave more dishonestly and selfishly if the situation requires it (Declerck, Boone, & Kiyonari, 2010). It seems that, rather than promoting morality, oxytocin might play a role in focusing our attention on social cues (Carter, 2014) – whether we use these cues to behave with greater or lesser generosity will be subject to the usual panoply of social and cultural influences, as well as our own personalities. In other words, claims that oxytocin can serve as a singular solution to humanity's macro-level problems seem to have been premature.

But by far the most common way in which behaviour is linked to biology must be through the brain itself. This has no doubt been aided by technological advances that allow us to produce images of how the brain functions and what its structure looks like. These images are typically produced using one of a handful of methods, such as computed tomography, or CT (a method for producing images of brain structure that involves taking multiple x-rays of the head from different angles), and electroencephalography, or EEG (a technique first used in the 1920s in which electrical activity is recorded from multiple different parts of the scalp, producing graphical representations of inferred underlying neural activity that can then be mapped to different parts of the brain). Probably the most significant imaging approach is functional Magnetic Resonance Imaging (or fMRI). fMRI records changes in blood flow in the brain using a very strong magnetic field, based on the principle that richly oxygenated blood will produce a greater magnetic pull than poorly oxygenated blood. By comparing sets of readings gathered in two different situations, researchers can compile images of the brain in which changes in blood flow can be highlighted. Images taken using fMRI have a very distinctive and recognizable format – against a black background floats a predominantly grey brain, with some areas vividly coloured in the style of a heat map in order to indicate the varying degrees of brain activity that have been recorded. These graphics have become somewhat iconic in our modern culture and are often employed figuratively in non-scientific contexts, such as advertising. However, it is worth bearing in mind that they are not real-time pictures of actual brain activity. In essence, they are computer-generated 'artist's impressions' of what brain activity might be taking place.

Indeed, the brain shown in such images is often not even the person's own brain. Rather, it is a standard template-brain onto which the person's brain activity has been projected. This requires various processes of digital image normalization, in which the recorded image of brain function is warped and smoothed to fit the generic brain template. A second point to remember is that the coloured highlighting is not based directly on recordings of brain activity. Rather, it is based indirectly on recordings of changes in brain blood flow (indeed, it is truly based on recordings of shifts in magnetic fields within and around the brain). As such, the validity of fMRI scanning is dependent on the assumption that brain blood flow faithfully reflects neural activation in both location and intensity. Thirdly, the interpretation of fMRI images is itself dependent on the assumption that neurons activated at a given moment serve some or other relevant and useful purpose. It could be that some neurons in the brain are *de*-activated during certain cognitive processes, or that the intensity of their activation is not directly correlated with the pertinence of their role. Fourthly, while the classic fMRI image suggests vivid but isolated brain activity taking place against a background of relative tranquillity elsewhere, this is not actually what the colour coding shows. The highlighting does not signify absolute activity, but activity relative to a prior set of measurements. The extent to which a particular brain area lights up will depend on the extent to which it was comparatively *less* active when the benchmark data were recorded. This means that an area of the brain might be shown as being 'active' simply because an increase in inward blood flow has been detected. There may be several other parts of the brain *more* active at that particular moment, but they are not highlighted as special because they were also highly active during benchmarking. And finally, even with the technological advances that have made fMRI a feasible tool for research in neuropsychology, the technique remains strikingly limited in granularity: the coloured highlights are based on the averaging of millions of neurons at a time, each of which is capable of firing hundreds of impulses per second (Fine, 2010). Imaging is an undoubtedly profound technology, but there are several technical caveats to bear in mind while gazing at those fascinating pictures of the human brain in action.

These caveats are not always appreciated. In fact, brain images are usually seen as being highly persuasive in their own right, especially by undergraduate students. In one study, students rated research articles containing fMRI scans as exhibiting better scientific reasoning than when the same articles were presented with the brain images expunged and replaced by mere diagrams (McCabe & Castel, 2008). Even though it could be said that fMRI images are themselves diagrammatic, it is as though readers take them instead to be somehow photographic, and attach to them the *seeing-it-with-my-own-two-eyes* weight that photographic evidence is typically presumed to carry.

Brain imaging research faces other limitations too. As mentioned in Chapter 3, brain imaging studies are typically based on very small samples of participants (this is partly due to the substantial expense of using imaging technologies). Mathematically, studies with small samples will be unable to detect genuinely true effects unless they are atypically pronounced in a given dataset. This means that reports of such effects will very likely be exaggerated, an outcome we can refer to as effect-size inflation. Perhaps even worse, it is also mathematically certain that small sample sizes will lead to a high number of non-effects being falsely heralded as statistically significant. As described in Chapter 3, one analysis of 461 neuroimaging studies found their average statistical power to be just 8 per cent (Button et al., 2013). This means that for every 100 non-null effects that exist to be discovered, these studies can be expected to find just eight of them. The fact that virtually all the studies reviewed reported statistically significant findings suggests that the vast majority of their results (92 out of every 100) were, in fact, false positives. Effect-size inflation means that those few which were not (the remaining 8 out of every 100) were probably exaggerated. The statistical problems do not end there. Even if we were to take the reported findings at face value, other reviewers have pointed out that neuroscience papers very often interpret statistics incorrectly. A review of over 500 papers from the five top-ranking neuroscience journals found that, for every paper correctly describing a statistical interaction, another paper reported such an interaction wrongly, drawing a false conclusion about its statistical significance (Nieuwenhuis, Forstmann, & Wagenmakers, 2011).

At a broader level of inference, critics of neuroscience have argued that brain imaging encourages psychologists to assume that discrete cognitive processes are linked to discrete brain areas. In reality, the neural nature of the brain is characterized by mass interconnectivity: while one area of the brain might be activated, it will be densely connected to literally millions of other areas at the same time, some or all of which might also be relevant to the psychological phenomenon being studied. The temptation to think of cognitions as being located in discrete localized sections of the brain, as if in the style of a modern form of phrenology, may be misleading (Uttal, 2001).

Such shortcomings might all be classified as technical matters, problems that in due course will be attenuated by greater awareness and better methodological rigour. However, there remains a fundamental conceptual challenge to brain imaging research: the graphical depiction of brain function is a *descriptive* activity, and generally not an *explanatory* one. It provides an account of what brain function is happening and where. The problem is that we already know that *some* brain function must be happening *somewhere*. This is because we know that psychological processes cannot occur independently of specific brain activity. When a person undergoes fMRI during a particular psychological process (such as a recognition memory task) we can predict that the resulting brain image will indicate some brain activity that might then be assumed to underlie that process. The fact that it does should be no surprise at all. What *would* be a surprise – a real shock, in fact – would be if *no* corresponding brain activity could be found. Likewise, if we know that psychological processes cannot occur independently of specific brain activity, it then follows that different processes must be derived from different specific brain activities. Thus, if fMRI scans of people engaging in different cognitive tasks failed to show differences in brain activities, this too would be grounds for astonishment. However, all too often the mere presentation of brain activity, or the identification of different patterns of brain activity that correspond with different cognitions, is greeted as a scientific breakthrough. We might better consider such findings as adding to the taxonomic classification of brain regions, a supplement to the knowledge we already have that the pertinent psychological phenomena themselves exist.

Let us briefly consider an example. In one well-cited study, researchers used fMRI to scan the brains of a sample of men and women who were 'in the early stages of intense romantic love' (Xu et al., 2011). They found that when participants were shown pictures of their lovers, parts of the brain normally associated with reward and motivation became highly active. Moreover, activity in the superior frontal gyrus (a part of the brain's frontal lobe) was associated with higher relationship satisfaction when the participants were followed up a year and a half later. However, let's consider the points made above. Firstly, it would be a significant problem if elicited emotions of romantic love were found to be associated with *no* brain activity at all. This would suggest that feelings of infatuation were occurring independently of the brain, which would then raise the question of how our consciousness of such emotions could otherwise be explained. Therefore, the fact that brain activity was observed in this study is not only unsurprising, it obviates a mystery. Likewise, the fact that largely similar brain activity was observed across all the participants is equally unremarkable: they were all shown pictures of their lovers, and so were made experience similar feelings. Further, the finding that this brain activity was similar to that associated with other types of elation is, again, highly reasonable. It suggests that romantic love is similar to other emotions that lead to elation. Finally, the fact participants with active superior frontal gyri turned out to have more stable relationships simply correlates with the likelihood that these participants had different patterns of behaviour and cognition as well. These findings do not explain what *causes* romantic love – where it comes from, when it might be expected to occur, what makes it last, or how to respond when it evaporates. They simply confirm that feelings of romantic love involve particular brain activity (which we assumed to begin with), that it is similar to emotions that we already recognize as being similar to it (which is a tautology), and that the kind of romantic love which leads to relationship stability is different from that which doesn't (which is yet another tautology). This type of study also suffers from a falsifiability problem. Had the researchers found a different part of the brain (other than the superior frontal gyrus) to be associated with relationship success, they would simply have reported so. In other words, there was no hypothesis to be falsified, which renders the research descriptive rather than explanatory. (We might generously pass over the fact that, with a sample of just 18 participants, the study's statistical power was so low as to be fatally debilitating.)

To extend the point slightly, let us imagine that being shown pictures of a lover would make a person blush. Blushing is caused by vasodilation of facial veins, where capillaries of the skin become exaggeratedly profused with blood. In humans, unless there is an alternative medical cause (such as the condition known as idiopathic erythema), blushing will almost certainly occur simultaneously with a strongly felt and socially relevant emotion. However, were we to observe consistent blushing in a group of 18 college students presented with pictures of their lovers, we would not likely declare that capillaries in the face were thereby implicated as a causal factor in human romantic love. It might well be interesting to know that people blush in this situation, and to speculate as to why. However, the mere observation of this phenomenon would allow us to do little more than catalogue it as yet another physiological aspect of the experience of emotion. With brain function, we are in a similar situation. The occurrence of brain function simultaneous to romantic emotions is essentially a correlation. Attempts to impute it with causal relevance amount to interpretational overreach.

Virtually all brain imaging studies in contemporary psychology can be critiqued from this perspective. Studies that show the brain activity of substance users to be different from that of non-users (Motzkin, Baskin-Sommers, Newman, Kiehl, & Koenigs, 2014) simply corroborate our expectations that (a) addiction involves the brain, and (b) addiction is different to non-addiction. The fact that the part of the brain so affected is also known to be active when rewards are processed is consistent with the already obvious idea that, for the addict, the addictive behaviour is rewarding. Similarly, research that shows pathological over-use of the internet to be reflected in characteristic brain structure (Lin et al., 2012) simply confirms for us that patterns of behaviour which are sufficiently conspicuous for society to declare them abnormally exaggerated will involve the brain. We would hardy assume otherwise. And when religious believers answering questions about religion are shown to experience different brain functions to non-believers (Harris et al., 2009), should we be astonished? We know these things before any brain images are produced: thoughts, feelings, and behaviour involve the brain; *ergo* differences in thoughts, feelings, and behaviour involve differences in the brain.

Conclusion

None of this is to say that brain imaging is scientifically unimportant, or holds no promise for psychology. Like all areas of science, brain imaging research can be pursued well or badly. Furthermore, taxonomy is an important activity and a feature of many sciences (such as botany or entomology). However, while the technology required to produce modern brain images represents the cutting edge of radiography, it does not necessarily follow that research based on its use will represent cutting-edge psychology.

Biological reductionism is limited by its inherent divorcing of psychology from the situational, social, and cultural contexts that envelop the human condition. To the extent that it does so, such reductionism can lead to pseudoscience. The necessary complexity of evolutionary, biochemical, and neuroanatomical approaches brings with it significant barriers to parsimony, a failing shared with many pseudosciences. Similarly, while a tolerance of imprecise measurement is a common feature of pseudoscience, it is something of a fact of life in brain imaging research. But perhaps the biggest challenges relate to falsifiability. It is a cliché to state that evolutionary theorists face a difficulty in being unable to directly observe evolution across historical epochs. That said, reasonable hypotheses can be formulated concerning the impact of evolution on organisms alive today. Problems with falsifiability arise when an evolutionary basis is assumed to override other factors. Take the idea that your choice to kiss somebody will derive from a desire to maximize the fitness of your offspring: at the level of such an individual behaviour, it is impossible to conceive of evidence that would truly contradict this explanation. Similarly, the mass-scale complexity of the biochemical and neurological aspects of behaviour heightens the risk that hypotheses about the impact of hormones or brain function will fall short on granularity. As such, when researchers encounter unexpected results, they can reasonably choose not to reject their theories, but instead to modify them. This might be excusable in some instances, but in others it might amount to a basis to ignore the need for hypotheses to be falsifiable. By providing a way to bypass the exacting requirements of scientific rigour, a slippery slope to pseudoscience is opened up.

Any diminution on falsifiability elevates the risk of confirmation bias. Conclusions drawn from research end up reflecting the researcher's prior assumptions, rather than any compelling implications of the data. Claims that men and women's brains differ in structural ways that are of real psychological relevance often far exceed whatever certainty any research findings have to offer. Indeed, they can appear quite far-fetched when you consider the fact that men and women's brains are so physically similar, it would be impossible for an anatomist to determine gender if presented with an individual specimen. Similarly, the idea that gender differences are the result of evolution are not insignificantly complicated by the fact that all men and women have an equal number of male *and* female ancestors; individuals are not clones of their parents, but combinations of two differently gendered half-clones. Likewise, studies of hormonal influences on morality, or of brain function and addiction, can end up reflecting more the value-systems of the investigator than the evidentiary value of the data (we will return to this impact of personal belief on research in Chapter 8). And the drawing of firm conclusions from tiny studies that are virtually devoid of statistical power hints at a level of overconfidence that some observers might consider flagrant. Biological aspects are undoubtedly inherent to the human condition, and the study of biological psychology is essential if we are to make sense of psychology as a whole. However, we should not get carried away with the idea that *biological* is synonymous with *scientific*. When reductionism leads scientists to persistently overlook critical limitations and draw empirical conclusions regardless, the boundary with pseudoscience cannot be far from view.

Chapter 7

Examples from the Mainstream: What Some People Say about What They Think They Think

Introduction: Why not just ask?

So, if biological reductionism is a problem, then what might be the solution? According to many psychologists, these obstacles are best overcome by circumventing biological paradigms altogether and reverting to a more common touch. If biological approaches are divorced from the individual human perspective, then the solution must lie in restoring the link: simply put, if you wish to explain the human experience, then go talk to some humans.

The idea that psychologists can best explore human psychology simply by talking to people takes many forms. It also rests on many assumptions. First is the basic assumption that the human experience is made up of discrete thoughts that occur within the human mind, most likely in the form of an inner narrative or 'live commentary' that comprises the individual's autobiographical monologue. A second assumption is that thoughts are available to the person who has them, and can be called to consciousness both readily and on demand. A third assumption is that most people will be willing to describe these thoughts should a researcher ask them to do so. And a fourth assumption is that experiences so described will inform psychologists in their attempt to extract generalizable principles about how humans think, feel, and behave. On these premises rests a huge portion of psychological research conducted over the past century, in both pure (e.g., social, developmental, personality) and applied (e.g., health, occupational, educational) domains.

However, each of these assumptions can be questioned in elementary ways. The first is perhaps the most obscure, but also the most fundamental. Whether or not human minds generate discrete sequential thoughts is very much open to doubt. Far more likely is the prospect that minds continuously generate a multiplicity of simultaneous thoughts, in the form of parallel processing. The idea that there is an inner monologue constantly representing our psychological state at any given moment is unlikely to fully account for the way we think. This also means that the second assumption – that people can directly access thoughts in a manner that supports their reporting them to a researcher – is mechanically questionable. Thoughts are not produced in such a way as to present themselves neatly to the person in whose mind they are formed. As described in Chapter 4, people often use obtuse reasoning to produce logically indefensible conclusions, and frequently derive erroneous impressions of what they remember. Thoughts not afflicted by heuristic error will nonetheless be subject to self-serving bias. The generation of ideas in the human mind is not ordered, apparent, and easily parsed; on the contrary, it is fuzzily sporadic, camouflaged by mental shortcuts, and virtually automatic. Such automaticity makes it parlous to assume that people even know what it is that they think.

A commonly cited example of cognitive automaticity occurs when we engage in a well-learned behaviour, such as driving a car. Driving requires a person to monitor their perceptual environments on a moment-by-moment basis, to continuously make a series of *if-then* type decisions regarding this input, to initiate actions (such as steering, signalling, and braking) as and when they are required, and to engage in several other related cognitive processes. When they are well practiced, these activities are achieved as though they were instinctive. Motorists are most likely to not even notice the detailed thinking they engage in as they glide along the roadway. Indeed, so submerged are their cognitive elaborations that drivers often feel their minds are sufficiently unencumbered that they can concentrate adequately on other things, such as conversing with a passenger or listening to the radio. Only in the event of a sudden interruption, such as a near-miss encounter with another vehicle, will specific driving-related cognitions intrude into conscious awareness. (One way to illustrate this point is to recall what happens when such behaviours are *not* well learned. For example, when you are taking part in your very first driving lesson, the sheer complexity of the required cognitions will probably *overwhelm* your consciousness to an extent that will make the task of driving seem a terrifying ordeal.) The point is that people who drive cars do so without having an up-to-the-second awareness of each and every thought required for the job. Asking them why they have driven in a particular manner might not be the best way to find out what they were thinking at the time.

Of course, an interview with a motorist about specific acts of driving might not make for the most compelling psychological research. Yet cognitive automaticity is not confined to banal or routine activities. Take, for example, the cognitions of a person attempting to console a distressed friend. When offering support to someone who is distraught, we try to choose very carefully not only what to say but also how to say it. To ensure that we react sympathetically to their responses, we monitor and recalibrate our tone as we go. We will approach the interaction having taken account of our own previous experiences, both direct and vicarious, as well as any relevant advice that we can remember receiving from others. Our choice of words will be influenced by our assessment of the various available options as we imagine and understand them, and these in turn will be based on the extent of our vocabulary and the conditions under which we learned each phrase and idiom. And our overall approach will be influenced by our own emotions at the time, including the sadness we will feel when witnessing our friend's distress. In short, our task will be inordinately multi-faceted and heavily laden with nuance. Nonetheless, it is one which we will probably pursue in an intuitive, almost instantaneous, fashion. We go with the flow, we open our mouths, and things come out. We won't deconstruct all the considerations we have to make: all the *if-then* choice points, the individual inferences about our friend's facial reactions and vocal pitch, all the momentary readjustments of our own demeanour and our efforts to act with appropriate empathy. If called upon afterwards to explain each and every choice we made (*Why did you use that tone of voice? Why did you choose that phrase? Why did you stand like that?*), we would have great difficulty telling a researcher the exact thoughts that led us to our decisions. As we will see in Chapter 9, such helping behaviour is of great interest to psychologists: many studies have examined the various informal ways we all try to support each other through distressing experiences, and many more have sought to identify the best way of doing so in a systematic therapeutic context. However, given the extent to which the critical details of these effectively automatic processes lie beyond the reach of our awareness, efforts to ask people about their thoughts when navigating these interactions seem almost pointless.

And then we have the third assumption – that people are willing to describe their thoughts when asked to do so. We have already considered how people moderate their utterances, engage in self-censorship, and embellish what they say for various reasons. When relaying events, witnesses frequently exaggerate some details in order to highlight what they feel is pertinent. Alternatively, they might choose to omit other details for the same purpose. The result of either approach is the same: an impression conveyed to listeners that is flawed in terms of accuracy. But perhaps of more concern is the way people's reports are affected by social motivations that arise from modesty and vanity. The number of respondents who will deliberately cast themselves in a negative light, or even run the risk of doing so, can be presumed to be very low. Indeed, very many people will expend effort to cast themselves in a deliberately *positive* (as opposed to neutral) light. For these reasons, when considering the reliability or accuracy of what people tell us, we must factor in the way they are trying to manage our impressions of them. This is hardly an unknown requirement: the problem of social desirability bias is one of the most discussed by research methodologists in psychology. However, the fact that psychologists readily recognize the issue, and ponder it at length, does not guarantee that they have any certain way of circumventing it.

With social desirability bias comes an inevitable interpretation paradox. Given the flimsy reliability and objectivity of anecdotal evidence, we can presume that researchers would prefer to use other methods should they be available. So the main reason researchers end up directly interrogating their participants (or giving them questionnaires, which amounts to much the same thing) is that they have concluded that no other such method *is* actually available. The subject matter is not something that can be observed using other approaches. The very fact that we must go to the trouble of asking people to tell us something indicates that their views on the matter are not ordinarily available for public scrutiny. We are inquiring about subject matter that the person has chosen not to broadcast, private thoughts that are hidden from view, presumably for a reason. Moreover, once disclosed these thoughts cannot be made private again. The very irreversibility of disclosure will likely make many respondents quite reticent. Thus, whether used to measure participants' feelings, attitudes, or past behaviours, self-report methods are most often targeted at *precisely* those subjects about which participants are cautious. In short, self-report is most likely to be used when social desirability bias is most likely to be a problem.

As an obvious example, take the issue of sexuality. Because human sex is primarily a private behaviour, nearly everything we know about it is filtered through the lens of social desirability bias. (In this regard, it is tempting to refer to a popular online parody video news reel, which describes the results of a new Teen Sex Survey: 'Nearly 100% of boys, ages 12 to 15, report that they have sex *all the time* and are definitely *not* virgins.'; 'Teen boys losing virginity earlier and earlier, report teen boys', 2014.) It can be argued that any subject matter with an emotional aspect will fall foul of similar difficulties. Emotions are experienced privately and so, to establish their existence (or measure their intensity), we must rely on their being communicated second-hand by the person experiencing them. The problem is that human emotions are blatantly communicative: people regulate their expression in an attempt to control what they reveal. As a result, often what is communicated is not precisely what is felt. In one study conducted at my own laboratory, a group of college students were asked to perform a logic task on a computer (Hughes, 2007). Immediately afterwards, some of the participants who had performed quite well on the task were misled into believing that they had done very poorly (we told them that their scores were as bad as the weakest competitors in their age group). In essence, we fooled them into thinking they had flunked the task. When asked how this made them feel, the group reported no more distress than other students in the study. However, all the students were having their blood pressure monitored, so it was possible to examine other aspects of their state of mind. The cardiovascular data showed that these students exhibited large spikes in blood pressure, akin to those of people undergoing acute mental stress. Students who were given neutral or positive feedback showed no such changes in blood pressure, suggesting that their feedback caused them little or no distress. Overall, therefore, the story told by the cardiovascular data cast doubt on the validity of the self-reports: students who thought they had flunked were unwilling to tell us they were disappointed. It is worth recalling that this was all in the context of a modest laboratory experiment at a university, involving anonymous participation in a five-minute computerized mental rotation task. Nonetheless, social desirability served to undermine the value of self-report in this banal context. It is hard to imagine that participants would have been particularly *more* forthcoming had they been asked to talk about their political opinions, personal values, altruistic tendencies, mental health, hygiene habits, condom use, intentions to quit smoking, or any of the real-life topics addressed by studies in which self-report methods are typically employed.

The fourth and final assumption is no less of a minefield: the idea that the testimony of a given group of participants can be used as the basis to make broad inferences about people in general. This is the straightforward enough issue of sampling validity, where researchers must think about whether the people they are studying are sufficiently representative of the population they wish to describe. The problem of ensuring sampling validity is not unique to studies that employ self-report methods. However, it is an especially important consideration for these studies, not least because of the vast range of ideas and opinions that human beings can have. It is easy to imagine that even a large sample of participants in a particular research study might hold views that are different to those of people not involved in the study. Indeed, because of the social desirability problem, the very fact that some participants are willing to participate in a study at all can mean that the views they report reflect a frame of reference that other people may not share. Sexually conservative people may not wish to participate in a study of sexual attitudes; emotionally reserved people may not wish to participate in a study of mental health. This conflation of sampling with subject matter is an acute problem for studies where the subject matter is sensitive. And as is argued above, self-report methods are usually resorted to for precisely those types of subjects.

In culmination, it can be sentimentally attractive to advocate the view that psychologists who wish to learn about the human condition should simply talk to people. However, the various questionable assumptions underlying such a strategy mean that the data collected will be hampered by a number of qualifications. Rather than yielding direct insights about *what people think*, it is safer to consider such data as comprising *what some people say about what they think they think*.

Our survey says...

Most self-report data are gathered through questionnaire-based methods. Sometimes this involves an attempt to solicit open-ended testimony, such as when participants are asked to describe their feelings about a specific topic. On other occasions, questionnaires are carefully prepared using a number of statistically-grounded methods that together are referred to as *psychometrics*. Psychometric tools usually require participants to supply a series of ratings or other quantitative responses to sets of rigidly structured requests. Technically, this latter style of instrument is not always a true 'questionnaire', in the sense that its various items may not be phrased as questions. Instead, the instruments might instruct participants to rank a list of options in order of their preferences, or to declare which of two (or more) statements best describes their feelings. A key advantage of these tools is that, by tightly structuring the format of participants' responses, it becomes possible to make reasonable assessments of response patterns using statistics. For example, it is possible to establish the statistically average response to a particular question, to compute numerical summaries of the respondent's own response tendency or style, and to identify profiles or patterns of responses across large numbers of questions in ways that then facilitate comparisons with other people. With this overall approach, psychologists have been able to develop relatively brief instruments that seek to measure a huge range of psychological variables, including attitudes, emotions, and mental health symptoms. By and large, these tools have proven to be valuable.

For example, when used in mental health, such psychometric instruments enable clinicians to form a confident view as to whether a client's self-reported feelings are close to the average of those reported by other similar people, or whether they are sufficiently far from the average as to warrant clinical attention. In the same way as a person's blood sugar or bodyweight can be classified as either normal or clinically high by comparing them with measures taken from healthy and unhealthy people, similar protocols can be developed using psychometrics to gauge a person's level of depression, anxiety, or (say) obsessive-compulsive thinking. Similarly, an occupational psychologist can employ a statistically-based psychometric tool to assess whether a candidate's attitudes resemble those of good managers, of effective innovators, or of incorrigible procrastinators. In these contexts, the strength of the psychometric approach lies in the ability to compare an individual participant's responses with statistical benchmarks derived from normative population samples. Without such data (including, critically, the benchmarks), practitioners would be left relying entirely on personal judgement to decide whether a client's psychological functioning was sufficiently distinct as to require comment or intervention.

Such tools have also proven useful in pure research contexts, by allowing psychologists to develop systematic ways of studying differences across people. A field where this has very clearly been the case is the study of personality. By using psychometrically developed questionnaires, psychologists have refined the way we understand how human personality is expressed, its consistency over time, and the way dispositions are inherited by children from their parents. Statistical techniques such as factor analysis have derived strikingly stable patterns in personality questionnaire data, across time and across cultures. For instance, virtually a century's worth of data now supports the view that human beings around the world can usefully be described in terms of a dimension ranging from introversion to extraversion. Every person can be scored as being extremely introverted, extremely extraverted, or somewhere specific in between. Moreover, the data shows that wherever a person is ranked on the dimension between the extremes, it is highly probable that they will continue to be ranked at roughly the same point whenever they are assessed in the future. It is even possible to statistically establish that introversion–extraversion is around 54 per cent heritable (Bouchard Jr & McGue, 2003). This means that across a group of people, more than half of the variation in this trait will be attributable to genetic, rather than environmental, factors. By comparison, this is around the same level of genetic heritability as has been established for body-mass index (Silventoinen, Magnusson, Tynelius, Kaprio, & Rasmussen, 2008) and cardiovascular disease risk (Fischer et al., 2005), two health characteristics that most people recognize as being biologically ingrained. As well as introversion–extraversion, psychometric research has found humans to be conspicuously characterized by other traits in ways that are similarly consistent and genetically heritable, such as their levels of emotional stability (which range from extremely stable to extremely unstable). In all, the number of major traits that have been identified as universally applicable to human personalities is believed to be relatively small, with the consensus largely favouring the idea that it is no more than five.

In sum, the psychometric approach to questionnaire research has been found to be useful. However, it is always worth remembering that this usefulness is almost entirely derived from the statistical nature of psychometrics. The scores produced by psychometric tools are contrived to be statistically consistent. Tools found to show poor reliability – that is, to yield scores that are not stable across repeated administrations – are deliberately overhauled or else simply abandoned. And when psychometricians modify a questionnaire, they do not dwell on the content of the questions; instead, they retain or drop questions on the basis of whether their responses exhibit the desired statistical stability. In this sense, the actual content of the question is almost irrelevant. Take the case of a psychometric questionnaire designed to measure a person's competence as a manager. If a particular question – for example, *'Do you enjoy watching television?'* – consistently shows a high degree of statistical association with managerial competence (acclaimed managers always give one particular response, while notably poor managers always give a different one), then this question will be retained even though its phrasing appears to have little or nothing to do with management ability. On the other hand, if a question consistently shows a *low* degree of association with managerial competence – for example, *'Do you enjoy being a manager?'* – then it will be discarded even though its phrasing appears, at face value, to be *very* relevant to management ability. The fact that the content of the latter question *appears* to be more related to management has no bearing on its usefulness in psychometric terms. The former question is more useful because, even though its wording concerns other things, good managers answer it in a way that makes them statistically discernible from bad managers. It is this statistical consistency that makes such a psychometric instrument actually work.

The big problem with questionnaires relates not so much to the method itself, but to the way casual observers view the field. To casual observers, the statistical underpinnings of psychometrics are typically invisible. They do not readily recognize that so long as a correlation is established between response patterns and outcomes, it hardly matters if the questions being responded to are semantically linked with those outcomes. Very often, people tasked with designing a questionnaire (such as psychology students embarking on a research project) become very invested in the content of the questions, and assume that the intrinsic value of the tool can be gauged from the way each is worded. They overlook the fact that usefulness is in fact determined by the comparative linking of responses with independent information concerning the construct that is to be measured (such as whether particular responses are consistently returned by good managers, but not by bad ones). It is only through examining statistical associations and comparisons with benchmarks that meaningful conclusions can be drawn. Casual observers tend to make the false inference that questionnaires, in their own right, are inherently useful in face-value terms. They feel that asking managers about management simply must be more informative than asking them about television, regardless of any amount of statistical data showing that what people say about television is more closely correlated with managerial prowess. Of course, when we refer to casual observers, we are not exclusively talking about amateur social scientists or lay readers of the psychological literature. It would appear that many mainstream researchers, including some who make questionnaire studies their life's work, observe the psychometric side of things quite casually indeed.

Ignoring the statistical potential of psychometrics, and instead relying on the semantic content of questionnaires to interpret responses, reduces the resulting data to the status of *what some people say about what they think they think*. It means that researchers are ignoring the various qualifications that make self-report testimony unreliable, and are holding firm to the four naïve assumptions that inform the idea that simply talking to people will generate useful insights about human psychology. This then leads to the problem of research findings that cannot be substantiated or, worse, the claim that particular findings are the result of scientific research when in fact they are based on evidence that is little more than anecdotal. Let us consider just a few illustrative examples of recent research appearing in major journals of psychology.

Example #1. How afraid are you of terrorists?

Surveying is a common way to attempt to gauge people's views on major political or cultural circumstances of the day. In one study, researchers used survey data to investigate whether British people's levels of perceived terrorist threat was a factor in determining their levels of social prejudice (Greenaway, Louis, Hornsey, & Jones, 2014). The researchers also aimed to assess whether people's sense of control over life moderated this link, based on the theory that feelings of control serve to nullify the way threat leads to paranoia. Having analysed data from over 2,000 UK citizens, the authors concluded that their theory was borne out. However, the measurement of all relevant variables was clouded by the use of self-report methods.

Take the measure of perceived terrorist threat. This was based on responses to the following question: '*Do you think a terrorist threat somewhere in the UK during the next 12 months is [not at all likely/not likely/ likely/very likely]?*' At the very least, this question is quite vague. What, for example, might be meant by the term '*terrorist threat*' in this context? An actual terrorist attack, or the mere existence of some unknowable and unrealized risk? Are we to take it that a *threat* is only deemable as *highly likely* if it is inevitably going to be followed through? Or is it legitimate to say that a *threat* is *highly likely* when there is some terrorist somewhere in the community, who has some vague but as yet unconsummated intention to consider action in the future? It is certainly conceivable that different people will attach different meanings to such a question, which then makes it difficult (if not impossible) to interpret the responses given to it. Secondly, take the measure of social prejudice. This was based on answers to a number of questions, including: '*Would you say it is generally good or bad for the UK's economy that people come to live here from other countries?*' Once again there is a level of vagueness to such a query. For one thing, it is possible to take a purely economic approach to the issue without being influenced by social prejudice at all (in other words, it is possible to hold a generalized view about the impact of transnational migration on economic growth in tariff-bound territories, without being addled by racism). At the other extreme, responses may be economically uninformed but driven entirely by xenophobia. Therefore, it is impossible to know whether a person answering such a question in the negative is socially prejudiced or not. In this case, the problems are undoubtedly compounded by social desirability. We can easily imagine that participants might be reluctant to admit to opinions that others may perceive as prejudiced; we can even expect that people's *honest* utterances will not fully reflect the degree to which social prejudices are implicit in their worldviews (Greenwald, McGhee, & Schwartz, 1998). In short, even though 2,000 Britons have reported their attitudes to the survey-takers, it appears wholly unsafe to treat their responses as *de facto* measures of the variables under investigation.

Example #2. How afraid are you of death?

In a second study, which addressed somewhat overlapping themes, researchers sought to investigate the way people's attitudes to mortality affect their willingness to become political martyrs (Orehek, Sasota, Kruglanski, Dechesne, & Ridgeway, 2014). For attitudes to mortality, the researchers employed a previously standardized psychometric tool, which was certainly a strength of the study. However, for martyrdom, the researchers took the approach of simply asking their participants questions that, at face value, concerned the relevant subject matter. Specifically, they asked the participants to rate their agreement with the following two statements: '*If faced with circumstances that required as much, I would sacrifice my life for a cause that was important to me*' and '*I would not sacrifice my life for a cause highly important to me*' (responses to the latter question were reverse-weighted when being combined with responses to the former). It can be noted that the 119 participants were all students at a North American university, and so were unlikely to have had day-to-day contact with political martyrdom. Nonetheless, even as an experiment dealing in hypotheticals, the reliance on self-report once again produced anomalies.

As before, a major source of confusion concerned the vagueness of the questions – in this case, the use of the phrase '*sacrifice my life*'. For some people, being willing to *sacrifice one's life for a cause in circumstances that require as much* might involve becoming a suicide bomber. However, for others, it might involve something far less malign. For example, it might involve refusing to move from your civilian home even though you know your enemy is about to launch a deadly air strike (in fact, it may be reasonable to speculate that the latter interpretation would apply to more people than would the former). It can further be argued that answers to questions on martyrdom will inevitably overlap with those on mortality attitudes. Both sets of questions ask participants to indicate their willingness to die; the statistical association of the responses more likely reflects this overlap in meaning than it does the existence of two discrete cognitive constructs, one causally driving the other. Notwithstanding the effort invested in presenting these questions as part of a systematically structured research study, it seems doubtful that cognitions relating to martyrdom can be gauged simply by asking participants to report them.

Example #3. How violent are you? (And how much do you eat?)

In the next example, researchers attempted to apply self-report methods to younger participants. In a study of teenage girls, a team of researchers sought to establish whether a history of violent behaviour was associated with a history of weight-loss dieting (their theory drew on previous studies suggesting links between adolescent weight-control and aggression; Shiraishi et al., 2014). Over 9,000 girls completed the necessary questionnaires. Based on statistical analyses, the researchers concluded that past engagement in weight-loss dieting was indeed associated with higher levels of violence towards both people and objects. However, as all variables were quantified on the basis of self-report data, the actual meaning of the statistical result is terribly unclear.

The main problems here relate to the fact that both target variables are socially sensitive. It is hard to imagine that all teenagers will be equally forthcoming in providing accurate information about their history of past violence. Many participants will inflate reports of such behaviour because of teenage bravado; others will do the opposite, especially if they fear that admissions would expose them to a risk of criminal prosecution (a concern that reassurances about confidentiality are unlikely to fully assuage). Likewise, it is doubtful that reports of dieting history will be straightforward either. For one thing, participants with extremely disordered eating habits will be reluctant to report them to others, and some participants will be unable to admit their full extent, even to *themselves*. In fact, there may be a logical reason for self-reported violence to be associated with self-reported weight-loss dieting in such datasets. It reflects the fact that people vary in the degree to which they are inhibited when asked to reveal risqué aspects about themselves: most people who are shy about reporting one of these behaviours will likely be shy about reporting the other as well.

Example #4. How lazy are you?

In the final example, researchers attempted to apply self-report methods to children. In this case, the researchers were interested in finding out whether physical activity (as opposed to sedentary behaviour) was associated with academic success in primary school (Haapala et al., 2014). Having analysed data from 186 school pupils, they concluded that children's physical activity levels were predictive of both their literacy and their numeracy: the more physical activities they engaged in, the better the children were at reading and counting. However, while academic abilities were derived from (presumably objective) school tests, the information on physical activity was based on self-reports. There are a number of reasons why such an approach might be unreliable.

Firstly, the children were very young – they were aged between 6 and 8 years old – and so may have lacked the concentration or conceptual understanding required to provide accurate answers to the several questions asked (for example, the children were asked to report separately the extent of their engagement – in minutes per day – in supervised exercise, organized sport, organized non-sport exercise, unsupervised physical activity, physically active commuting, physical activity during school breaks, and so on). Secondly, because they were so young, their parents helped them to complete the questionnaires. While this may have improved the accuracy of some responses (such as reports of physical activity in which the parents were involved), it may have had an adverse influence on others. Indeed, the social desirability of the parents may have influenced the answers of the children (after all, few parents would wish to be perceived as failing to encourage their children to be healthy). Ultimately, the fact that *self-reported* physical activity was found to be correlated with literacy and numeracy may reflect the possibility that children who were less good at explaining and less good at counting answered the questions differently from their peers.

Considering individual examples helps to illustrate the ways in which research that is reliant on self-report will always face certain limitations. These limitations correspond to the underlying assumptions: assumptions relating to the formation and accessibility of thoughts, the willingness of participants to share their thoughts, and the degree to which such thoughts truly represent the discrete constructs being investigated. These problems are not isolated to a few recent (and hand-picked) cases. When the major journals of social psychology are scrutinized, similar methods appear prominently and frequently. Of articles describing empirical studies that appeared in the *British Journal of Social Psychology* during the last decade, the top five most cited included the following: a study linking self-reported health habits to self-reported self-esteem (Verplanken, 2006); a study linking self-reported social identification with self-reported social support and self-reported life satisfaction, but not self-reported stress (Haslam, O'Brien, Jetten, Vormedal, & Penna, 2005); and a study linking self-reported cultural stereotypes with self-reported in-group favouritism (Cuddy et al., 2009). During the same period, the top five most cited empirical papers in the *Journal of Personality and Social Psychology* included: a study linking self-reported positive emotions with blood pressure, but also with self-reported benefit-finding during negative situations (Tugade & Fredrickson, 2004); a non-blinded intervention study where self-reported positive emotions were linked with self-reported social support, self-reported life satisfaction, and self-reported depression (Fredrickson & Cohn, 2008); and a study linking self-reported political orientation with self-reported values and moral judgements (Graham, Haidt, & Nosek, 2009). It is clear that self-report survey methods are heavily relied upon in certain areas of psychology. However, whether the problems associated with the underlying assumptions are ever properly dealt with, or are simply ignored, is less clear.

Researchers will often argue that surveys are the only feasible method with which certain variables of interest can be examined. Certainly, many variables – such as self-esteem, political orientation, personal value systems, intentions, life satisfaction, and depression – are difficult to examine without consulting participants directly. Private behaviours – such as sexual activity – are also hard to examine objectively, but for other reasons. However, these variables are rarely ever *impossible* to examine in an objective way (a person's political orientation might alternatively be inferred from their behaviour; and, while still presenting a minefield, a person's self-reported sexual activity might be corroborated by consulting their partner). As such, in academic debates it occasionally appears as though such defences of self-report are ritualistic rather than fully thought through. For example, in one high-profile article defending self-reports in health behaviour research, two prominent authors offered this observation:

> It is virtually impossible to obtain objective measures of some health related behaviors (e.g., condom use), and for many others (e.g., exercise, physical check-up) objective measures are expensive and time consuming. (Ajzen & Fishbein, 2004, p. 432)

However, in the *very same paragraph*, the authors went on to argue the following:

> In some behavioral domains, such as condom use (Jaccard, McDonald, Wan, Dittus, & Quinlan, 2002)...self-reports are found to be quite accurate... . (Ajzen & Fishbein, 2004, p. 432)

Note the authors' point about condom use: they simultaneously describe it as *virtually impossible to measure objectively* and yet *amenable to quite accurate measurement by self-report*. But how exactly can they know that self-reports are 'quite accurate' if it is 'virtually impossible' to produce objective measures with which they can be compared? The paper they cite to support their statement inferred accuracy of self-reported condom use from the fact that such self-reports are often highly correlated across individuals (Jaccard, McDonald, Wan, Dittus, & Quinlan, 2002). In other words, the merit of one person's self-report is to be determined by its resemblance to *another* person's self-report. This appears very similar to the basis on which conspiracy theorists attach credibility to claims that the Loch Ness Monster actually exists.

Put another way, the idea that self-reports are self-verifying exposes research to precisely those pitfalls of anecdotal positivism that the scientific method was intended to avoid. The fact that some researchers simply find it difficult to imagine other ways of examining their subject matter does not, in and of itself, make pure self-report – the kind intended to be read at face value, with no psychometric benchmarking against population norms – any less weak.

Correlation, causation, conflation

In Chapter 1, we noted how difficult it can be to interpret correlations. When X is correlated with Y, it can mean a number of things. It can mean that X caused Y, that Y caused X, that Z caused *both* X *and* Y, or that X and Y happened at the same time by coincidence. When an observed X–Y correlation has been inferred from *what some people say about what they think they think*, a further scenario arises: maybe X and Y are actually *the same thing*. In other words, maybe the labels 'X' and 'Y' are simply two ways of referring to the one underlying concept. This is a particular risk with anecdotal data because, psychologically, we cannot verify whether or not those entities people think are discrete are in fact discrete. When people choose to discuss a subject, its elements and associated thoughts exist first and foremost in their minds; the fact that they talk about things as though they were separate does not actually guarantee that they are separate things at all. What is so often referred to as a problem of correlation and causation could actually be one of conflation.

Imagine a study where a researcher is investigating 'happiness in general' and 'happiness with daily life'. The researcher duly starts by asking participants some questions about happiness in general. After this, the researcher asks questions about happiness with life. And thus the study takes place. However, although it is possible to concoct different questions for each concept, it is far from clear that the two are in fact authentically separate – it seems much more likely that they are the same underlying construct being referred to using two different terms. But the mere fact that it is possible to set aside different *questions* for each means that the researcher can compile separate *data* on each. This is of course problematic: when the researcher asks about the first concept, the answer given will convey information about both; then, when the researcher goes on to ask about the second, the next answer will be similar. In the end, it should be no surprise when a statistical test shows the responses to be correlated. It isn't that there are two separate variables (an X and a Y) that are statistically intertwined; it is that what are being treated as two separate variables are not actually two variables at all. Happiness in general and happiness with daily life are the same thing: happiness. While nuances of vocabulary enable separate questions to be asked, this does not mean the concepts themselves can necessarily be separated.

The above variables are so overlapping that it may seem pushy to use them to illustrate the point. However, this too reflects some nuances of vocabulary. An alternative term for 'happiness in general' is *positive emotion*, while another term for 'happiness with life' is *life satisfaction* – and there are very many studies indeed to have investigated the links between positive emotion and life satisfaction. Virtually all of them are based on data gathered by asking people how they feel. One way to appreciate the extent to which such variables are conflated is to consider the questions used to conduct the research. For example, in one typical study (Cohn, Fredrickson, Brown, Mikels, & Conway, 2009), researchers measured positive emotion by asking participants to rate their feelings of ten different sentiments, four of which were *contentment*, *gratitude*, *joy*, and *pride*. Meanwhile, they measured life satisfaction by using a scale (Diener, Emmons, Larson, & Griffin, 1985) that asked participants to rate their agreement with five statements, including '*The conditions of my life are excellent*', '*So far I have gotten the important things that I want in life*', and '*If I could live my life over, I would change almost nothing*'. The problem is that the statements on life satisfaction directly relate to the precise feelings used to assess positive emotion (namely, *contentment*, *gratitude*, *joy*, and *pride*). Thus the first measure is targeting the same feelings as the second and, with semantically equivalent items, both scales are measuring the same, single, thing. The fact that the researchers show the separate measures to be statistically correlated is as unsurprising as their separation of them is misleading. After all, in mathematical terms, every variable is correlated with itself.

Sometimes the overlap between concepts is partial rather than whole. But a partial overlap is an overlap nonetheless. The study mentioned earlier concerning attitudes to mortality and martyrdom is a case in point. A person's willingness to die and their willingness to become a martyr are aspects of the same underlying sentiment. Saying that a willingness to die is *associated with* a willingness to become a martyr implies that they are separate sentiments that have co-occurred in a way that reveals something we did not know before. It is as odd as saying that being a frog is *associated with* being an amphibian. Not all amphibians are frogs, and not all people who are willing to die are willing to become martyrs. However, being an amphibian is a necessary aspect of being a frog, and it would seem logical to conclude that being willing to die is a necessary aspect of being a voluntary martyr. Because every variable is correlated with itself, these partial conceptual overlaps will then be reflected in inevitable statistical correlations. But such correlations do not at all imply that the variables from which they are mathematically derived are conceptually distinct.

This use of a research method to create the impression that a multiplicity of entities exists where there is in fact but one is akin to what philosophers refer to as an *analytic truth*: the claim to truth arises from the process of scrutiny, rather than from reality per se. An example of an analytic truth is the generalization that 'all triangles have three sides'. It is a matter of definition that triangles are three-sided; therefore, the conclusion as to the status of 'all' triangles can be inferred from the very fact that it is triangles that are being spoken of. The truth of the statement depends only on the meanings of the terms in the statement. In much the same way, the truth of an assertion like 'positive emotion is correlated with life satisfaction' can be derived from the meanings of the terms *positive emotion* and *life satisfaction*. Referring to real world data is not necessary to demonstrate this, and going through the motions of doing so generates no more than an illusion of empirical corroboration.

One area that has been particularly identified as falling foul of these problems is the field of *social cognition modelling*. Social cognition models are theories that try to explain the way people take account of various different considerations when making decisions about their behaviour. For example, when thinking about whether to take a particular action (such as whether to wear a seatbelt, to go jogging, or to carry an organ donor card), a person might evaluate the difficulty of the action, the consequences of the action, and how other people might feel about the action. Any one of these considerations on its own might fully determine whether or not the person proceeds to act; alternatively the choice might be influenced by two or three of the factors, or none of them. The ultimate aim of social cognition modelling is to allow psychologists to identify those factors that determine human choices, ideally in order to suggest ways of changing people's behaviour patterns. Accordingly, most of the interest in this approach comes from applied fields, such as health promotion, where changing people's behaviour is of major concern.

By far the most prominent such model is the Theory of Planned Behaviour (TPB; Ajzen, 1985; 1991; 2011). This model posits that decisions about a particular behaviour are a function of several discrete cognitions, including: people's sense of control over the behaviour (whether they find it easy, or feel that circumstances allow it); their attitudes towards the behaviour (whether they think it leads to positive outcomes, and that such outcomes are likely); and their perceptions of what others think about the behaviour (whether other people would approve, and whether this actually matters). While there are some complexities around the sequencing of these cognitions, it is essentially true to say that according to the TPB, people's decisions about behaviour are formed on the basis of a handful of discrete perceptions. Most importantly, according to TPB researchers, it is possible to investigate these perceptions by simply asking people to report them. The TPB has been a major influence on health psychology research for over three decades (Sniehotta, Presseau, & Araújo-Soares, 2014), with many hundreds of studies appearing in the scientific literature. Government agencies across the world have invested extensive budgets in research designed around the TPB, and public awareness campaigns based on its findings have been rolled out in several healthcare domains. It would not be an exaggeration to say that, in many instances, policymakers have entrusted people's lives to this particular theory.

However, the TPB is severely hamstrung by the conflation problem. Take for example a study investigating people's decisions about healthy eating. When researchers measure cognitions about personal control, they typically ask respondents to answer questions like '*How easy will it be for you to avoid sugary snacks in the future?*' But in order to assess the impact of such cognitions on ultimate behaviour choices, the researchers then ask the respondents to answer such questions as '*How likely is it that you will avoid sugary snacks in the future?*' (e.g., Masalu & Astrom, 2001). The rub here is that the two variables are conflated (Ogden, 2003). A respondent who declares that they will find it 'extremely difficult' to avoid sugary snacks will, by logic, have to report that their likelihood of doing so is going to be relatively low (for example, it will certainly be lower than for people who find such avoidance 'extremely easy'). To take the matter to its plausible limit, respondents who report that they find this behaviour *impossible* will be compelled to report that their likelihood of engaging in it is *nil*. Even if only accounting for a small subset of respondents, this inevitability alone will tilt any dataset in the direction of a pattern, which in turn will emerge as a statistical correlation. In other words, the phrasing of the researchers' questions makes the resulting statistical association unavoidable. Given that the correlation can be confidently predicted without consulting the data, the association is an analytic truth.

When TPB researchers measure variables like perceived control, they are usually also gathering information about behavioural intentions. The belief that there are two statistically correlated variables here is illusory. Insisting on treating one variable as if it were two is an over-complication that stands in stark contrast to the parsimony demanded by scientific methods. The fact that the variables so flimsily reflect underlying realities brings this work into the realm of vague measurement. A sanguine attitude toward parsimony or measurement accuracy is more characteristic of pseudoscience than of science. It is therefore perhaps unsurprising that the TPB, when scrutinized thoughtfully, has been found to be of pretty limited practical usefulness, despite its widespread presence in health psychology. At a bare statistical level, the theory fails at explaining the empirical evidence that has been gathered across the plethora of studies that make up the TPB literature (Sniehotta et al., 2014). There has also been some concern around the way researchers seem to habitually employ inappropriate statistical procedures in their attempts to shake coherent conclusions out of TPB datasets (Weinstein, 2007), as well as their practice of reconfiguring the theory, instead of refuting it, when it becomes obvious that the data are not lining up in its support (Ogden, 2003). But ultimately the core problem is conceptual: the TPB falsely assumes that all behaviour is consciously 'planned', takes no account of the possibility that behavioural choices are influenced by emotions or other intangible factors, and is reliant on a hodgepodge of conflated variables that can only be assessed by self-report.

Moreover, the main outcome targeted by TPB researchers is not *actual* behaviour but *intended* behaviour – or to be more specific, it is intended behaviour as *reported by the respondents*. Now it doesn't require a great deal of insight to observe that people's intentions are very often far removed from reality, especially when they concern good behaviour. Saying you are going to eat wholesomely, drink less frequently, and take more exercise does not necessarily mean that you will ever follow through on becoming a paragon of virtuous health. In the TPB literature, this discordance between self-reported aspiration and real-world action has become known as 'the intention-behaviour gap'. Indeed research into the intention-behaviour gap has itself become something of a spin-off field of study in its own right. However, although pithy in its own way, the term is basically a jargonistic euphemism for 'the poor validity of our methods' or 'the extent to which our research fails'. The fact that there *is* a gap between (self-reported) intentions and (actual) behaviour is a non-trivial matter: the gap is not so much a tricky problem to be borne in mind when interpreting TPB research, but more a critical methodological flaw arising from the nature of self-report.

A number of articles critical of the TPB have appeared in high-profile journals, with some editorials now openly calling for the theory to be abandoned forthwith (Sniehotta et al., 2014). Nonetheless, these position papers remain far outnumbered by empirical studies that persist in using the TPB. It appears as if some psychologists feel empowered just to ignore the criticisms, as though the extant volume of TPB research creates a protective consensus. This in turn may relate more to professional expedience than epistemological confidence. According to a recent editorial in the journal *Health Psychology Review*:

> Three decades later, the TPB has lost its utility. It does not help practitioners to develop helpful interventions. It does not lend itself well to experimental tests and it does not provide explanatory hypotheses that would differ in a meaningful way from other prevalent theories ... Moreover, the TPB fails in the primary function of a theory: it does not accurately communicate accumulated empirical evidence. The TPB has become an empty gesture to tick the box that science should be theory-based. (Sniehotta et al., 2014, p. 4)

In other words, researchers seem to be using the TPB because they are extrinsically reinforced for doing so: the belief that theory-based science is better than atheoretical science encourages them to employ any theory they can lay their hands on, even if lacks construct validity. It is a little ironic that the theory itself offers a useful framework within which to describe this behaviour. Rather than choosing a methodological approach that has clear-cut scientific merit, researchers instead employ the TPB because (a) they are influenced by a belief that such studies are easy to design and conduct, (b) they anticipate that using the TPB will lead to positive outcomes, such as publications and grants, and (c) they strongly feel that that doing so conforms with the social norms apparent in their peer group.

Qualitative approaches and psychology

Any inquiry into the merits of doing research by talking to people about the human condition will inevitably lead to the topic of qualitative methods. While variously described, qualitative research methods are essentially those that attempt to capture, and focus on, the elements of subjective conscious experience. Typically they involve a number of techniques where there is direct engagement between researchers and participants, including (but not restricted to) in-depth interviewing, focus group discussion, on-site observation of participants engaging in daily life, discourse analysis, and the analysis of biographical narratives written by participants. One of the reasons qualitative researchers focus so much on subjective experiences is because they feel that, since each of us constructs a unique personalized understanding of the world as we go through life, every human being perceives things somewhat differently. Consequently, the information gathered in such research is almost always heavily intertwined with respondents' own particular predicaments. Qualitative methods attempt to directly track those nuances and subtleties of human psychology that more traditional scientific research methods seem to avoid. The data in qualitative research are usually very rich and detailed, being carefully gathered from first-hand testimony or observation, and their analysis can be intensive and laborious. Usually, this will involve the extraction of pertinent themes by the researcher from texts or transcripts, with an overall aim of gaining insights into the ways particular respondents have made sense of their circumstances in a given context.

It is only fair to immediately point out that the distinction between qualitative and non-qualitative approaches is a source of not inconsiderable tension in psychology. Qualitative methods are wedded to the principle of interpretivism. This is the idea that research observations must be decoded by researchers, rather than simply recorded by them. In short, qualitative methods require participants to think about, interpret, and then report upon their perceptions, after which it is required that researchers think about, interpret, and then report upon what it is that the participants have said. This layering of interpretation upon interpretation brings qualitative research very quickly into the realm of *what some people say about what they think they think*, and so to the perennial concerns that arise with such data. In psychology, qualitative approaches have been known to generate serious consternation, not least with regard to their potential impact on the reputation of the field. Non-qualitative researchers often construe qualitative methods as the deliberate infusion of subjectivity into research, and thus as flying in the face of the principle that objectivity is, in and of itself, an epistemological virtue. It is clear that the mainstream scientific method explicitly champions objectivity, and considers subjectivity to be a denigration of the scientific enterprise. As such, according to some observers, the use of qualitative methods in psychology might be seen as a dilution of psychology's claim to be a science, and a threat to psychology's standing both in universities and in wider society (Morgan, 1998).

Qualitative researchers are firm in defending their approach. According to them, objectivity is a noble, but futile, aspiration. Following in the tradition of critics of logical positivism (as discussed in Chapter 1), they propose that qualitative research should be applauded for acknowledging the inherent subjectivity of scientists. They suggest that research that aspires to be objective is doomed to failure, and that it is in fact the traditional scientific attachment to objectivity that threatens to undermine psychology. According to the qualitative worldview, human experience is personal and therefore all knowledge that flows through our species's narrative community is, at root, subjective and relative. The merit of one person's utterance simply cannot be tested against an external reality, because there is no non-subjective way of verifying the nature of that (purported) reality. In essence, qualitative researchers feel that subjectivity is worthwhile because objectivity is naïve. A second defence of qualitative research is that it actively seeks to avoid reductionism. Qualitative researchers regularly make the point that the human condition is highly complex, that its elements defy quantification, and that it is wrong to assume that humans are so generically alike as to facilitate the extrapolation of experiences from a single unique individual to other people. Other, more superficial, defences are also mentioned. Qualitative researchers often argue that their listening orientation is more appropriate to the therapeutic philosophy of (clinical) psychology, and – as we saw in Chapter 6 – might better reflect gender-specific aptitudes in ways that serve to neutralize various institutional and societal injustices.

Such defences are not without their own shortcomings. The idea that objectivity is futile is, in one sense, an analytic truth: the very meaning of the term objectivity, if taken literally, implies a divorce from the perspective of any one person, and yet we are all – scientists included – individual persons. It follows, then, that our perceptions are, by definition, subjective and that objectivity is, by definition, unattainable. However, this approach to the notion of objectivity seems unhelpfully pedantic – is it really what scientists refer to when they use the term? More likely, when scientists (or politicians, or journalists, or lawyers, or parents, or children) refer to 'objectivity', they do so to recommend that we remain vigilant against bias, skewness, and other related sources of inaccuracy and imbalance. The scientific emphasis on objectivity is not a call for perfection; it is a recognition of human fallibility. As such, a research approach that dismisses objectivity as some kind of positivist affectation would appear to lack an important safeguard. Of course, it could be that critics of objectivity do not intend to be so literal as to dismiss even the effort to avoid bias in research. Instead, they could be arguing that there are always limits to scientific objectivity, especially when scientists turn their attention away from molecules and chemicals and towards human behaviours and feelings. However, while this is undoubtedly reasonable, it is not in itself an effective argument in favour of *subjectivity*. If anything, subjectivity is much more susceptible to the biases of human perception, and extravagantly so. If critics really do fear that human researchers will never be able to fetter their prejudices, then it seems very strange indeed to recommend, as an antidote, a set of methods grounded in subjective interpretation.

The notion that qualitative methods avoid reductionism by assuming hypercomplexity is not unrelated to an idea we have encountered previously: the claim that human consciousness is special and transcends the science conducted by mere mortals. The counterargument, of course, is twofold (see Chapter 3): the complexity of human systems is not in fact inordinate when compared to the subject matter of other sciences; and even if it were, this would not in itself be a barrier to ordinary scientific scrutiny. The associated claim that human beings are so unique as to defy generalization seems strangely blind to the presence of any cross-species similarities at all. When we say that 'everybody is different', we hardly mean that 'everybody is different in every conceivable way'. Just as there are recognizable patterns in the shapes and sizes of human bodies, so too are there likely to be patterns in the way people think, feel, and behave. This is why conducting research on some people will help shed light on our understanding of others. It is not assumed that *all* aspects of experience can be studied in this manner, but simply that many aspects can be. In any event, as above, such criticisms of non-qualitative research fail as arguments in favour of qualitative alternatives. Just because individual human beings are distinct in multiple complex ways does not guarantee that complicated research methods will succeed where simpler ones fall short. Maybe they will both fall short.

In one sense, the very existence of qualitative research in psychology stands as evidence against its own core assumptions. For example, if researchers are addled by subjectivity and, thus, incapable of making inferences that can be treated as objective truth-statements, then presumably this applies to qualitative researchers too. If so, then how do we gauge the truth-value of their statements about the limits of objectivity? And if not, then how have qualitative researchers been able to bypass these problems? And if *this* is the case, then does it not demonstrate that such problems *can* be bypassed? Secondly, if it is impossible to extract generalizable principles from observing people amidst all their idiosyncratic complexities, then how do qualitative researchers become so convinced of the merits of each other's shared scepticism towards traditional scientific paradigms? If one person's subjective perspective is so unique as to defy extrapolation to other people, then how come qualitative researchers agree so much with one another about epistemological matters? In reality, of course, qualitative researchers (whether they realize it or not) are thoroughgoing positivists: through their words and actions, they exhibit an unquestioning reliance on the existence of external reality, on the patterned nature of observation, on the comparability of people's experiences, on the utility of generalized statements that describe multiple agents, and on the possibility of being wrong.

That last point – the possibility of being wrong – is important to mention. This is because the relativist underpinnings of qualitative methods imply that, in qualitative research, it might actually be *impossible* to be wrong. This in turn arises from the assertion that human beings are so complex and unique that their subjective experiences cannot be generalized. If true, then a problem of replication arises. Take, for example, a qualitative study of the experiences and feelings of medical students, in which researchers identify an obsession with death to be a dominant theme in the students' testimony. According to the principles inherent in the qualitative approach, such a theme cannot be presumed to reflect the views of other medical students not included in the study. Therefore, if a second group of researchers were to conduct an identical study on a different sample of students, they might find a different theme (say, an obsession with bodily appearance) to be dominant. In this case the second study has failed to corroborate the first. In science, a failure to replicate a finding usually stands as a warning that one of the studies is flawed. It could even mean that both studies are flawed. It will certainly be taken to imply that the overall findings cannot be relied upon, nor can they be seen as informative. However, when the research is qualitative in orientation, then a failure to replicate is not seen as a problem. In fact it is seen as utterly reasonable, something that is entirely par for the course. Neither study is 'wrong'. But if it is impossible to be wrong, then how is it possible to be right? Such a scenario epitomizes what is meant by non-falsifiability.

In these discussions, traditional scientific methods are usually referred to as 'quantitative'. It is striking how readily commentators assume that quantitative methods are exclusively those for which it is required that concepts be rendered using numerical variables and then tested statistically (Trafimow, 2014). Among other problems, this often unhelpfully leads to observations concerning purported gender differences in mathematical ability, with the associated claim that quantitative (i.e., mainstream) methods are somehow unfriendly to female researchers. As we have seen in Chapter 6, such an anxiety is unwarranted given the lack of empirical evidence for the alleged gender difference. However, it is also poorly grounded in conceptual terms. The term 'quantitative' refers to something that has a magnitude or an extent, which in effect means that it refers to a construct that is externally verifiable: it is possible to observe its status and, if necessary, for this observation to be independently corroborated. A distinction between 'something that is present' and 'something that is absent' is a quantitative distinction, as is the distinction between 'same' and 'different'. When a bird returns to its nest and sees that one of its eggs is missing, it is forming a quantitative conclusion. When a mouse becomes wary of mousetraps because past experience has shown them to be dangerous, this conclusion too is based on quantitative reasoning. Nobody would seriously suggest that such decision-making requires birds or mice to be statistically numerate, or to be adept at null-hypothesis significance testing. Quantitative constructs are researched in their own right precisely (and only) because they are objectively verifiable; the use of statistical methods when doing so is a secondary matter.

Qualitative and quantitative approaches coexist in psychology, albeit somewhat uncomfortably at times. That being said, it should be noted that by far the majority of psychology research is of the traditional, scientific, 'quantitative' kind. The presence of qualitative methods owes much to psychology's scholarly proximity to fields like sociology and anthropology, from which most qualitative approaches have been borrowed. In this sense, qualitative approaches are not native to psychology but rather have immigrated into it, a process that has involved the various challenges of assimilation, integration, incorporation, cooperation and so on (Bhati, Hoyt, & Huffman, 2014). However, while many working methodologists have called on psychology to simultaneously absorb *both* qualitative *and* non-qualitative methods into the field, and even into individual research studies (Howe, 1988), there are some significant challenges to doing so. Mainstream scientific methods descend from philosophical traditions of positivism and post-positivism, while qualitative methods descend from relativism and constructivism. At their core, these traditions are philosophically contradictory. Qualitative approaches are premised on assumptions that directly conflict with those of ordinary scientific research. Mainstream science tries to achieve reliability and validity, while qualitative research considers such aims to be futile and misguided. Further, the extent to which qualitative research embraces anecdotalism, eschews (if not rejects) parsimony, defies falsifiability, and valorizes subjectivity might eventually create a resemblance to pseudoscience that would be a poor fit for psychology's scientific aspirations.

In the end, qualitative research is either useful or it is not. There are very few ways to evaluate usefulness in research other than in terms of whether the knowledge generated is found to be applicable outside the study from which it arose. In qualitative research we have problems with the interpretation of data, but also with the data themselves. Interpretation is hampered by a number of likely distortions, not least those resulting from the cognitive heuristics discussed in Chapter 4 (Paley, 2005). However, the main difficulties in qualitative research relate to the extent to which its core data comprise a *valid* representation of that which is intended to be studied. If we are concerned about people's fears, aspirations, or personal sense of identity, then we need information about these very constructs. The reliance on personal testimonies, from interviews and suchlike, is problematic, because the best that can be said about them is that they relate to *what some people say about what they think they think*. This core problem occurs at the level of the data, and so will remain a problem no matter how meticulously the data are subsequently scrutinized.

Conclusion

One way to consider these methods is to note that that, rather than stand at a distance from that which is being studied, they in fact represent psychological interactions that are *themselves* worthy of study. When a person is asked about their attitudes to death, it is naïve to consider whatever they say to be purely distilled from the very milieu that sees them being asked such questions to begin with. The ways people hear and understand such questions, and how they choose to answer them, themselves constitute complex and subtle psychological interactions between respondents and questioners. This is not to recall the idea of reflexism, that trope of the anti-science movement which alleges that psychologists cannot study any behaviour at all because of some fear that the act of studying will interfere with that which is being studied. It is merely to point out that any interpretation of *what some people say about what they think they think* is going to be far from straightforward. Rather than take such utterances as face-value statements of fact, it would almost be safer to assume that they are *anything but* face-value statements of fact. Claims that they constitute some kind of objectively verifiable data seem to be very tenuous, and to completely ignore the benefits of objectivity itself (if not indeed the very meaning of the term). Championing such communication as raw material for science seems akin to the valorization of anecdotal evidence.

Such concerns are compounded by the way self-report methods struggle with conflation. The possibility that people might refer to single entities in multiple ways, whether in survey-based or qualitative research, exaggerates the likelihood of inferring the presence of associations that aren't actually there. Moreover, it presents our understanding with unnecessary, and unverifiable, elements of description. In the end, no amount of survey-taking, questionnaire-filling, or in-depth interviewing will allow a researcher to establish whether a participant's perception of their happiness with daily life is caused by the same underlying emotion as is their happiness in general. As such, whether such research could be expected to achieve parsimony (in the sense of containing only that about which we are sure) is quite doubtful.

The most important limitation of research based on *what people say about what they think they think* is its problems with falsification. While psychometrics offers a way of identifying robust patterns in how people return answers to tightly structured questions, it effectively does so by ignoring what a respondent is telling us: it simply computes the relative frequency with which certain responses are returned, and the extent to which such frequencies are distinctively associated with other variables of interest. This is how a respondent's attitudes toward television might end up being used as a statistical indicator of their management abilities. But research that asks people questions *and then seeks to interpret their responses at face value* quickly runs into several minefields. Simply put, it is extremely difficult to use such interpretations to conclusively test the accuracy of an associated claim. If a person responds to a question on martyrdom by declaring a willingness to die for a political cause, can this declaration be relied upon? Nobody knows, of course. So it is difficult to see how such declarations could ever be used to challenge the accuracy of claims that, for example, a particular category of the population is really a group of suicide-bombers-in-waiting.

But science is supposed to be self-correcting. Indeed, this is one of its most powerful attributes. The scientific method evolved as a way of checking the soundness of claims against verifiable observations, such that false claims could be exposed and robust claims – even if they are unpopular or threaten the status quo – could be retained. Any subfield that finds falsification difficult will have difficulty being scientific. A subfield that, as a matter of dogma, rejects the value of falsifiability altogether, is very far removed from the scientific enterprise. Such research may well have an important place in academia because, after all, not all important scholarly endeavours are sciences. But rejecting and avoiding the twin objectives of refuting claims and replicating findings cannot be described as a way of doing science. In this sense, research that relies heavily (if not exclusively) on studying *what some people say about what they think they think* might find it difficult to retain the claim that it is part of a scientific discipline at all. Purporting to be science while rejecting scientific standards is how pseudosciences are born.

As has previously been noted, psychology is a very popular subject area. It is perhaps mostly for this reason that researchers take epistemological risks. The extent to which some areas of psychology approach the boundary with pseudoscience might reflect the pace at which research is being pursued. In a sense, sheer enthusiasm for the discipline is leading some areas of psychology to attempt more than can reasonably be achieved. Psychologists are trying to run with idealistic all-answering paradigms, rather than walk with mundane incremental methodologies. In an effort to scramble towards useful findings and humanly meaningful conclusions, they end up cutting corners around parsimony, accuracy, and falsifiability. This of course is not exclusively a problem of studies that rely on questionnaires or interviews; as we saw in Chapter 6, it can also arise in research that is usually construed as being very scientific indeed. One of the main challenges psychology faces is precisely this enthusiasm factor. The requirement for scientists to be objective and dispassionate at all times clashes with the more humanitarian or sociopolitical concerns that underlie psychologists' own motivations when they pursue their profession. Personal motivations and value-systems have a very strong bearing on the quality of psychological science, and greatly affect the way it is received and understood by wider society. In the remaining chapters we will consider the extent of these distorting influences, and whether anything can (or needs to) be done about them.

Part III

Psychology and Pseudoscience in Context

LEABHARLANN
CO. CILL DARA

Chapter 8

Biases and Subjectivism in Psychology

The problem of values

Science is supposed to be value-free, whereas people who are value-free are (rightly) considered psychopaths. Frequently, psychologists find this incompatibility frustrating: requirements to be dispassionate – to leave one's feelings at the door – often jar with the empathic impulses that first led them to the field. The arising tension has a number of problematic effects. Firstly, it resuscitates concerns about whether science really is the best way of studying psychological subject matter. The fundamental subjectivity of human experience leads some observers to conclude that psychology, as a field, must also be fundamentally subjective. The assertion that objectivity falls short in psychology is endorsed not only by critics of psychology's status as a science (see Chapter 3), but also by psychologists themselves when they support claims about quantum consciousness (Chapter 5) or the use of qualitative methods (Chapter 7). But ultimately, the critique is self-defeating. The challenge of subjectivity – which undoubtedly exists – is unlikely to be dealt with by resorting to *more* subjectivity. In fact, identifying subjectivity as a challenge resonates strongly with the ethos of *objectivity* to which scientific methods aspire. Rather than naïvely assuming human perfection, aspirations for objectivity arise from maturely recognizing the fact that human judgement is typically undermined by *imperfection*. This is why the majority of research-active psychologists unhesitatingly advocate the use of orthodox scientific approaches.

A second problem arising from the collision of scientific principles with personal values is that supposedly objective findings can be contaminated by bias. Because it is impossible for psychologists to not have their own views, what transpires as knowledge in psychology will, by necessity, be filtered through the perspectives of its human creators. And as many value systems are communal, scientific psychology will be pushed and pulled by social preferences as well as individual ones. Thus, when psychologists conduct empirical research, not only must they set aside their own beliefs and expectations, they must also identify the extent to which their data are framed by arbitrary theoretical frameworks and presumptive methodologies. Such deciphering is much easier to advocate than to achieve.

181

Difficulties escalate when data pertain to moral or ethical matters. Psychologists who strongly oppose a particular public policy on moral grounds may find that the available empirical data serve to undermine their case. They are then faced with something of a trilemma: do they describe the data objectively (and so risk being perceived as an advocate for a view they abhor), do they describe the data subjectively (and so risk being accused of dishonesty), or do they simply avoid describing the data at all (and so risk being accused of selective reporting, or even censorship)? And what if psychology research is conducted amorally but ends up informing the morals of others? For example, should psychology make assumptions about what is normal and abnormal, given that these might be seen as establishing what is correct and incorrect?

In this chapter we will consider ways in which biases intertwine with the expectation of objectivity in psychology. Some biases are held explicitly – psychologists will have preferences about what outcomes they would like to see in the research that they conduct. The problem is that, while psychologists will be conscious of such views, they typically underestimate their effects on their scientific objectivity. The example we will consider concerns politics, and the way psychologists hold either liberal or conservative views. Other biases are held explicitly but arrived at inadvertently – psychologists will hold unquestioned assumptions about the nature of human behaviour. The problem here is that, while such assumptions might appear superficially banal, they are often quite profound. The example we will consider is when psychology attempts to distinguish normality from abnormality. A third set of biases are those which are held implicitly rather than explicitly. These biases undermine objectivity in a way the psychologist is not aware of. The views underlying such biases might be so implicit as to reflect contentious social values – even prejudices – that the psychologist would otherwise disavow. We will begin this chapter by considering one such example, which refers again to the masculinist bias historically embedded in wider society. We have already seen how poorly scrutinized biological tropes regularly lead to erroneous views about gender differences in cognition and behaviour. In this chapter, we will consider whether the weight of social gender bias might hamper the entire enterprise of science, the resulting implications for psychological research, and whether this bias extends beyond professional interactions to the very logic of psychology, thereby warranting the adoption of a new, feminist, form of epistemology better suited to the field.

Sociopolitical biases in psychology: Masculinism

The masculinist society

It is an uncomfortable historical fact that just about all human cultures were founded on a presumption of sexual inequality. For centuries, in almost every imaginable context, erudite people assumed that men were dominant over women, and that all human affairs should be approached with this expectation in mind. Even in the very earliest democracies – including those established in the Ancient Greek city-states that gave the system of 'democracy' its name – it was believed that an entitlement to vote in elections could only ever extend to male citizens. The idea that women might be allowed to vote was considered simply ludicrous, and female suffrage was virtually unheard of anywhere in the world until the middle of the 19th century. Countries whose egalitarian uprisings shaped the historical narrative of Western democracy were no exception. For example, while the French Revolution took place in the late 18th century, it was not until after the Second World War, in the autumn of 1945, that French women were allowed to vote. The United States passed its Nineteenth Amendment in 1920, some 140 years after its republican revolution, while Britain, which codified the rights of citizens in the Magna Carta of 1215, waited seven centuries before allowing women to help choose the government, in 1928. In the patriarchal frame of reference, it was simply illogical that a woman could vote given her subservience to men.

For hundreds of years, societies around the world adopted a norm of coverture, in which the rights of married women were subsumed into those of their husbands, such that effectively only men could acquire property, sign contracts, pursue an education of their own choosing, or keep whatever salaries they earned. In essence, women were *owned* by men. It is little surprise that virtually all the accoutrements of human civilization – its political systems, social conventions, economic activities, and organized religions – developed structures that were derived from, and that therefore reinforced, this starkly sexist reality. Only in the past century or so – and only in some parts of the world – has the notion that women should be presumed equal to men become something of a mainstream idea. And in many of these places, the concept seems ruggedly stuck at the ideas stage, only slowly, if at all, making a practical difference to social habits, civic practices, or the law.

The fact that patriarchal dominance has been the norm for the majority of human history is often easy to overlook. When students are taught about historical epochs and events, the turning points of civilization and the watersheds of sociopolitical upheaval, they are rarely reminded that the venues for such developments were societies in which women were considered (at best) second-class citizens. Effectively half of the human population were confined to a social grade below that required to be mentioned by name in history books. Arguably, the legacy of such gender apartheid continues to be seen throughout modern society. Most attempts to compare men and women's earnings suggest that men extract greater remuneration from their economic activity. Men's salaries are often simply higher than women's, and the difference is further compounded by disparate pay across occupations held with different frequency by men and women. Often the pay gap reflects the fact that in the corporate world (as in public service and politics), men appear in disproportionally greater numbers among the senior, more influential, echelons. Society's downgrading of women also extends to matters of basic security: women are far more likely to be victims of spousal abuse, domestic homicide, and sexual assault, while men are far more likely to perpetrate these crimes. Notwithstanding this, rape is considered to be one of the most under-reported violent crimes in industrialized countries (Allen, 2007), and in Britain fewer than 20 per cent of those rapes that are reported result in the perpetrator even being charged with a crime, never mind convicted of it (Casciani, 2014).

Many of these discrepancies are argued to have plausible alibis, in that there exists a repertoire of common arguments to be cited in the effort to explain them away. For example, many observers question the authenticity of the gender pay gap by pointing to the fact that women often *voluntarily* choose jobs that attract lower salaries than the jobs chosen by men (Ceci & Williams, 2011). However, this overlooks the possibility that so-called 'women's jobs' attract deflated salaries precisely because they are held, in the main, by women. A number of research studies have shown that managers simply recommend lower salaries to women job applicants even when their male counterparts submit identical CVs (Davison & Burke, 2000), suggesting that the gender of job-holders may indeed influence the way salaries evolve over time. The relative scarcity of women in senior management or in politics is often said to reflect the non-availability of suitably qualified or willing female candidates for such positions, rather than sexist prejudice in the minds of promotions boards or the electorate. Even if true (which is difficult to substantiate), this rationalization merely shifts the question downstream, in that the factors contributing to gender-imbalanced candidate pools themselves require explanation. Even the claim that sex crimes are most commonly perpetrated by men on women is disputed by some, with arguments centring on the definition of rape, the compilation of crime figures, and the validity of individual crime reports (Matchar, 2014). It is notable, however, that commentators seeking to minimize the problem of sexual assault on women are often prone to misquoting the relevant statistics (Grether, 2014).

It would appear that humanity considers males more valuable than females. One subtle way of inferring this is to consider patterns of parenthood and marriage. Across the world, decisions of parents upon having daughters instead of sons, or sons instead of daughters, appear to conform to regular patterns. These patterns are difficult to discern from personally observing the societies in which we live, but are relatively vivid when considered in the form of large statistical datasets. For example, parents who have two daughters are much more likely to have a third child than parents who have two sons (Halpern, 2007), parents who have three daughters are much more likely to have a fourth child than parents who have three sons (Landsburg, 2003), and so on. The only aspect of this pattern which varies internationally is its scale: wherever data have been gathered, the direction has been the same. Parents of girls are more likely to become separated and divorced than parents of boys, and divorced parents who have custody of daughters are less likely to remarry than divorced parents who have custody of sons. In fact, when divorced parents with daughters do remarry, those marriages are less likely to succeed than the second marriages of divorced parents who have sons (Dahl & Moretti, 2008). Perhaps most intriguingly, when unmarried pregnant women undergo ultrasound monitoring to discover the sex of their unborn child, they are less likely be married by the time they give birth if the child is female than if the child is male (Dahl & Moretti, 2008). While all such patterns are open to interpretation, one way of explaining them is to hypothesize that parents find families with sons to be a more comforting proposition than ones with daughters. They are more likely to create such a family (by getting married and/or maximizing the probability of male offspring by extending fertility) and to maintain it (by not getting divorced). Of course, the subtle statistical patterns only serve to corroborate other, more macabre, trends that have been seen historically throughout the world. Specifically, the demographic make-up of many societies suggests that patterns of abortion and infanticide are also selective, in ways that imply a valuing of sons over daughters (Halpern, 2007).

In summary, while explicit equality norms inform the framing of law and policy in modern Western democracies, these are superimposed onto cultural norms and extant practices that have historically espoused anything but gender equality. It seems difficult to believe that the weight of this history could have no effect on our current circumstances. So our question is this: if society at large is burdened by the legacy of an inherent masculinist bias, then shouldn't we assume that science (and specifically scientific psychology) is also so afflicted? Does the profession or practice of science, or its accumulated corpus of organized knowledge, betray a legacy of patriarchal dominance? And if it does, then in what way are such biases manifested?

Masculinism, science, and psychology

We have already considered how social assumptions regarding gender differences have skewed the discourse around research into men and women's psychology, with most of these assumptions casting women's behaviour in a negative light. When sloppy reasoning (verging on pseudoscience) consistently produces the same errors – unwarranted claims that women are more timid, more demure, less mathematically-minded, less self-assured, and so on – then such errors seem loaded rather than random. Therefore, insofar as these gendered assumptions are persisted with despite being contradicted by empirical evidence, we might well conclude that they reflect a masculinist bias in the way knowledge is filtered.

This bias can directly affect the professional and cultural institutions in which science takes place. Historically, women have been discriminated against in science to no less an extent than in wider society (Lee, 2013), and this is as true of psychology as it is of any other discipline (Gross, 2009).

For example, women psychologists such as Mary Calkins (1863–1930) and Margaret Washburn (1871–1939) are substantially less famous than many of their male contemporaries, despite making significant empirical and professional contributions to the development of scientific psychology (Gross, 2009). Up to one hundred years ago, women who conducted scientific research were seen as not merely intruding on male territory, but as effectively ceasing to *be* women. A total of 149 of the 4,000 'men' listed in the 1906 edition of *American Men of Science* were, in fact, women (Rossiter, 1982). This diminution of women's roles as scientists reflected the linguistic conventions (and implicit social attitudes) that prevailed in early 20th-century society. However, its influence on the writing of histories has culminated to produce a distorted impression that women have traditionally had no place in science.

Statistics suggest that women are nowadays reluctant to consider science as a career. Some efforts to address this seem only to exacerbate the problem, highlighting the way (male) scientists and policymakers continue to view women as exotic beings unconnected to their world. When in 2012 the European Commission organized a campaign aimed at encouraging women to pursue careers in science, their launch video attracted so much ridicule that it had to be withdrawn within days. The video featured a dour bespectacled and white-coat-wearing male scientist working carefully with a microscope, only to be distracted by the sudden appearance of three giggling female scientists who spend their time dropping lab equipment and playing with make-up. Against an electro soundtrack reminiscent of a high-street fashion boutique, the three females stride back and forth, occasionally posing, as if on a catwalk. Eventually the video ends with a caption claiming that science is 'a girl thing', in which the letter 'I' is replaced by a stick of lipstick. As one specialist on gender equality in education described it, the video was 'so shocking that the EC had to [officially] deny that it was an attempt at irony' (Rice, 2012). Unsurprisingly, the gender pay gap affects professional science as much as any other occupation; and experimental studies have shown that scientists are just as likely as other managers to recommend lower salaries for women job applicants than for identically qualified men (Moss-Racusin, Dovidio, Brescoll, Graham, & Handelsman, 2012).

However, some critics suggest that recommending lower salaries for women and interpreting data in sexist ways are relatively superficial forms of masculinist bias in science. They argue that the very epistemological conventions of mainstream science reflect styles of scholarship that were developed to suit men, and assumptions about knowledge that betray an incorrigibly gendered rigidity. This rigidity is said to place a question mark over the outputs of all science. As a body of commentary, the stronger forms of this critique have collectively become known as *feminist epistemology*. This sub-field of the philosophy of science queries the conceptualization of knowledge, the impact of perspective, and the context-dependency of research. Its primary assertion is that all knowledge is 'situated', in the sense that it reflects the perspective of the people who possess it. As science has historically been written from the perspective of male scientists, its 'situated' nature ensures that claims to knowledge are thus undermined by bias. For example, consider how scientific psychology conceives of the relationship between minds and bodies. It is a matter of fact that male bodies are different to female ones. As such, depictions of first-person perspectives on inhabiting a body will be different for men and for women, at least in some respects. Insofar as psychology explores the way people inhabit their bodies (a concept that could be said to encompasses such topics as body image, sexuality, neatness and hygiene, physical freedom and constraint, self-consciousness, attitudes to death, and even children's play), choices around what research questions are more important and what conclusions are more convincing might be made differently by men and women. The fact that for most of its history research psychology has been conducted predominantly by men may then have influenced which theories of minds and bodies psychology has deemed compelling (Young, 1990).

Some of the most prominent feminist epistemologists have argued that, despite their practitioners' ostensive allegiance to scientific objectivity, *all* sciences operate within frames of reference that reflect gendered worldviews. One famous example of this critique concerns behavioural biology. American physicist and feminist theorist Evelyn Fox Keller (1983) argued that it was masculinist bias which prevented scientists from properly understanding the behaviour of slime mould cells (*Dictyostelium discoideum*). Slime mould is the gelatinous scum that accumulates on or near dead plant material. In everyday life, humans see such mould as dirt, or possibly as a fungus, but it is actually composed of a slowly expanding congregation of individual single-celled organisms. Despite having no brain or centralized nervous system, and, thus, no apparent cognitive resource, this amoeba-like creature is able to form clusters, shape-shift, and navigate complex environments with apparent intelligence. It even seems able to spontaneously vanish (it actually disaggregates into its individual unit cells, which are invisible to the human eye), only to reappear nearby at a more useful location. In laboratory studies, clusters of slime mould have been trained to navigate complex mazes (Nakagaki, 2001). This elaborate behavioural repertoire, in the absence of a biologically plausible cognitive system, confounded scientists for decades. Keller argued that her peers had been over-reliant on explanatory concepts that reflected masculinist assumptions about group behaviour. Traditional attempts to account for the movement of slime mould sought to invoke theories of leader-cells and follower-cells, but these were never able to properly explain the complexity of what slime mould was able to do. Eventually, biologists developed newer theories in which cells were hypothesized to work collaboratively with one another. These collaboration theories proved to have superior explanatory power. Keller pointed out that leadership and followership were conventionally male interests, and that rigidly assuming such influences to be important served to hamper science's ability to understand slime mould. Only when scientists considered collaboration, which Keller identified as a female interest, did they make progress in this area. According to Keller, this was but one example of the insidious way masculinist bias affects everyday science.

In a similar vein, Belgian-born French philosopher of science Luce Irigaray (1982) questioned the accuracy of Albert Einstein's famous mass-energy equivalence formula. Her argument was that $E = mc^2$ is a 'sexed equation', in which the squaring of the fastest entity, the speed of light c, is based more on a masculine fascination with speed than on its arithmetic success. Irigaray also bemoaned the state of fluid mechanics, blaming its relative lack of intellectual development on a male obsession with solids, at the expense of feminine softness and lability. Others have questioned why palaeontologists assume that all ancient carvings of the female form must be fertility symbols (Russell, 1998), and why evolutionary biologists are prone to frame their theories of prehistoric life in the form of heroic narratives (Haraway, 1989). Likewise, theories of leadership and of intelligence are often accused of arbitrarily defining their subject matter in ways that incorporate (allegedly) male habits, such as single-mindedness. And we have already seen, in Chapter 4, how concepts such as self-esteem can be arbitrarily gendered. As Carol Tavris (1993) pointed out, when men rate themselves more self-flatteringly than women, the dominating view in psychology is not that men are relatively more conceited, but that such response patterns show how women, in relative terms, lack basic feelings of self-worth.

Several feminist epistemologists have argued that even the way scientists draw their conclusions reflects styles of thinking rooted in the male perspective. They assert that when scientists employ tight argumentation or atomistic analysis, they are exhibiting distinctly male styles of thought that owe their priority to men's dominance of science. These theorists propose that feminine ways of making one's point, such as deploying a narrative to excite a listener's imagination (e.g., Keller, 1985), are just as valid and effective as the conventional, masculine approaches. Whether or not men and women really do *think* differently – that is, whether men tend towards argumentation and women towards narrative – is very far from clear. But the depiction of men as adversarial and of women as affiliative is a very common stereotype, common enough perhaps to imbue these different types of reasoning with gender-specific symbolism in ways that appeal differently to men and women (Rooney, 1991). Therefore, science's preference for 'masculine' methods may lead to an unfair priority being attached to certain types of research at the expense of others, and thus may end up skewing the way knowledge is ultimately accumulated (Keller, 1985). In a classic analysis, the American philosopher Sandra Harding (1986) asserted that mainstream science's prioritizations of reason over emotion, of objectivity over subjectivity, of the abstract over the concrete, and of the general over the particular, each reflected a bias toward androcentric modes of thought; and that the practice of science would be generally improved by incorporating more feminist cognitive styles, such as those involving emotion, reflexivity, and social values (cf., Longino, 1990; 2004).

To some extent, these arguments overlap with those relating to biologically reductionist gender differences (in defining female thought as substantively different from male thought) and qualitative research (in questioning the feasibility of objectivity). As such, they are subject to the same shortcomings as those fields. In the first instance, the idea that men and women practice or prefer gender-specific styles of perception, reasoning, and inference is far from supported by empirical evidence. It is equally unclear whether science synthesizes data in ways that are more compatible with masculine stereotypes of competitiveness than with feminine stereotypes of collaborativeness. Even if it did so, it is mere speculation to construe this state of affairs as reflecting a gendered provenance: in reality, the practice of replacing old theories with new ones on foot of objectively corroborated data might have come to be preferred simply because it *works better*, rather than because, historically, most scientists were men. As previously emphasized, the claim that a failure to be objective systematically leads to unreliable results is not in itself an argument for new forms of subjectivity. Just because purportedly feminine styles of research are traditionally neglected does not make them any more preferable than their masculine equivalents, because ultimately there is nothing to suggest that they would be any more useful. If objectivity is contaminated, then alternative forms of subjectivity seem a poor antiseptic.

Indeed, alternative forms of subjectivity can end up repeating unwanted histories. For example, when Taylor formulated the new name for her theory of the female stress response (see Chapter 6), it is notable that she chose a pair of verbs – *tend* and *befriend* – that, when alluded to as descriptors, reflect positively on the people being described. *Tending* and *befriending* are virtuous behaviours. As such, the terminology lacked concordance with the earlier jargon it was intended to complement: Cannon's account of the (male) pattern had referred to two proclivities – *fighting* and *fleeing* – that were vices rather than virtues. In reality, all these behaviours are neutral. If she wished, Taylor could have chosen terms that corresponded with the negative tone of Cannon's original expression (perhaps replacing '*tend-and-befriend*' with '*pamper-and-ingratiate*'). For that matter, Cannon could have avoided the dual stigmas of aggression and cowardice had he described his '*fight-or-flight*' theory as one of the '*engage-or-evade*' response. The point here is that, presumably inadvertently, Taylor's choice of terms served to convey a gender distinction that cast women in a positive light in contrast to men. This is hardly an advance on masculinist theories that, equally inadvertently, do the opposite.

It is clear that sexism exists in society, and must therefore exist in psychology. Sexism constrains scientific fields by influencing the priorities given to particular research questions, the use and framing of terms, assumptions regarding gender itself (including assumptions about what constitutes 'normal' sexuality), and of course the salaries of female scientists. In the interests of empirical accuracy, it is always useful to be vigilant against assumptions that are arbitrarily framed in non-neutral ways, such as the idea that male self-flattery represents confidence instead of conceitedness. And, in the interests of ordinary human dignity, it is important to eliminate any research practice that treats women with contempt, or which reinforces sexual stereotypes without adhering to the standards of good science. However, it is much less clear whether a masculinist bias truly affects the underlying logic of psychology. Many feminist epistemologists dispute the idea that ordinary scientific reasoning is corrupted by bias and so should be discarded. Indeed, the idea that there exist 'feminine' forms of thought, or 'female' metaphysical perspectives, or 'feminist' approaches to knowledge-generation, would appear to impose strongly defined norms of femininity on women psychologists. The imposition of such norms by society is usually seen as problematic. These ideas also fall into the relativism trap that ensnares qualitative psychology. If objectively reliable knowledge is so elusive, then how are we to trust the commentators who warn us that this is so? How can we be assured that their identification of this elusiveness is objectively reliable? In addition, such claims undermine any view that gender stereotypes are empirically unjustified: attempts to refute prejudice with evidence will be rendered pointless if the conventions by which evidence is produced are declared to be flawed.

By aspiring to objectivity, to verifiability, and ultimately to accuracy, the scientific method stands as an effective way of addressing prejudices and inequalities. For this reason, many feminist epistemologists prefer to see themselves as 'doing science as a feminist' rather than as 'doing feminist science' (Longino, 1987). As psychology is a science, the comparison can duly be rephrased to articulate strong arguments in favour of 'doing psychology as a feminist'. This, however, raises its own paradoxes. Specifically, can a value-free pursuit such as science justify advocating for any particular value system, no matter how ethically laudable it happens to be?

Sociopolitical biases in psychology: Liberalism

Politically subjective knowledge

At least doing psychology as a feminist involves *consciously* and *explicitly* supporting a particular value system. It is of course possible for psychologists to support value systems both *unconsciously* and *implicitly*. In fact, according to some epistemologists, researchers are *always* influenced by values without realizing it. One relatively abstract concept to bear in mind here is that of 'underdetermination', a notion conventionally associated with the American philosopher Willard Van Orman Quine (1951), among others. Very loosely speaking, the problem of underdetermination relates to the status of a theory that is supported by some evidence, but not to the extent that we can be absolutely confident about what we should believe. This is a pretty convoluted way of expressing a familiar idea: in the case of many theories, we are not totally sure whether they are true. According to Quine, we will *never* be totally sure, because multiple theories will always exist to fit any available data. This means that we will always need to use intellectual judgement to make the leap from data to interpretation.

Underdetermination of theory by evidence is often described as causing problems for science. However, it would be misleading to say that the problems caused are always fatal. After all, the entire scientific enterprise is based on the realization that there are very many things about which we are unsure. The very fact that theories are not conclusively supported is what drives our efforts to conduct more and more scientific research. Far from being an intrinsic problem for science, in a sense, underdetermination is one of its necessary preconditions. What makes it problematic is when scientists succumb to epistemological haste – when they disregard the fact that the evidence is insufficient, and proceed to believe the theory anyway.

With some science, underdetermination is not much of a problem at all. For example, when examining the anatomy of a newly discovered species of lizard, a zoologist may have strongly held expectations about the locations of the animal's various internal organs. These expectations may be borne out by the zoologist's subsequent observations, but they also may not be. The newly discovered lizard may have a different anatomy to that of similar species studied heretofore. If so, then the zoologist's original expectations would have been premature, and any assumptions made prior to examining the lizard are likely to have been erroneous. The reason this is not a problem is because the internal anatomy of a lizard can be defined, observed, and verified by objective third-party scientists. Our zoologist's observations can be double-checked and confirmed. In psychology, especially in domains such as social psychology, things are a little bit different. Because psychological phenomena can be difficult to define or observe, the interpretation of data often remains strongly influenced by the psychologist's initial expectations. This is true even when psychologists make every effort to be balanced and unbiased.

Consider, for example, a psychologist who wishes to investigate the causes of trait sociability. In an effort to infer why some children grow up to be sociable adults, the psychologist gathers data from families over a long period of time. Typically, data of this kind will show that children of gregarious parents are statistically more likely to themselves be gregarious. So what might the psychologist conclude from such a pattern? Well, if the psychologist had an *a priori* expectation that human traits are biologically ingrained and transmitted genetically through the generations, then the data might well be interpreted as supporting a 'nature' model of personality development. However, if the psychologist had an *a priori* expectation that such traits are shaped by environment and experience, then the same data could be used to support that view as well. This is because, by and large, children will grow up in the same home as their parents, or at least with very extensive parental contact. We might expect that sociable parents are likely to actively encourage their children to be sociable, and, through their own actions, to serve as strong role-models for sociability in the eyes of their offspring. As such, the children of sociable adults will likely be immersed in conditions in which sociability is seen as the norm, and so are themselves likely to grow up to be sociable people. Data showing that the children of gregarious parents are themselves more likely to be gregarious is just as consistent with this 'nurture' interpretation.

When it comes to sociability, the comparative merits of 'nature' or 'nurture' theories might be a matter of dogma for some researchers, but for many more it will be a matter of data. In other words, relatively few psychologists (and few readers of psychology research) will have passionately held views on the issue of sociability. However, now consider an alternative domain of research – specifically that of investigating the causes of criminality. The framing of this research proposition is almost identical to that concerning trait sociability, and the methods used to study the issue will be much as described above. Furthermore, similar findings are likely to emerge: many datasets will suggest that the children of criminals have a statistically significantly increased chance of themselves becoming involved in criminal behaviour. (Certainly not all datasets show this; but for the sake of argument, let us hypothetically assume that they do.) As before, if a psychologist has a strongly held *a priori* belief in 'nature' arguments, then such data are likely to be interpreted as supporting this narrative of biological transmission. And if a psychologist has the opposite view, then the role-modelling of parents and the shared parental/childhood environment can be highlighted instead. The key point here is that, unlike sociability, criminality is a hot political topic, one for which opinion divides largely on political grounds. Classically defined, social conservatives will lean towards the 'nature' side of these debates, in which people (rather than circumstances) are to blame, and where unsavoury behaviour is incorrigible and should be punished as such. On the other hand, social liberals will tend towards 'nurture' arguments, which highlight the imperfections of society at large, and which hold out the prospect of altering criminal behaviour through community development and rehabilitation. As such, when the data are ambiguous, when the theory is underdetermined by the evidence, it is subjective opinion that drives interpretation. And unlike a zoologist dissecting a lizard, a psychologist cannot easily resolve uncertainties by getting a colleague to confirm the observations.

Instead, psychologists are likely to seek confirmation from further data. Those who favour nature-based theories might point to children of criminals who were separated from their parents and raised in other families: if these children also demonstrate a tendency towards crime then, conventionally at least, the nature argument is strengthened. However, even in such cases, the picture is far from clear. For one thing, the data are intertwined in real-life complications, including the very fact that being separated from one's parents is itself a stressful experience. Even the most fortunate of these children are likely to face several disadvantages in life, which on their own would only bolster the 'nurture' position. Only if criminals' children were raised in entirely privileged environments – and only if they were kept oblivious to their parentage and personal histories – could they be considered appropriate test cases for 'nature' theories. Moreover, given the likelihood of situational variability and measurement error, very large numbers of such children would need to be studied. In reality it is doubtful that a meaningfully sized sample of such cases has ever been identified, never mind studied. Meanwhile, 'nurture'-oriented psychologists might try to bolster *their* positions by citing additional data on, say, specific behaviours that occur in specific families. For example, if it can be shown that in families where criminal parents were violent towards their children, those children were themselves more likely to be violent in later life, then this might be taken as suggesting a link between experience and behaviour. But these data too are ambiguous. The specific acts of violence perpetrated by these parents could be said to have resulted from the parents' own genetically heritable violent tendencies, and their children's later violence could be said to be the result of the same genetic inheritance. The fact that violence occurred in two successive generations does not show that the first event caused the second any more than it shows that both generations were affected by the same genes.

This confounding of parental style (nurture) with parental genes (nature) causes much ambiguity in developmental and social psychology. It is difficult to avoid the feeling that many interpretations in the literature reflect more the political preferences of authors than the strength of data. Indeed, nature–nurture confounds are just one source of ambiguity in such studies. Other sources include the fact that socially relevant topics attract a good deal of social desirability bias. Another is that they typically arise in highly complex real-life ecologies, where it is impossible for researchers to take account of all relevant variables. As such, data gathered in these contexts are almost archetypally ambiguous: it is always possible to conceive of multiple explanations for them. In choosing which explanation works best, some investigators will refer dispassionately to objective criteria. However, others will feel their way towards a decision, ultimately settling on the particular interpretation that to them seems most intuitively compelling. Interpreting ambiguous data in ways that support a preferred theory is not particularly scientific; in fact, it is the definition of confirmation bias. And as mentioned in Chapter 3, confirmation bias is a hallmark of pseudoscience. According to many observers, psychologists examining politically relevant subjects nearly always exhibit this exact type of decision-making.

Politically subjective psychology

This might just about be excusable if it were the case that political bias was evenly distributed in psychology. However, it has been observed that academic and scientific psychologists (in Europe and North America) are predominantly biased in one political direction. Specifically, in the relevant surveys, a large majority of these psychologists self-identify as political liberals. The prevalence of liberal political orientation in psychology has been gradually increasing over the course of its history. According to recent polls, some 84 per cent of US-based psychologists self-identify as political liberals, with only 8 per cent declaring as conservatives (Duarte, Crawford, Stern, Haidt, Jussim, & Tetlock, 2015). This is very far removed from the demographic pattern across the US population as a whole, where national surveys show liberals to have been outnumbered by conservatives for many years (Gallup, 2014). The challenge of liberal bias in psychology was highlighted in 2011, when US social psychologist Jonathan Haidt of New York University delivered a keynote lecture at the annual conference of the Society for Personality and Social Psychology (Duarte, Crawford, Stern, Haidt, Jussim, & Tetlock, 2015). During the course of his speech he asked his audience to indicate their political views by raising their hands. Only three audience members – from a total of over 1,000 – raised their hands to say they were social conservatives (compared with over 800 of the audience who did so to indicate they were liberals). Surveys conducted with international social psychology groups have emulated this overall pattern of pro-liberal consensus (Inbar & Lammers, 2012).

Notwithstanding the commitment to objectivity that individual researchers might hold, the risk is that implicit confirmation bias will arbitrarily drive certain types of data interpretation. A related concern is that, across psychology as a whole, liberal worldviews will influence the choice of which topics get researched and which get ignored. For example, American law professor Richard Redding (2001) observed apparent contradictions in psychologists' positions regarding adolescent competence. Redding argued that the consensus of psychology researchers appears to be that adolescent women are sufficiently competent to make independent medical decisions, including the decision to have an abortion. However, for other scenarios – such as whether adolescents should be tried in court as adults – the consensus in psychology is different: in these cases, adolescents are no longer seen as sufficiently competent to make independent adult decisions, and so should not be seen as truly culpable for their crimes. According to Redding, this tendency by psychologists to consider broadly similar findings in radically different ways highlights the impact of personal political philosophies on the research process.

The claim that psychology exhibits an unfairly liberal bias has been made with regard to a range of social issues, including affirmative action, ethnic stereotyping, gender segregated education, drug legalization, and even the nature of social conservatism itself (such as whether it is associated with low intelligence). A number of controversies have arisen around the study of family configuration, an issue that typifies the divide between liberals and conservatives. The traditional conservative position includes requirements that families be composed of married heterosexual couples and their children. Families which diverge from this convention – for example, those in which the adult couple are gay – are believed by many conservatives not only to be morally parlous, but also to be at elevated risk of adverse outcomes, such as family breakdown and child mental ill-health. Liberals, on the other hand, believe that diverse families are no more at risk of encountering such predicaments than are any other people, and that principles of social equality demand that no distinction be made (for example, in law) between traditional and non-traditional families. Whether or not certain types of family structure are more or less associated with well-being is very difficult to research, due to several complications (a small subset of which includes: the heterogeneity of such structures; case-level uniqueness; conflation of cause and effect; the necessity for longitudinal studies; and challenges in defining terms). Nonetheless, psychology research is frequently cited in public debates, and a number of psychologists' organizations have published policy statements to the effect that the weight of scientific evidence unambiguously supports views that child adjustment is unrelated to parental sexual orientation (e.g., American Psychological Association, 2012). It may well be that such research interpretations are correct and the liberal position reflects the true impact of family configuration on outcomes. However, in reality, the complexity of the research question, coupled with the dominance of social liberalism among psychologists, makes it difficult to rule out the possibility that confirmation bias is propelling the consensus.

Beyond the choice of research questions or interpretation of data, political bias might express itself in more unsavoury ways. The very nature of politics is often such that advocates of strong views will not only disagree with their opponents, but disapprove of them as well. Socially conservative psychologists have written extensively of hostility received from liberal colleagues (Redding, 2012). Others have suggested that conservatives are customarily discriminated against in a manner similar to the persecution of ethnic and religious minorities (Jussim, 2012). Indeed, recent surveys suggest that many psychologists feel it would be entirely appropriate to discriminate against conservative researchers by, for example, rejecting their grant applications and journal paper submissions, declining to invite them to seminars, and not hiring them when they apply for jobs (Inbar & Lammers, 2012). A number of common rationalizations used by liberal psychologists to explain the scarcity of conservatives in their midst seem, to say the least, insulting. Such explanations include claims that conservatives just don't like academia and so voluntarily opt out of it, that conservatives are less intelligent and so flounder in their attempts to pursue academic careers, and that the conservative worldview is incompatible with (if not anathema to) the open-mindedness required for productive intellectual work. In passing, we might observe that these rationalizations closely resemble those used to explain the scarcity of women and ethnic minorities from a variety of prestigious professions. But beyond this resemblance, empirical studies would appear to cast doubt on their accuracy. For example, research shows the links between intelligence and conservatism to be very unclear, and certainly not of a kind that would explain the predominance of liberalism in academia (Duarte, Crawford, Stern, Haidt, Jussim, & Tetlock, 2015). If it is true that the profession of psychology presents a hostile climate for people who hold socially conservative views, then the degree of true choice exercised by those who decide to 'opt out' of such careers can be questioned. Ultimately, psychology's political monoculture may turn out to be scientifically undermining: processes that narrow the range of perspectives are likely to quash intellectual creativity and diminish the rigour of peer-review (Redding, 2013).

All this being said, we should remember that political categorizations are highly relative. A person described as liberal in one society might appear conservative in another. Although psychology is most commonly critiqued as demonstrating a liberal bias, this might serve to distract us from the overwhelmingly hegemonic nature of a white, middle class, middle-aged, male academic field shaped by a century of Euro-American dominance. For example, it seems inconsistent to be simultaneously worried about liberal bias *and* masculinist bias in psychology. After all, if psychology really was ultra-liberal, then we might not expect much concern about so illiberal an orientation as chauvinism. Similarly, psychology's attitude to sexual minorities has long come in for criticism (Gross, 2009). The way psychology textbooks approach homosexuality often seems quite coy, especially when compared with other academic disciplines or with popular culture at large. While there may be a liberal-sounding consensus concerning issues like gay parenting, it would be an exaggeration to say that homosexuality is seen as mainstream by psychology. Overall, the apparent pro-liberal bias of psychology may relate most to the particular hot button issues that characterize the liberal–conservative schism seen in public politics. For most other issues – those concerning 'private politics' – psychologists are as liable to display a generally conservative worldview as anybody else.

While value systems can affect science, should science affect value systems? One day, scientific research may indeed establish that criminality is the result of nurture instead of nature, that affirmative action is stressful for some people, or that children in non-traditional families find it slightly harder to make friends. The broader question is whether any of this matters when deciding what moral choices should be made. It is erroneous to assert that values should be based on empirical observations of the world. Sometimes doing the right thing will cause inconvenience, upset, or even harm; in fact, given that for every injustice there is a party whose interests are being served, then addressing that injustice will cause at least *that* party some inconvenience, upset, or harm. The interested party might be a privileged segment of society. For example, it is undoubtedly the case that data will show that permitting women to drive cars leads to more women being killed in car crashes. However, countries which prevent women by law from driving are typically seen (by people in other countries) as unacceptably regressive, even totalitarian. The fact that scientific findings can be cited to confirm chauvinistic claims that women would be safer if prohibited from driving is neither here nor there. Values should be meritorious, and worth advocating, in their own right. Indeed, if a value requires the findings of a research study in order to convince people to hold it, then we might question whether it really is a value at all.

The belief that the natural state of things in the world, as established by science, should determine what humans should or should not value is known as the *is–ought problem*. It is a fallacious belief because natural occurrences are not subject to any guiding moral influences, and so cannot be assumed to be indicators of morality. The wish to infer *what ought to be* from *what is* is intertwined with the human instinct to classify things as either normal or abnormal. Such classifications are a lot more complicated than they sound.

Sociopolitical biases in psychology: Presuming normality

The nature of normality

It is striking that the discipline of psychology contains a major subfield called *abnormal psychology* when in fact no agreed definition of the term 'abnormal' really exists. Indeed, many of the behaviours studied in abnormal psychology are, in some senses, far from abnormal. Lots of people experience depressed mood, anxiety, phobias, obsessive tendencies, sexual dysfunction, substance use dependency, stress, age-related cognitive decline, and so on. It seems strange to classify these experiences as lying beyond the realm of the ordinary. Insofar as the average human being stands a good chance of encountering mental health challenges during at least some stages of his or her life, it could be said that abnormality is actually quite normal.

In fact, it is possible to quantify just how normal abnormality really is. If we follow the mathematical logic of continuous probability distributions, we can see that the commonness of most traits is 'normally distributed': a person's score for a given trait will fall between two extremes ('lowest' and 'highest'), and the probability of a particular score will increase the closer it is to the mid-point between the extremes. In probability theory, normal is typically defined as the mid-ranging 95 per cent of all cases, or those which have a score within 1.96 standard deviations of the mean of all scores. This means that, for each human being, there is a 0.95 probability of having a given characteristic within its normal range of intensity.

Interestingly, human beings are made up of very many characteristics. In Chapter 4, when discussing conjunctions, we saw that for multiple independent events, the total probability of all of them happening together is computed as the product of the probabilities of each individual event (i.e., all the individual probabilities multiplied together). This means that, were we to consider just 59 different human characteristics, the probability of scoring within the normal range on all of them would be 0.049 (i.e., 0.95 × 0.95 × 0.95 ... and so on 59 times). In other words, the probability of being *that* normal would fall outside the normal range. Therefore, while acknowledging that abnormality is actually quite normal, we should also note that normality is somewhat abnormal.

In reality, what makes a behaviour or feeling suitable for coverage by abnormal psychology is not that it is, per se, abnormal, but that is clinically relevant. Abnormal psychology is really the study of clinically relevant psychology. When abnormal psychology addresses depression, it is really referring to the type of severe depression that is destructive, biologically extreme, and deserving of clinical intervention. When abnormal psychology talks about obsession, it is not referring to the intense interest of a fanatic, but to the disordered preoccupations and unwelcome intrusive thoughts that for some people are truly paralysing. By considering the extent to which particular behaviours and feelings interfere with happiness, productivity, relationships, and so on, abnormal psychology has pressed ahead without ever fully having resolved the matter of what is, and what isn't, abnormal.

That is not to say that abnormal psychology textbooks avoid raising the question of defining abnormality. Indeed, they would seem to be very well practiced at it: the basic themes used when considering the issue today very closely resemble those discussed in psychology journals nearly a century ago (e.g., Moore, 1914; Skaggs, 1933; Wegrocki, 1938). Such themes include the idea that a behaviour or a feeling is potentially abnormal if it is statistically infrequent, if it causes distress to the individual concerned (or to other people), or if it is debilitating. The statistical frequency approach has certain limitations, as outlined above. In addition, not only is the 95 per cent mathematical threshold itself arbitrary, but the fact that traits vary in two directions (high *and* low) can sometimes create anomalies. For example, while extremely high levels of substance use dependency would appear to be conspicuously problematic, the same could hardly be said of extremely *low* levels of such dependency. A further point is that many behaviours and feelings are statistically rare without causing difficulties. As such, any statistical approach will need to be supplemented by other considerations. Therefore, the notion that distress is pertinent is intuitively appealing: a behaviour or a feeling might be considered abnormal if it causes significant distress. However, some people experience disruptive behaviours and feelings without exhibiting obvious distress. A person may have regular hallucinations, be incapable of empathizing, or have no feelings of guilt, shame, or embarrassment, all without ever experiencing any distress whatsoever. Some attempts to define abnormality try to accommodate this by taking account of the distress caused to *other* people in the person's life. However, this is reliant on the assumption that suitable other people exist, and that they are capable of making consistent assessments of the situation (for example, some friends might find a person's lack of empathy pathological, while others may not). Accordingly, consideration is often given to the idea of debilitation: if behaviours and feelings undermine a person's productivity, relationships, and ability to participate in society, then they might be classified as abnormal. But a central dimension to this will be the person's own wishes: some people may not *want* to hold down jobs, or enter stable relationships, or be part of mainstream society. So, in order to assess the degree of debilitation, we must first decide whether a person's *wishes* are normal or abnormal, which brings us back to square one.

A further set of approaches to the definition of abnormality (and, thus, normality) concentrates on the *compatibility* between person and environment. Most commonly, behaviours and feelings are presumed to be abnormal if they contradict or disregard the dominant conventions of the society in which the person lives. This is premised on the idea that normality is, in a sense, 'normative': it can only be discussed with reference to the norms established at a cultural level within broader society. When people fail to adhere to such norms, then it can be anticipated that they are likely to encounter a range of personal difficulties, which include the possibility of distress for themselves, distress for others, and debilitation. This does not entirely avoid the need to assess the normality of a person's wishes, but such a cultural approach would imply that people who consciously withdraw from society are indeed behaving abnormally. It is therefore perhaps obvious that the biggest problem with this approach is its inherent arbitrariness: it is entirely dependent on the cultural norms prevalent in a particular place at a particular time. This means that behaviours that are normal in one place, or at one time, might be abnormal elsewhere, or at another time. For example, in some cultures it is socially acceptable for people to express their emotions very vividly, whereas in other cultures the etiquette is to be reserved, especially in public. If somebody moves from one culture to another, then their behaviour might be seen as abnormal in their new setting. Vocalizations, gestures, and intrusions that mark orthodox expressions of distress in one place might come across as near-lunacy in the other; while the stern, expressionless austerity that the latter culture requires might be seen as verging on psychopathy elsewhere. The question of how these differences in cultural norms might affect the way a person's behaviour is classified is not a trivial one. It could lead to members of cultural minorities being inappropriately diagnosed as psychiatrically ill, and may partly explain why in many societies ethnic minority status seems to be a risk factor for mental ill-health (Halpern, 1993; Missinne & Bracke, 2012).

Normality as socially constructed

It should be noted that most diagnostically relevant psychiatric conditions emerge quite similarly in different cultures. People who are clinically
depressed or who have schizophrenia are very likely to be diagnosed as
such wherever they live. Also, very many psychiatric diagnoses are related
to known *physical* defects in the brain, including conditions that are secondary to physical brain injuries. Diagnostic inconsistency arises mainly
in relation to less extreme situations, where there is no clear brain injury
or defect, and when a person's symptoms lie at the (perceived) threshold
between normality and abnormality. The fact that inconsistency arises
shows us that the true definition of psychological normality, rather than
being objectively definable, is actually dependent on a social consensus. If
everyone in society agrees that a behaviour is normal, then it is defined as
such; if everyone agrees it is abnormal, then this is the diagnosis. In past
centuries, female sexual desire was considered so socially unacceptable as
to be a sign of mental disease, with one prominent 19th-century physician
declaring that around a quarter of all women were sufficiently amorous as
to warrant psychiatric attention (Briggs, 2000). Other Victorian authors
described a condition of suffocating nervousness associated with various
advances in technology, which was said to be especially pronounced among
ethnic minorities and the poor. In his classic book, *American Nervousness:
Its Causes and Consequences*, neurologist George Beard (1881) identified the steam engine, the telegraph, and the daily newspaper as particularly dangerous in this regard. Another 19th-century American physician,
Samuel Cartright, described a condition known as *drapetomania*, a mental
illness seen among African-American people held as slaves, the main symptom of which was an objection to slavery and a wish to escape captivity
(Guthrie, 2004). It might be clear from these examples what the problem
is: cultural definitions of normality and abnormality may represent an offshoot of social totalitarianism, where the wishes of controlling interests are
protected, and where dissenters are dismissed as insane.

All told, our consensual definitions of normal and abnormal often reflect
our underlying social, cultural, and political worldviews. In other words,
they may represent yet another example of value-laden psychology. To the
extent that our conventions of normality reflect our value systems, then our
study of normality could be said to be tainted by bias. Certainly this type
of problem is not seen across all areas of scientific research. When values
intrude on scientific inquiry, the accuracy and objectivity of whatever data
are measured is called into question, the inclusion of arbitrary assumptions as tenets threatens the requirement for parsimony, and the risk of
confirmation bias is heightened, thereby undermining the principle of falsification. There is certainly a dilemma here: psychologists feel driven to
force through a working definition of abnormality not because they enjoy
toying with epistemology, but because they want to provide help to people

who are psychiatrically ill. They do not feel it is appropriate to wait until some philosophically satisfying resolution can be found to the paradoxes presented by the culture-bound nature of normality. They can see that vulnerable people require psychological therapies, supportive interventions, and (maybe) psychiatric medication right *now*. The fact that proceeding to treatment involves compromising on scientific principles like objectivity, parsimony, and falsifiability seems a small price to pay. We will consider further the implications of this ethos of helpfulness for the scientific standing of psychology in Chapter 9.

Not all distinctions between normal and abnormal relate to clinical applications. Many theories in psychology imply a presumed normal state of things, which can be used to predict how people function in everyday life. For example, some major theories of human personality specify a standard template for majority human nature, with exceptions arising from sheer improbability or situational quirks. The American humanistic psychologist, Abraham Maslow (1954) developed a very famous theory of personality, which specified a hierarchy of needs by which, it is said, all people are motivated. According to the theory, people are motivated to satisfy needs in a sequence, ranging from basic lower-order needs (such as the physiological need for food) to intrinsically human higher-order needs (including the humanistic need for self-actualization). Theoretically, people will only be motivated by higher-order needs once their lower-order needs are satisfied. However, while this theory has been very influential in the areas of self-help and life coaching, its specified needs hierarchy has been found to be very arbitrary indeed. It is now well accepted that people *can* strive for higher-order needs without all their lower-order ones being satisfied, that for some people the hierarchy is ordered differently, and that for many people needs are strived for in no particularly consistent hierarchical order at all. In other words, the theory now has as many caveats as it has features. A major problem with Maslow's theory of universal human personality is that he developed it after studying a very narrow group of people at a pretty unique time – namely, Californian teenagers attending (one) American college during the 1950s. The fact that his theory reflects one particular worldview that turns out *not* to be universally shared should perhaps be no surprise. Indeed, the theory's focus on personal self-actualization might be said to reflect a type of individualism that would not be at all valued in more collectivist societies.

In the end, psychological normality is a social construction. This means that its definition is arrived at not by individual objective observation of the world, but through a process of collective and subjective discussion. Such a distinction sounds a lot like that between science and non-science: while science focuses on objective, accurately measurable, and replicable empirical observation, non-science takes account of other things, such as conventional dogmas regarding how things are in the world, subjective judgement by experts, and anecdotal advice. In this sense, a rigid adherence to the study of the psychologically *abnormal* as a subject in and of itself (as distinct from the study of 'depression' or 'phobias') requires a tolerance for non-science. Our human drive to frame the world as consisting of things that are normal and things that are not may be just another cognitive heuristic, a strategy for making environmental complexities feel agreeably comprehensible. It may also reflect a very useful instinct for expecting nature to conform to order and regularity, an important appreciation that helps us think in scientific ways. But not all notions that pop into our minds authentically reflect order and regularity in nature. Splitting humanity into the categories 'normal' and 'abnormal' is not the same as identifying 'males' and 'females'. Rather, this forcing of a dichotomy is little more than the rhetorical tactic of false bifurcation. A reasonable case can be made that the terms 'normal' and 'abnormal', when used as descriptors for human beings, are inaccurate by definition and so should be avoided, especially in science.

Should psychologists have values?

Many criticisms of psychology relate to its proneness to social constructions. Unlike the subject matter of chemistry or physics, the concept of the 'human condition' is something that our species has projected into existence, by virtue of our being able to reflect on our own affairs. That is not to say that the human condition doesn't exist. But while it certainly exists, it exists as an inferred entity, a collection of clearly detectable person-level occurrences (thoughts, feelings, behaviours, memories, and traits) that, because of their self-referencing and interdependent nature, cohere into a discrete package of concepts: a freestanding 'thing' in its own right. Psychology is not the only discipline to deal in such 'things'. Entities such as 'the economy', 'climate', 'the immune system', or even 'history' are in essence social constructions. Such topics can be complex to study. As well as comprising a multiplicity of components, their identification (and, thus, the ability to distinguish them from *other* entities) relies on the availability of socially agreed definitions. But just because specification is dependent on convention does not make a concept useless or, for that matter, unscientific.

Some critics argue that all of psychology – and indeed all of science – is afflicted by an overwhelming relativism. Relativism is the idea that because truth is a function of perception, all of the things ever discussed by humans are, in fact, socially constructed. Taken to its extreme, this position logically argues for solipsism: the idea that one has no basis to believe in the existence of anything outside one's own conscious experience. As noted in Chapter 7, the observation that real objectivity is impossible because humans cannot step outside their own subjectivity is an analytic truth: it is true by virtue of the very definition of the terms 'objectivity' and 'subjectivity'. Furthermore, a genuine solipsist would have to conclude that the entire universe might in fact be a product of their consciousness, and that their impressions that other things (and other people) exist are hallucinatory. When applied to science, the position requires that all claims to objective knowledge be viewed with suspicion: scientific findings are constructed by humans rather than revealed by empirical research, and as such cannot be assumed to correspond with reality (whatever that is).

The Austrian philosopher, Paul Feyerabend (1975) argued that, because knowledge is socially constructed, no authentic distinction can be made between science and pseudoscience. In a series of anarchic treatises, he proposed that activities such as witchcraft and astrology were as legitimate a means of establishing true knowledge as was any scientific method. According to Feyerabend, scientific claims to objectivity are not only misguided, they also serve to camouflage scientists' ulterior motives and vested interests. Feyerabend's legacy is seen in contemporary epistemologies whenever attempts are made to declare the standard approaches of science obsolete (for example, when it is floated that orthodox science should be supplanted by a so-called 'quantum' paradigm). While attaining fame as one of the world's most prominent philosophers, Feyerabend ultimately failed to convince his peers that orthodox science was so addled by constructionism as to be no more worthwhile than pseudoscience. Technological medicine had proved far more effective than witchcraft, and, when it came to predicting events in the universe, astrophysics had a much better track-record than astrology. Ironically, Feyerabend made little secret of the fact that his own philosophical positions were imbued with sociopolitical intent: he wished to argue that science was morally objectionable because he perceived it as serving the interests of Western imperialism. Perhaps influenced by his belief that objectivity was futile, Feyerabend unashamedly assembled his philosophical arguments in ways that reflected his political views.

In reality, scientists frequently acknowledge that their work is shaped by sociopolitical values. Many choose their entire field of study on the basis of a moral preference: they wish to use their scientific career to help make the world a better place. As such, they choose fields that enable them to conduct research into climate change, the alleviation of poverty, treatments for diseases, and assistive technologies. The aim to enhance the world not only influences people's career choices, but also directs the questions that are addressed by their studies. This is why there is much more research on ways to treat cancer than on, say, ways to fold handkerchiefs. Of course, such influences are not always optimal: diseases prominent amongst the world's affluent populations attract much more research attention than those common in developing countries. Nonetheless, in most cases, the shape and scope of science is certainly influenced by its practitioners' good intentions. Science is itself an endeavour founded on values: scientists endorse a particular value-set when they call for transparency, integrity, and open communication in research (Derry, 1999). Furthermore, doing psychology as a feminist, adopting 'nurture' rather than 'nature' perspectives on human development, and being sceptical toward notions of psychological abnormality are all examples as much of ethos as of epistemology. In short, scientific practice can often be described as socially constructed and informed by social values.

When considering how subjectivity influences the science of psychology, it is important to distinguish the *course* of knowledge-generation from its *content*. Personal values may well influence how investigators choose research questions, but values should not influence how researchers interpret otherwise ambiguous data. When contributing to public debate, psychologists may well have their own preferred positions. The conventions of democracy suggest that they are perfectly free to advocate for particular policies. While some may wish to simply let the data do the talking, more will offer themselves as arbiters of scientific information, honest brokers of policy alternatives, or even advocates for particular contested views (Pielke Jr, 2007). Strictly speaking, subjectivity is not scientifically problematic in these contexts. Subjectivity becomes problematic when explicit biases are hidden and implicit ones go undetected. It then interferes with assumptions, distorts inferences, and generally wrecks the epistemological principles of science. In other words, while *scientists* can afford to be subjective, *science* is supposed to remain fundamentally value-free.

Chapter 9

Religion, Optimism, and Their Place in Psychology

Looking for a brighter future

Not all value systems relate to prejudice or politics. In fact, very many people are *neither* prejudiced *nor* political, but are still distinguished by an array of strongly held personal beliefs. Nearly everyone has values relating to the importance (or unimportance) of life, the order and nature of entities in the universe, what constitutes justice, how people should treat each other, and the difference between what is welcome and what is not. Most people will instinctively differentiate good from evil. Of course, not everyone will hold the same views about these matters, but they will hold *a* view of some kind regarding each. This is so universally true that we often fail to recognize such views as constituting 'values' per se – that they represent enduring *opinions* about what states of affairs are to be considered preferable – and instead assume them to be free-standing axioms that come with life itself. As nearly everybody is influenced by such concerns, we could say that, by definition, such value systems are humanly unavoidable. Therefore, insofar as they interfere with science, values are likely to present perennial complications for psychologists. This is highlighted by the fact that, as described in Chapter 8, scientists (and psychologists among them) typically choose to enter their careers because of a personal wish to make the world a better place. In other words, they choose to do so because of an attachment to a particular value. Far from being value-free, their psychology is value-dependent.

The desire to improve the world is psychologically interesting because it rests on a number of assumptions about the nature of humanity itself. Firstly, it assumes that human beings are themselves important, and that striving to improve their predicaments is an activity of inherent merit. This in turn implies that people will automatically welcome the efforts of others to improve their lives, and that the entire helping enterprise will be greeted universally as a good thing. For many, these views will emanate from a general belief in a higher power that transcends the ordinary mortal aspects of the universe, a belief that itself has profound implications for a person's understanding of what it means to be human. Secondly, the desire to improve the world is informed by an assumption that it is actually possible to do so, and by a related assumption that simply *trying* to do so is itself worthwhile. In other words, there is a common view that striving for improvement is not only more likely to succeed than not, but also inherently worthy in its own right. It is worthy even when it does *not* succeed. As a result, one of the worst failings a human being can be said to exhibit is disregard for the welfare of others, and one of the most damning criticisms to be accused of moral turpitude (with perhaps the second most damning to be accused of moral ambivalence).

In this chapter we will consider those ways in which the aspiration to create a brighter future can impact upon the challenge of doing psychology as a science. In considering optimism and the ambition to do good in psychology, we will examine three strands that focus on the issue of helping and improvement: we will look at 'social support', the benefit that people are said to derive from the assistance of others; we will look at the case of psychological therapies and how they are conceived of and understood; and we will look at 'positive psychology', an approach to psychology that explicitly seeks to orient the field towards a mission of human improvement, both personal and social. But before examining these strands, we will begin by considering one of the most profound – and controversial – classes of all value-systems, a set of moral paradigms that human beings in their billions have adopted for centuries. This is the perspective that the existence of humans is subject to that of a transcendent higher power – a celestial authority that not only facilitates optimism and morality, but *mandates* them too.

Religion and psychology

Religiosity and the social context

It is perhaps an obvious point, but of all the philosophical or ideological paradigms within which human beings can ponder their own existence, religion has endured as by far the most common for many thousands of years. Part of its endurance is its flexibility in the face of social change. While a small number of major religions account for the majority of religious people in the world, they comprise a milieu of denominations, congregations, churches, sects, and other spin-off movements, which emerged and evolved in response to shifting demographics, historical events, and advances in human thought. Religion endures also because of its customizability at the level of the individual: broadly speaking, persons who declare themselves affiliated to a particular religion may not adhere to all of its teachings. They may subscribe to only a subset of their church's ideas, rejecting or ignoring those that do not sit so comfortably with them. Accordingly, reported religious membership may not always reflect enthusiasm for religious belief. In one Canadian opinion poll, 33 per cent of Catholics and 28 per cent of Protestants reported that they did not even believe in the existence of a deity (Ipsos, 2011). According to the most recent official Canadian census data, this equates to around 23 per cent of the total population of Canada, substantially exceeding the proportion of Canadians – 16.5 per cent – who formally describe themselves as not religious (Statistics Canada, 2005). In other words, the survey suggested that the majority of Canadian atheists are, in fact, Christians.

The proportion of EU citizens reporting themselves as belonging to a religion is 75 per cent. This includes 66 per cent of the population of the United Kingdom (European Commission, 2012). Unlike the Canadian study, the standard EU survey does not ask those who classify themselves as 'religious' whether they actually believe in deities. Moreover, such polling data are usually drawn from the adult population. This is relevant because most religions allow parents to designate their children as members of a church long before the children themselves are in a position to form views on the matter. As such, official figures regarding the number of members in each religious grouping are often complicated by the counting of children along with adults. While children are very likely to eventually adopt their parents' views on religion, it is not exclusively the case that they will – and, on a cognitive level, it is difficult to argue that a child's views on the existence of celestial beings are truly comparable to those of an adult. This further suggests that data on religious adherence can often be inflated. Referring back to Canada, we can note that around 16 per cent of that country's population is aged 15 years old or less: if this proportion is shared equally across the religious groups, then it can be inferred that while Catholicism represents the largest religion in that society, only around *half* of Canadian Catholics are actually fully grown adults who believe in the existence of a god.

Overall, while the observance of religious belief might well be different today compared with the past, the holding of structured views on the special nature of human life and its association with a realm beyond the limits of our earthly existence remains a highly prevalent influence on popular thought. This far-reaching aspect of religious influence ensures that, in just about every walk of life, significant numbers of people will declare an allegiance to an organized religion, or will hold religious or quasi-religious beliefs about the nature of the universe. This might have little direct relevance in many professions, but in the case of science – the profession that seeks to *study* the nature of the universe (and all the things in it) – it is liable to create paradoxes. After all, religious belief-systems are typically derived from teachings handed down through generations, rather than from empirical research studies. Most religions have recourse to a holy text or sacred history said to comprise the unerring message of an omniscient god (or gods), and many religions will attach special authority to the views of a formally ratified clergy. For most of human history, it was through these channels – and not through scientific journals or research reports – that explanations of the universe were conveyed. Given that people who become scientists are just as likely as anyone else to have been acculturated in such an environment, it will be little surprise to learn that scientists (and thus psychologists) are just as subject as anyone else to the influence of religious thinking.

As well as individual scientists, entire academic subjects can be affected by acculturation. It is often forgotten that the late-19th century universities into which modern scientific psychology emerged were deeply enmeshed in a religious culture (Reuben, 1996). With very few exceptions, major universities up to the late 19th century offered what was essentially a seminary education, implicitly designed to prepare young men for careers as religious ministers. Bible studies was a core subject, along with ancillary domains, such as Latin, that were intended to facilitate students in deepening their appreciation of scripture. Even the study of modern languages originated as a way to combat Biblical criticism from German and French secular philosophy (Brown, 1969). Academic science began under the title *natural philosophy*: this positioned it as a complement to *natural theology*, the subject concerned with using reason to argue for the existence of a deity (and, by extension, as an adjunct to *revealed theology*, the subject concerned with citing the Christian Bible for much the same purpose). The sub-area of philosophy that considered psychological subject matter was referred to as *mental philosophy* or *mental science*. It was this specialism which pre-existed and eventually gave way to the new scientific *psychology* (Fuchs, 2000). Some famous written histories of psychology describe this transition as one in which a generation of impatient and disruptive laboratory-educated psychologists blazed onto campuses worldwide, aggressively rejected the theologically informed mental philosophy of the past, and set about replacing godly paradigms with ones citing evolution, psychophysics, and biology (e.g., Boring, 1929). However, in reality, the move from mental philosophy to new psychology was quite convivial. Contemporary writings show how mental philosophers – by and large – took a keen interest in the new psychology and willingly provided it with an academic home in the university system; moreover, the new generation of laboratory psychologists – by and large – retained their personal religiosity and structured their new textbooks and courses along the same lines as those of their predecessors (Fuchs, 2000).

If anything, the nexus between science and religion has attracted more scholarly controversy in modern times. Some commentators assert that religious discourse can and should inform scientific worldviews (Peterson, 2003), such as in the case of fields like intelligent design and creationism. Others argue that moral reasoning is an empirical matter, and that religious approaches to ethics should defer to scientific positions on what is and is not valuable (Harris, 2011). However, perhaps most argue that science and religion simply occupy different epistemological zones (Frazier, 2003). Within this frame, some observers see science and religion as non-overlapping pursuits that should easily coexist without controversy (Gould, 1999) – but for others, science and religion are necessary foes: the success of one is seen as absolutely requiring the demise of the other (Dawkins, 2006; McGrath & Collicut McGrath, 2007). In this latter view, science and religion are perceived as subverting one another's objectives. Science is accused of trivializing the human experience in ways that ultimately lead to the commodification of people, and of undermining the very basis of morality by insisting on philosophical materialism. Religion is accused of deluding audiences into falsely believing in their own exceptionalism (and even in their own immortality) in order to cultivate unquestioning mass support for hegemonic and destructive social hierarchies, as well as of being paradoxically ambivalent towards the suffering of non-coreligionists. Such perceived incompatibilities may account for an apparently lower level of religiosity seen in academic scientists relative to the general population (Ecklund & Scheitle, 2007).

But not all scientists are equally irreligious. Indeed, one large-scale survey – of 2,000 academic scientists working at US universities – suggests clear patterns of variability across sub-disciplines (Ecklund & Scheitle, 2007). The pollsters found that around one in ten psychologists (10.8 per cent) declared themselves to be convinced of the existence of a deity. This was similar to the number of undoubting theists in chemistry (10.9 per cent) and economics (10.4 per cent), but much higher than the numbers seen in physics (6.2 per cent) and biology (7.4 per cent). On the other hand, while one in three psychologists classified themselves as undoubting *atheists*, the same was true of only one in four chemists. Physics possessed the largest subgroup of clear-cut non-believers (40.8 per cent), and chemistry the smallest (26.6 per cent). To complicate matters further, religious belief did not mirror religious behaviour: while similar numbers of psychologists and economists believed in deities, nearly twice as many economists (8.5 per cent) as psychologists (4.9 per cent) were weekly church attenders. In short, in terms of all-round religiosity, psychology was found to be a generally middle-of-the-road scientific field. (It should of course be pointed out that much of the preceding discussion relates to the situation in Western countries, where the dominant historical religion has been Christianity, albeit with significant subgroups such as Islam and Judaism. There has been much less study of the relationship between religion and science in non-Western countries. However, in the main, religion seems to be held in generally higher regard in such countries, and so we might reasonably speculate that the numbers of psychologists there who hold religious views is similar to or greater than what is seen in the West.)

In summary, when it comes to religion, psychologists are not unlike other people in the world. Very many of them hold religious views regarding spirituality, the afterlife, the origins of the universe, the nature of morality, and the existence of deities; very many of them are members of a major organized religious grouping; and very many of them were raised to be religious by their parents. Notwithstanding all this, very many of them hold only a subset of those formal beliefs set down as dogma by the religious group they belong to; and very many of them hold and express their religious views in a moderate way, with only a minority being sufficiently driven as to engage in zealotry. Of course, very many psychologists are none of these things at all: atheism, agnosticism, and secularism are common in psychology as they are in most other sciences, which is to say more common than in mainstream society. Nonetheless, this is relative: the prevalence of religiosity in psychology is still very high, with survey data suggesting that only a third of psychologists hold views that could be described as genuinely atheistic. All in all, if a religious ethos is truly incompatible with a scientific one, then the prevalence of religiosity among psychologists raises practical questions about how psychology deals with the fundamentals of cognition and behaviour – and whether it can address the more challenging, unconsoling, or even unpalatable aspects of the human condition with authentic impartiality.

Religiosity and scientific epistemology

Just about all religions seek to enhance the quality of life as it is lived by all humanity, to promulgate doctrines of individual moral behaviour, and to strive for human order, harmony, and peace. In a sense, that is the easy part. Were such intrinsically noble ambitions the totality of what religions sought to represent, then religion itself would be a wholly uncontroversial topic. What makes religion a fraught subject – in some cases, the pretext for centuries of intergenerational sectarianism – is that religions tend to do more than just promote aspirations for moral living. Religions don't just talk about how life should be lived; they also offer a view on what life actually *is*, how it came to exist, and what it actually means. In essence, each religion offers a distinctive theory of the factual universe, a set of teachings explaining how things are in nature and how they came to be that way. Such metaphysical theories help religions imbue their moral codes with the status of the extraordinary: a code becomes compellingly persuasive when it originates from an entity more powerful than humanity itself. Thus, moral teaching becomes intertwined with the existential, and the represented order of nature becomes cited as the reason *why* certain behaviours are moral and others not. All this culminates in the intermingling of ethics with observance: proscriptions on theft, violence, and killing are placed on an equivalent moral footing as those on eating certain foods, shaving one's beard, or working on certain days. Admirable principles referring directly to the prevention of human suffering are listed alongside arbitrary ones referring to signifiers of a particular religion's customs and traditions. Morality is bundled with mores.

The way religions seek to explain the origins of the universe and the nature of humanity has many implications for scientific psychology. Consistent with the all-encompassing scope of religious narrative, these implications span a wide range of ideas and conceptual contexts. For conciseness, we will consider the following contexts in turn. First, we will consider the *trivial*, the *supernatural*, and the *existential* – three contexts that relate to relatively cosmetic aspects of the universe, including its history and biology. Then we will consider the *metaphysical*, the *moral*, and the *epistemological* – three contexts that relate to somewhat more intangible notions. In each case, there are both direct and indirect implications of religion for psychology; in many, it is the indirect implications that pose the subtler, and therefore greater, challenges.

The trivial, supernatural, and existential contexts. Contexts might be said to be *trivial* when they arise from contestable assertions regarding the narrative of historical events considered factual within a particular belief system. For example, within Abrahamic religions, it is asserted that the first human being was a man known as Adam, while the second human being was a woman known as Eve, who was created by the deity from one of Adam's ribs in order to serve as his helper. Many committed religious adherents view such depictions in mythological or metaphorical terms, but for others they are seen either as contracted descriptions of actual events (Madueme & Reeves, 2014) or as unambiguously factual events in themselves (Ham, 2012). Insofar as they are asserted to be true, such depictions broadly clash with the scientific view of psychology. This is because the notion that human beings were formed as a complex organism *ab initio*, as opposed to having evolved from other species – not to mention the idea that the female human body form is a derivative of that of the male – runs counter to the scientific understanding of evolutionary biology that informs modern psychology in several ways. Another historical assertion seen in several religions is that, in ancient times, a global flood wiped out all life on earth except for certain animals and humans who were protected by a supreme deity in order to subsequently repopulate the world. Again, such an event would have altered the course of human evolution in ways that contradict the current scientific consensus. However, while these various stories are argued to be historically authentic by some proponents, probably the majority of religious adherents understand them to be allegorical rather than literally true, or to reflect a plausible (but now obsolete) scientific understanding of the world that existed at the time they were first recorded. Therefore, as these examples clash with science at a very superficial level, critics who labour such points of factual incompatibility might rightly be accused of pedantry.

Something is said to be *supernatural* if it defies scientific explanation or exists in a way that is not subject to the laws of nature – literally, if it stands 'above' nature. In this sense, then, virtually all religions promote supernatural concepts, by proposing the existence of deities. This *supernatural* context of religion presents both direct and indirect challenges for scientific psychology. A direct challenge relates to the fact that many deities are held to be omniscient. The idea that a single being can know everything about the universe challenges the standard depiction of knowledge as the product of psychological experience and a function of brain-based cognition. There can hardly be a more fundamental aspect to scientific psychology than the proposition that knowledge is formed within minds, over time, through experience. Moreover, the accumulation of personal knowledge involves the human brain: certain types of neural damage result in memory loss, while other types will prevent the formation of new memories. Claims that certain humans can perceive thoughts directly from other people's minds, or can perceive distant events in ways other than through their sensory organs, are generally considered to be pseudoscientific (see Chapter 5). Thus, the claim that a supernatural being can do all that and more must certainly be challenging to scientific psychology. An indirect challenge for psychology arises from the implication that there exist entities that, because of their supernatural standing, are inexplicable. If it is argued that inexplicable entities exist, then no scientific account of any phenomenon can ever be said to be trustworthy: this is because it will always be possible to offer an alternative account by invoking inexplicable entities. For example, while a meteorologist might offer an empirically based explanation for bad weather, a proponent of supernaturalism can always insist that the bad weather was, in fact, caused by something other than nature. If it is agreed that supernatural entities are possible, then this latter explanation can never be refuted. In essence, by rendering *all* propositions irrefutable, the accommodation of inexplicability renders *all* science futile.

The *existential* context refers to how religions account for the emergence of the universe and of life within it, and the way special status is attributed to human life relative to that of other species. In other words, it refers to the import of human existence. One direct challenge this presents relates, again, to the framing of scientific psychology within a standard biological model of evolution by natural selection. This broad-ranging paradigm not only informs the trajectory of biological, cognitive, and clinical psychology, it also shapes the assumptions of much developmental, social, and personality psychology. Of course, as discussed in Chapter 6, it is untrue to say that all areas of human psychology can be explained using one biological theory. However, the assumption that human beings are, at root, animals subject to biological impulses, and that human mental functions have been shaped by the history of the species, is always part of the picture. As a result, assertions that the human form (including its mind) was in fact framed by an external strategically-focused intelligence, such as a supreme deity, undermine the basic explanatory paradigm of psychology.

Of course, it should be acknowledged that, nowadays, the hierarchies of most major religions declare themselves comfortable with Darwinian theories of evolution, and to see no contradiction between natural selection and religious views on the development of species across time. However, such declarations are usually based on a slightly modified version of evolutionary theory and so remain scientifically controversial. Specifically, religious depictions of evolution retain the idea of a divine blueprint for humanity's ultimate form, towards which the processes of natural selection are said to be directed. Human bodies and minds are held to conform to an archetype laid down by divine will. However, the idea that there exists an ultimate form towards which evolution is directed is simply not part of evolutionary theory – in fact, it flagrantly contradicts it. Evolution by natural selection postulates that species gradually become distinguished over time when successive generations of organisms genetically inherit small physical variations that helped their ancestors to survive and procreate. Such variations occur in each generation of a species as the result of random genetic mutations (in the same way that children physically resemble their parents but are never identical to them) and so it can be said that evolution is a never-ending process. We can therefore assume that the human species continues to evolve with each passing generation and, in the distant future, will look and behave very differently to the way it does now (Buller, 2005). There is no blueprint or ultimate final form for humanity, or indeed for any particular species. Depictions of evolution that attempt to posit an ultimate final form, or to incorporate a role for an external agent either as architect or engineer, are simply not the standard form of evolution upon which the assumptions of evolutionary biology, and thus scientific psychology, are based.

The indirect challenges posed by the existential context relate to the way the standard religious depiction of human exceptionalism conflicts with broader scientific knowledge. The idea that human beings are special within nature, and indeed are superior in value (if not in all-round competence) to any other species, seems difficult to support in the context of what is known scientifically. In reality, humanity represents a very tiny detail of the known universe. In the context of our own planet alone, humans are but one of nearly eight million different species of animal (Mora, Tittensor, Adl, Simpson, & Worm, 2011). With only seven billion humans alive today, we are a relatively small species in terms of population. There are, for example, twice as many live chickens as there are humans in the world (even allowing for the fact that millions of chickens are slaughtered every hour). Bacteria are by far the most populous animal, with more than 4 quadrillion quadrillion (i.e., 4×10^{30}) in existence. In Chapter 3, when discussing people's tendency to exaggerate the complexity of their own species, we put that mind-boggling statistic in perspective: we noted that each of us – individually – has more independent living bacteria inside our digestive tracts right *now* than the total number of human beings who have *ever* lived (Dethlefsen, Huse, Sogin, & Relman, 2008). The combined biomass of all humans is less than that of all ants (although, to be fair, there are 14,000 different species of ant), and two-thirds that of cattle (of whom there is one species, *Bos taurus*; Groombridge & Jenkins, 2002). While our planet has sustained living organisms for 3.6 billion years, humans have existed for a mere 200,000 of these, equating to less than 0.01 per cent of the history of life on Earth. And the Earth itself is hardly that significant in cosmic terms. Our planet may be one of eight orbiting the sun, but it is joined by more than 100 billion others, of varying sizes, strewn across our galaxy, the Milky Way (Swift et al., 2013). Were you to shrink our nearby sun to the size of a single white blood cell, and then shrink the rest of the universe to the same scale, the Milky Way would still be the size of the entire continent of Europe (Fact Stream, 2011). In other words, our planet seems irrelevantly small when compared to the size of the galaxy in which it is situated. But there's more: even that galaxy seems negligible when you consider the size of the universe as a whole. The Milky Way is merely one of over half a *trillion* (i.e., 500 billion) galaxies known to exist. If the Milky Way were shrunk to the size of a single teardrop (say, one millilitre of fluid), and the rest of the universe reduced to scale, then all the remaining galaxies combined would occupy the space of two *hundred* Olympic-sized swimming pools (and that ignores the space *between* galaxies, which exceeds that occupied by the galaxies themselves).

Overall, the human species is but a fleeting biological moment in the history of a minuscule planet tucked away in some imperceptibly small corner of an unimaginably vast space. The idea that humanity enjoys some kind of special privilege in nature, and that the entire universe exists in order to accommodate our destiny, seems somewhat detached from the broader scientific perspective. Conferring this special status upon humanity within nature is likely to be misleading, and unlikely to materially assist our ability to use empirical methods to properly understand human psychology.

The metaphysical, moral, and empirical contexts. The three remaining contexts relate to somewhat more abstract notions. Once again, each is characterized by both direct and indirect implications. The *metaphysical* context refers less to entities that exist in our universe and time, and more to the very nature of 'being' itself. Virtually all religions posit some form of life after death (such as reincarnation, purgatory, or a heavenly or hellish hereafter). This reflects a fundamentally dualist position on psychological matters: the view that each person's human existence transcends the physical realm. For many people the idea of an afterlife seems more compelling than any other religious concept. One large-scale British cohort survey of 42-year-old adults found that more believed in an afterlife (49 per cent) than in a deity (31 per cent; Sullivan, 2012). It may be that, by bringing with it a form of immortality, the idea of an afterlife is just too attractive to dismiss. It may even be that the thought of *no* afterlife is so terrifying as to be virtually unbearable, and thus avoided automatically in order to stave off despondency (Goldenberg, Pyszczynski, Greenberg, & Solomon, 2000). However, there are other reasons why large numbers of people might entertain the possibility of life after death. According to some psychologists, the common human intuition that there exists an afterlife, the entry to which is contingent on near-term good behaviour, is a natural consequence of the way human minds evolved to ensure that we regulate our behaviour even when we are alone (Bering, 2011).

Whatever it is that sustains a belief in an afterlife, the dualist nature of the concept poses a direct challenge for psychology. To say that human life continues after death requires that human thought take place in the absence of a functioning brain, an idea that, however seductive or consoling to people who worry about the death of their brains, runs very much counter to mainstream psychological research. As noted in Chapter 6, everyday observations of the way drugs, alcohol, and brain injury can create profound changes in human thoughts, feelings, and behaviours help to illustrate how human consciousness flows from brain function, rather than the other way around. Similarly, the sometimes mechanical nature of consciousness that is revealed by empirical research (see Chapter 5) belies the idea that humans are possessed of ethereal spirits that transcend the physical realm. The metaphysical context also poses some indirect challenges. One relates to the way presumptions of an afterlife can influence people's decisions in the here and now. When people anticipate a post-death existence of infinite heavenly comfort, a prospect that is defined in dogma as unimaginably more fulfilling than the mortal life experienced here on Earth, their motivation to live worthwhile lives in the present may become skewed. In the context of a heavenly hereafter, all current earthly problems – such as human poverty and social inequality – can be seen as ephemeral: we need not become despondent by their intractability in this life, because the world's poor and downtrodden will be rewarded in the next one. For psychologists trying to effect positive change in clients' lives, or in society at large, such distortion of people's personal decision-making horizons can be a significantly complicating factor.

The *moral* context arises from the claim that morality and moral behaviour are god-given and so reliant not only on the existence of a deity, but also on a belief in one. Psychologically, this would mean that the ability to reason in moral ways does not arise naturally in humans, but rather is conferred upon them, from outside, by a god (or gods). The idea that it is necessary to believe in gods in order to be moral has been mooted for centuries, and notwithstanding philosophical rebuttals, is today widely held around the world. Of 39 countries recently surveyed by the Pew Research Center (2014), 22 had clear majorities who believed that theism was necessary for morality. In countries such as Ghana and Indonesia, this belief was expressed by 99 per cent of those surveyed. However, it is notable that the global distribution of this opinion is patterned. Specifically, the view that belief in a god is necessary for morality is more apparent in poor countries and less so in rich ones: the position was supported by at least 75 per cent of the population of all six African countries in the survey, but by no more than 49 per cent of the population in any of the nine European countries studied. In the UK, the survey showed that only 20 per cent of the population held this opinion. Based on GDP per capita, the correlation between national wealth and belief in a theistic basis to morality was statistically significant, at –0.82 (the United States was the obvious outlier: despite being the richest country surveyed, 53 per cent of its population saw religious belief as required for morality). The direct challenge to psychology here is that psychological theories of altruism, moral reasoning, and prosocial behaviour are all premised on a developmental or learning model, in which people come to be morally literate through reason and experience (Killen & Smetana, 2006). Insofar as aspects of moral reasoning are innate, they are seen as reflecting adaptations that help to promote personal well-being and social cohesion. Philosophically, it has been noted for centuries that in order for a deity to teach humanity what is moral, it is required that morality exist independently of that deity; therefore it is possible for people to comprehend morality without reference to the deity. Similarly, in order for humans to choose which aspects of religious teaching to take seriously (such as the prohibition on killing) and which to dispense with as anachronistic (such as the Christian Bible's apparent endorsement of slavery; Giles, 1994), it is required that they possess a personal capacity for moral judgement. In short, psychological (or philosophical) models of morality do not accommodate notions of an external moral code that is crafted by a non-human third party and strategically imprinted into human reasoning from outside, nor do they accommodate corresponding notions of dispositional *im*morality, such as original sin. As such, psychology's attempts to account for moral behaviour are premised on wholly different assumptions to those put forward by religions.

In Chapter 8, we considered how psychologists' social or political views might weigh upon their interpretation of research data. Religious views on morality may present a case in point. For example, we previously discussed how the neurohormone oxytocin has been promoted as the molecule underlying morality, despite the fact that its actual impact on moral behaviour is far from clear (see Chapter 6). We noted that the champion of this 'moral molecule' hypothesis, Paul Zak, was affiliated to Loma Linda University in California (in fact, as he highlights a number of times in *The Moral Molecule*, Zak lives in the city of Loma Linda itself; Zak, 2012). LLU is a fine educational institution with an undoubted reputation for academic excellence. It also promotes itself as adhering to a strongly held religious mission, namely that of the Seventh-day Adventist Church (Loma Linda University, 2015). This reflects the standing of Loma Linda as one of the world's main population centres of Seventh-day Adventism (the university's church boasts one the biggest Adventist congregations in the world). While LLU supports students and employees from many religious groups with diverse chaplaincies and equal opportunity policies, it is nonetheless institutionally committed to promoting Seventh-day Adventist principles. These principles include very clear teachings on the rootedness of personal morality in religious belief. Undoubtedly, Paul Zak is an objective scientist who approaches research in a very balanced way. For example, in *The Moral Molecule*, he records how elderly Loma Linda residents who are religious show *lower* levels of oxytocin than those who are not (Zak, 2012). Under

Zak's hypothesis, this would reflect how feasible it is for moral behaviour to exist *independently* of religiosity. The point here is not that scientists attached to religious universities are incapable of objectivity, but rather to highlight the fact that individual researchers will frequently find themselves studying phenomena about which their employers and neighbours will have strong prior assumptions. Whether it is ever possible to truly stand apart from such contexts when choosing research questions or interpreting data is difficult to surmise. Zak's research presents an optimistic view of the potential for all human beings to encourage generosity in their communities, as well as an argument for the promotion of personal morality as a target for public policy. Had his research produced contrary implications – for example, had it implied that morality had no advantages for society – then we can only speculate as to the predicament this might have presented for him.

The final context in which religion has implications for psychology is possibly the most fundamental one, at least in terms of how scientific psychology is done. This is the *epistemological* context. The reason it is possibly the most fundamental is that it relates directly to the demarcation between science and pseudoscience. We saw in Chapter 1 that science is the systematic use of empirical methods to resolve uncertainties in worldviews in an effort to establish independently verifiable, and thus factual, knowledge. It arose as an alternative to other claimed methods for obtaining knowledge, such as the consulting of authorities. Scientific reasoning pursues parsimony (the rejection of suppositions that cannot be independently confirmed) and avoids anecdotalism (the attaching of reliability to personal testimony or traditional consensus). It would not be an exaggeration to say that the scientific method is an explicit rejection of claims that true knowledge can flow from faith or conviction. In contrast, religion is premised on an undertaking to accept, as reliable, truth-assertions for which there is no third-party verifiability or objective evidentiary support. Religions require that adherents accept certain things as factual irrespective of their rootedness in evidence. They require that adherents place trust in claims that cannot ever be proven. Indeed, when it comes to core religious beliefs on metaphysical or existential matters, it is their very unprovability that makes them critical – adherents are called upon not to seek proof, or to question the veracity of what is being described, but to have 'faith' in the teachings set down by divine providence. It is this very 'faith' that defines them as truly observant of the religion to which they subscribe.

Therefore, when a psychologist holds religious beliefs, it involves the juxtaposition of two competing forms of epistemology. One the one hand, the person is committed to the general principle that assertions require verification in order to be considered useful or true. On the other hand, the person is committed to upholding certain unverifiable views not just as useful and true, but also as inspiring and profound. Many people, including many scientists, feel it is reasonable to compartmentalize their attitudes in this regard. They view the implications of their day jobs as far removed from those of their religious philosophies. A psychologist working in human cognition, social behaviour, child development, or mental health might feel that the specifics of these phenomena are utterly unconnected to the origins of the universe, the nature of morality, or the existence of holy spirits. However, the direct challenge that arises is that the two types of reasoning are, literally, contradictory. More specifically, the latter type of reasoning – the religious principle of holding certain beliefs to be true irrespective of their empirical status – trumps the former. The fact that a person's scientific work might appear far removed from their religious thoughts is not strictly a factor, because the concept of 'truth' is not tied to a particular domain of life. It is difficult to argue that truth-assertions require verification in one context when one is happy to accept at least *some* unverified assertions as true in another. Or at least it is difficult to be convincing while doing so. Insofar as psychology seeks to convince anyone about anything – and especially insofar as it seeks to dismiss pseudoscience as *un*convincing – then challenges will be presented.

One way to minimize these challenges is to consider psychology something other than a pursuit of truth. For example, individual psychologists might come to view their field as purely a practitioner-discipline, focused on helping people to thrive or on curing mental illness. In essence, this involves seeing psychology as a non-science. For the very many reasons espoused throughout this book, such a manoeuvre would involve a radical departure from convention and the abandoning of psychology as we formally know it. An alternative strategy would be to recast one's views on religion. It is possibly the case that for the majority of people, the most inspiring aspects of religion are not its embedded mysteries or epistemological challenges, but its attempts to guide humanity toward fulfilling lives and positive social values. The theological underpinnings of religion would likely be far less seductive were there not also this shared positive vision for humanity. As such, individual psychologists may reconcile their scientific and religious epistemologies by downplaying the theistic aspects of the latter. This tendency to emphasize optimistic themes ahead of theological ones might help psychologists to navigate several of the challenges presented by combining scientific psychology with religious adherence. However, as we shall now see, the invoking of *optimism* as a philosophical orientation is not without its own significant complications.

Optimism and the aspiration to do good in psychology

Psychology and the bright side

Optimism is one of the main ways in which human beings reliably differ from one another. People can be extremely optimistic, extremely pessimistic, or somewhere in between, but wherever they are on the spectrum of optimism, there they are likely to remain. People who come across as optimistic in one situation or on one day will generally do so again in other situations or on other days; those who have been pessimistic in the past are liable to be pessimistic in the future. So stable is the trait of optimism (or pessimism), it is regularly used as a dominant personality archetype with which to depict symbolic fictional characters. One of the most memorable literary pessimists is Eeyore, the gloomy grey stuffed donkey who befriends Winnie the Pooh in the classic books of A. A. Milne (e.g., 1926). When his friends say good morning, Eeyore replies that he doubts whether such a glib claim can be substantiated; when they find his lost tail, he explains how he imagines it is only a matter of time before he loses it again. Eeyore frequently complains that people are ignoring him, even though Pooh Bear, for one, regularly goes to the trouble of dropping by to see him. At one stage, Eeyore claims that nobody has spoken to him in 17 days, even though several of his friends have visited during that time. When someone compliments the weather, Eeyore points out that, for all anybody knows, there could be a blizzard the very next day: 'Being fine today doesn't mean anything', he says. In psychological terms, Eeyore demonstrates several different types of pessimistic thought: he frequently discounts the positive, magnifies the negative, uses all-or-nothing reasoning, over-generalizes adverse consequences, and engages emotion instead of logic when resolving cost-benefit analyses (Lee, 2011).

In contrast, probably the most famous fictional optimist is Pollyanna, the lead character in Eleanor H. Porter's (1913) novel of the same name. The eleven-year-old Pollyanna develops a secret strategy for dealing with the uncanny number of misfortunes that befall her, which is to reframe her perception of events in order to identify some aspect about which to be joyous. In what verges on black comedy, Pollyanna's strategy of incorrigible optimism coincides with an ever-deteriorating series of adverse life events, culminating in her being run over by a car and paralysed from the hips down. True to form, however, Pollyanna determines to focus on the bright side of paraplegia, and reflects on all the benefits of having *once* had functional limbs. She reframes the ability to walk as a variable rather than a constant, and expresses satisfaction at having lived a life in which, for a period at least, she enjoyed the use of her legs. Indeed, throughout the book, Pollyanna's optimism is seen as having an inspiring effect on other people, such that she becomes something of a heroine in her community. And ultimately, her heroism is rewarded: in a plot twist that perhaps reveals much about our culture's attitude towards optimism, Pollyanna walks again following treatment at a specialist hospital.

Notwithstanding our capacity to pick out pessimists and observe optimists, it can reasonably be argued that, in real life, far more people tend towards Pollyanna than to Eeyore. Reflecting the innate human penchant for self-aggrandizement, people are inclined to view the world around them with what has become known as an *optimistic bias* (Weinstein, 1980). For example, respondents to surveys typically report that they believe the future will be better than the past, even though – technically – there is little direct evidence to guarantee that this will be the case. In their global end-of-year poll for 2014, the survey company Gallup International reported that more than half of respondents expected the coming year to be better than the previous one, compared with only 15 per cent of respondents expecting it to be worse (Gallup International, 2014). Of course, it could just be that the world is genuinely becoming a better place. A large body of historical, sociological, and psychological research suggests that the sociocultural development of human societies serves to make them less inhospitable over time (Pinker, 2011). Nonetheless, people are consistently found to be optimistic about events that could not reasonably be predicted in terms of global human progress, such as their likelihood of having a gifted child or of enjoying their first job more than their peers (Taylor & Brown, 1988). In a related vein, people expect future events to tend naturally towards justice. In assuming that the consequences of a person's actions will eventually provide a morally fair outcome, most people exhibit a common logical fallacy referred to as the *just-world hypothesis* (Furnham, 2004). In short, they believe in the theory of just desserts: ultimately, virtue is rewarded and vice punished. Aspects of optimism are also core to the teaching of many religions. For one thing, in seeking to promote human generosity religions reveal an optimistic view of what it is humans are capable of. Further, as discussed above, of all possible hypotheses concerning our personal futures, it is hard to envisage one more optimistic than the prospect of a never-ending perpetually blissful post-death existence.

As described in Chapter 4, one benefit of optimistic bias is that it motivates us to pursue future happiness by making it seem more realizable. However, there are disadvantages too. Pessimism is a safety mechanism for inhibiting over-confidence; unmitigated optimism neutralizes its effects. When people unreasonably assume the inevitability of success, their efforts to prepare for failure may be diminished. Moreover, they may ignore the risk of adverse consequences, such as when smokers underestimate their statistical chances of developing cardiovascular disease. Likewise, when people assume that the world has a way of providing its own form of justice, they are more inclined to rationalize their neighbours' misfortunes as having somehow been deserved. This blame-the-victim tendency may shield people from the stress of survivor's guilt, but it also makes them less likely to offer help or sympathy to those who might benefit from it. And, of course, very often optimism involves holding views that are factually incorrect (such as the belief that your own success is truly inevitable) or logically parlous given the paucity of available evidence (such as the conviction that all you have to do to succeed is work hard). As outlined in Chapter 4, a proneness to erroneous or flimsy beliefs is unlikely to be constructive over the longer term.

So how does all this affect psychology? In one sense, optimism affects psychology in exactly the same way it does all human endeavours. For example, when researchers *expect* their research to be successful, they are less oriented towards critiquing the flaws in their work, leading to a proliferation of confirmation bias, self-fulfilling prophesies, overlooked alternative interpretations, and the like. But optimism affects psychology in more specific ways too. One cross-cutting feature of optimism is the view that our fellow human beings are inherently benign (as opposed to malign or neutral). A second is the expectation that positive future outcomes are inevitable – or, if not quite inevitable, then easily attainable. Both these orientations imply certain assumptions about human thoughts, feelings, and behaviour. For starters, they assume that human beings incline toward beneficence. Secondly, they assume that human happiness lies readily within reach. A third assumption is a consequence of the first two: that human beings' instinctive good behaviour will inevitably lead to their (and others') happiness. The problem is that all of these assumptions are questionable, at least in the scientific sense of the term. People may *not* be intrinsically benign, future outcomes may *not* all be positive, and altruism may *not* lead to benefit. To simply assume that these beliefs are true, in the absence of any firm evidence to support them, is to adopt an arbitrary starting position in one's research. We should recall that the scientific method invokes the principle of parsimony: uncorroborated assumptions should be avoided where possible, and the extent to which they intrude upon research is the extent to which that research is scientifically undermined.

In reality, quite a lot of psychology research rests on optimistically infused assumptions. An example is that set of studies focusing on how people treat each other. Another is the research that looks at how therapists treat their clients. Much of this work appears to assume that the fruits of human benevolence will automatically be worthwhile, and that the challenge for investigators is merely to examine the upper limits (if any) of therapeutic success and the precise contextual features that exploit the productivity of human interaction. A further example is the research literature on how people treat themselves. This includes an explosion of investigations into such things as emotional growth, benefit-finding, value, virtue, 'flow', mindfulness, happiness, and so on, as well as much research into how these paradisiacal states of being can most effectively be attained. None of these bodies of work seems to consider whether its optimistic assumptions are misplaced, or whether there are any downsides either to the behaviours being studied or to the research studies themselves. We will consider each area in turn.

Being supportive

Some research into how people treat each other concerns the many ways in which humans perform social helping acts or otherwise convey benefits to each other within their communities. Much of this work focuses on the concept of 'social support', a term that has acquired several formal definitions in psychological research on health and well-being. These definitions often appear pedantic in their refinement of distinctive meanings. For example, while one classic theory tells us there are four categories of support – '*appraisal*', '*emotional*', '*informational*', and '*instrumental*' (House, 1981) – another tells us there are four slightly different ones – '*companionship*' and '*esteem*' as well as '*informational*' and '*instrumental*' (Wills, 1985) – while yet another tells there are just three – '*emotional*', '*informational*', and '*tangible*' (Sarason, Sarason, & Pierce, 1990). Social support researchers have frequently added a further dimension to this train of thought by separating 'perceived' versions of support from 'received' ones (Barrera, 1986). 'Perceived support', reasonably enough, alludes to the idea that the value of the help a person gets from others is often best judged through the eye of the beholder. Attempts to measure it involve trying to gauge beholders' views as to the capacity of their social support networks to provide help should it be needed. By contrast, 'received support' is said to refer to those specific support interactions that have occurred in real life, rather than to hypothetical capacity estimates. One confusion in all this is that 'received support' is typically measured by self-report, meaning that its quantification rests on how respondents *perceive* their support histories, while perceptions of support-network capacity are often informed by the respondents' experiences of having *received* help in the past. In other words, 'perceived support' includes support that has been received, while 'received support' is a function of how support is perceived.

As if that weren't enough, social support researchers occasionally add even further layers, such as the separation of 'functional' support from 'structural' support (Wills, 1998), or efforts to take account of how the social role of support provider affects the way their support will be perceived (or received). For example, a hug from one's life partner will probably be seen as much more comforting than a hug from one's boss; and if you ask a police officer for directions to the train station, you are unlikely to be impressed if they refuse to answer your question but offer to pay for your ticket instead.

Much of the research into social support – whether it be informational or instrumental, received or perceived, structural or functional, or so on – falls foul of the many problems of self-report methodologies that were discussed in Chapter 7. The more nuances are added to the concept, the more researchers have to rely on respondents to answer questions about it in questionnaires. Whether respondents really do have access to support, whether they really have been provided with support in the past, or whether they just *say* these things are true (or believe them to be so) even though they are not, is all impossible to untangle. The only information questionnaires can gather on the matter is what some people say about what they think they think about it. If questionnaires are avoided, then the scope for subtle assessment is greatly lessened. Of course, such difficulties do little to prevent research from taking place. Notwithstanding the proliferation of divergent definitions or the absence of a standard methodology, social support research continues to be conducted and published in vast quantities. In almost every case, the context for doing so is an implied assumption that social support must be a good thing. And so we have findings cited throughout the literature describing the positive impact of social support (or social interaction, social integration, social capital, or whatever) on mental health, educational attainment, adherence to rehabilitation, recovery from illness, the reduction of blood pressure during stress, and even the delaying of death.

One hint that all might not be well in such research is the sheer diversity of methods by which social support is quantified by researchers (Hughes, 2008a). While many studies use questionnaires, these tend to be varied and short instruments or ones designed on an *ad hoc* basis for the particular study in question (and, as such, not comparable to measures used anywhere else). Many studies forgo formal psychometrics and instead rely on proxy measures of social support such as family size or marital status. Some studies attempt to record people's social activities, with a number inferring support from participation in social events (such as church attendance). Several studies examine social support experimentally, perhaps asking participants to perform a stressful laboratory task with or without an audience (sometimes as small as one person), inferring the audience group to be 'supported'. Other studies present all participants with an audience, but ask the audience to smile in the 'high support' condition and to remain dour in the 'low support' condition (Christenfeld et al., 1997). Other experimenters ditch the audience altogether and ask participants to simply imagine being supported during their performance of the task (Gramer & Reitbauer, 2010).

The biggest risk with social support research is circularity. It is relatively easy to show that healthy, able-bodied, disease-free, happy people are more embedded in social relationships than others. But this is because there is a very reasonable likelihood that people who are *not* healthy, able-bodied, disease-free, or happy will find it harder to acquire social relationships in the first place, or to maintain them if they have them. People who are ill just don't socialize as much as those who are healthy; participation in social activities is less a buffer against incapacitation, as incapacitation is a barrier to social activity. And depression, stress, and anxiety not only undermine relationships, they also colour people's perceptions of them. If (self-reported) low social support is found to be associated with negative emotion, it would be unsound to infer that low social support is the cause and negative emotion the effect. Likewise, experimenters who foist audiences on participants in laboratories might well find smiling spectators to be associated with lower physiological stress responses than poker-faced ones. However, they should not conclude that smiling audiences help participants to prosper. It could be that poker-faced audiences are sufficiently intimidating as to undermine participants' confidence and impede their ability to cope. Rather than show a benefit of being smiled at, such studies might instead be demonstrating the disturbing effects of being ignored. The core problem here is that social interaction is such a complex aspect of any individual's experience of the world, an aspect that is comprehensively bound up in the thickets of that person's ecological circumstances, it almost makes no sense at all to propose it as a single variable that can be quantified in a questionnaire, operationalized on the basis of census data, or replicated by research assistants. It could be that 'social support' is an illusion, a term liberally applied to a plethora of interconnected, but yet distinct, micro-features of daily life. Rather than bundle them all into one elaborate and multi-faceted unitary research concept (which, as discussed above, has led to definitional disarray), it may be more coherent to ignore the term 'social support' altogether and simply study the micro-features in and of themselves.

The impulse to package social support into a positive single commodity owes more to optimism than to empirical rigour. Indeed, what is referred to as 'social support' can create as many negative effects as positive ones. For example, much research suggests that the supportiveness of a social interaction is secondary to its reciprocal balance. Actors typically see helpfulness as a form of trade, and distribute support to each other in ways that ensure a fair exchange. As receiving support from another person creates a debt that requires future repayment, being helped by somebody you can't help in return may lead to sufficient indebtedness as to damage your relationship. Receiving social support in this manner produces a gain in the short term, but only at the cost of long-term distress. Another feature of social support is that it involves the transmission of assistance from someone who is capable of giving it to someone who is in need of it. This inherently implies a power differential between giver (who is capable) and receiver (who is in need). Many studies have shown that interacting with someone more powerful than you can be stressful if it draws attention to your own shortcomings or problems. In one memorable study, couples in marriage counselling were found to do poorly if they attended group sessions facilitated by people who themselves had healthy marriages (a format that many counsellors instinctively feel would be helpful). Instead they fared much better when they attended group sessions with people who had even *worse* marriages than their own, presumably because it made them reflect more on the relative merits of their own situation (Buunk & Hoorens, 1992). A related point is that people who are in need of social support often tend to receive more of it (either because they seek it out or because they exhibit such obvious distress that support providers come rushing to help them), whereas people who are doing just fine will attract far less support and may need none at all. This state of affairs suggests there should be an *inverse* correlation between support and well-being (as help will be offered more frequently to sad people than to happy ones), a state of affairs that contrasts sharply with the positive helping-happiness link predicted by so many studies. All these dynamics are encapsulated in several well-studied paradigms of social psychology, including social exchange theory, social comparison theory, and self-esteem theory, literatures that seem to remain conspicuously ignored by most social support researchers (Hughes & Creaven, 2009).

The claim that, in principle, support is a good thing is hardly controversial, given that the very definition of the word *support* implies the provision of assistance or benefit (that said, this does not prevent some researchers articulating the somewhat strangled notion of 'negative support'; Shiozaki, Hirai, Koyama, & Inui, 2011). The related claim that human beings enjoy the company of others, whilst a generalization, can be seen as largely reasonable given the history of human culture. However, such sweeping assertions as these do not automatically imply that human beings are inherently helpful, or that people always benefit from being assisted. They do not imply that social support is a resource that can easily be harnessed and deployed, at will and without risk, in any situation. The rush to valorize social support in research belies the amorphous nature of the term, the flimsiness of its operationalization, and those negative consequences of human interaction that disrupt the conventional, optimistically oriented, narrative.

Being psychotherapeutic

For some psychologists, the question of how human beings might best help others is particularly acute. These are the psychologists who focus their careers, and their reputations, on the provision of formal psychotherapy to people whose mental health needs require professional attention. Whether such therapies really do produce powerful effects for clients has been a target of scrutiny for many decades, and it is perhaps fair to say that the research has been controversial from the beginning. The fundamental question to be addressed is whether psychotherapy – defined by the BPS as 'the practice of alleviating psychological distress through talking rather than drugs' (British Psychological Society, 2015) – succeeds in producing for clients unique benefits greater than those which would arise in the absence of psychotherapy. One main complication is determining what the benchmark should be: what does the 'absence of psychotherapy' look like? In the case of surgery or pharmacotherapy, the 'absence of treatment' is quite straightforward: it means not being cut open by a surgeon and not taking drugs. However, the absence of psychotherapy is harder to define. Psychotherapy largely revolves around talking, but real life involves quite a lot of talking too. Psychotherapy involves talking with someone who is empathic, humane, supportive, and (often) non-directive and non-judgmental; but real life regularly provides exactly that type of experience. Some psychotherapies involve encouraging us to avoid focusing on the negative; but again, in real life, people often give us that kind of advice. Other psychotherapies prompt us to deliberately focus on the negative and to learn to accept it; once more, this is not unlike advice that people in real life might offer. Psychotherapy often encourages us to self-reflect, improve our decision-making, or think differently about our problems; however, yet again, these are strategies that other people might suggest to us, or that we might even suggest for ourselves. The point is that quite a lot of what goes on in psychotherapy resembles stuff that goes on in ordinary life. Rigorous empirical tests of whether psychotherapy 'works' need to capture and scrutinize that which makes psychotherapy truly different from what takes place otherwise. In other words, it needs to focus on that that which is unique about psychotherapy. This task is a lot more difficult than it sounds.

Simply looking at whether people in therapy recover from what ails them is not a particularly strong research design. After all, given enough time, many unwell people will experience spontaneous remission, a recovery that occurs in the absence of therapy. Indeed, some attempts at therapy might even *delay* a person's spontaneous remission. In experimental terms, the standard methodological workaround for this is to compare a treatment group (people who receive the therapy) with a control group (people who do not) in order to benchmark against rates of spontaneous remission in the latter. However, this approach is heavily reliant on comparability and consistency. The people receiving therapy need to be truly comparable to their counterparts in the control group in the various factors known to affect mental well-being (for example, diagnoses, personalities, life histories, environments, relationship statuses, medical treatments, and so on). Meanwhile the treatment being tested needs to be administered in a rigidly consistent way across all cases so that a fair conclusion can be drawn as to whether the observed effects can fairly be generalized to other cases. Needless to say, these ideals are hard to achieve in studies of psychotherapy. An even more robust research standard would involve a third group who receive a modified or alternative therapeutic intervention, to check if the very experience of being *in* therapy (rather than the specifics of the therapy itself) is important. The big challenge with this is to ensure a fair comparison. In drug trials, it is easy to give patients a placebo pill that is identical to real treatment in all appearances. But in psychotherapy research, any differences in treatment will be discernible to the person undergoing it, which introduces the risk that such perceptions of treatment differences will influence the outcomes.

A further challenge is the arbitrary nature of psychiatric conditions – and not least the fact that, over decades, their definitions and diagnoses are liable to be amended and updated with each revision of the relevant clinical guide. Throughout the world many mental health professionals (and state agencies) refer to the definitions listed in the American Psychiatric Association's *Diagnostic and Statistical Manual of Mental Disorders* (DSM) in order to diagnose mental health conditions. Up until the late 1960s, most DSM-listed mental disorders were described purely in Freudian terms, meaning that they had to be diagnosed using psychoanalytic methods. Subsequent revisions have seen many radical amendments. While a number of formerly recognized conditions are now no longer listed, several of the most recent re-writes mean that people who were mentally healthy in the past will now meet diagnostic criteria for a disorder (American Psychiatric Association, 2013). For example, the symptom timeframe for a diagnosis of bulimia nervosa has been reduced from six months to three, meaning that people who binge-eat weekly for four months (but no longer) can now be classified as bulimic. Similarly, an anxiety disorder diagnosis used to require that sufferers be aware that their experienced anxiety was disproportionate to their predicament. However, this requirement was removed from the latest DSM update: now any person exhibiting excessive anxiety can be diagnosed with an anxiety disorder, even if they are oblivious to their emotional excess. Such alterations to diagnostic standards have characterized each revision of the DSM. Given that the DSM has been updated approximately once per decade during the last half-century, this means that psychotherapy research based on formal clinical diagnoses may suffer from a problem of built-in obsolescence. If the diagnostic criteria for, say, anxiety disorders changes every ten years, then therapy studies conducted in one decade will include participants who may not even be diagnosed with anxiety in another. How confident can we be, then, in comparing studies of anxiety therapy that have been conducted in different decades?

Researchers have queried the empirical status of psychotherapies since early in the 20th century (e.g., Rosenzweig, 1936). The debate intensified in the 1950s and 1960s, when British psychologist Hans J. Eysenck (e.g., 1952) and German-American psychotherapist Hans H. Strupp (e.g., 1963) published strident exchanges. Their conflict can be summed up very briefly: Eysenck argued that the empirical evidence for psychoanalytic psychotherapy was almost entirely lacking, while Strupp argued the opposite. Unlike previous controversies, this one reached the attention of a very wide audience, even featuring in the pages of the *New York Times* (Wampold, 2013). Among Eysenck's criticisms of psychotherapy research were that it was uncontrolled, non-blinded, and non-standardized, and that it failed to control for spontaneous remission rates – which he said were higher than the rates of recovery seen in psychoanalysis. Notwithstanding Strupp's rebuttals, it was these methodological limitations and doubts about efficacy that fed the popular media curiosity about psychotherapy's empirical standing. However, in retrospect, Eysenck and Strupp's debate had its limitations. Firstly, while Eysenck was very focused on critiquing psychoanalysis, he simultaneously championed the merits of behavioural therapy without scrutinizing it as vigorously (Wampold, 2001). Meanwhile, Strupp's motivation in critiquing Eysenck's views stemmed less from a commitment to empiricism in psychological science, and more from a loyalty to the Freudian school of psychoanalysis (Wampold, 2013). Furthermore, Strupp's rebuttals to Eysenck were largely anecdotal rather than empirical (Barlow, Bullis, Comer, & Ametaj, 2013). In other words, this was less an academic debate about science than a competition between different theoretical stances in psychology. A further limitation relates to the shifting nature of diagnostic practices across time, as discussed above. Eysenck, Strupp, and their contemporaries spent much of their energies debating the relative merits of treatments for 'neurosis', a diagnostic classification that now no longer even exists.

Today, there is a generally wide consensus that, in terms of outcomes for clients, undergoing psychotherapy is distinctly better than not undergoing psychotherapy. However, this consensus is as nuanced as it is wide. For example, it applies in a very general way to a broad range of conditions, rather than in a specific sense for specific conditions. While research data can be marshalled to support claims that nearly 80 per cent of people receiving psychotherapy will enjoy better mental health than those receiving no treatment (Wampold, 2001), it is much harder to say that a particular therapy for, say, depression will be effective in any particular case. Meanwhile, although large multi-study datasets show how something like cognitive behavioural therapy can be effective, its demonstrable efficacy is restricted to relatively specific diagnoses: for example, for cannabis and nicotine dependence (but not opioid or alcohol dependence), for short-term psychotic symptoms (but not long-term ones), and for obsessive-compulsive disorder and social anxiety (but not so much for generalized anxiety disorder; Hofmann, Asnaani, Vonk, Sawyer, & Fang, 2012). Further, even though the research literature on CBT is extensive, if not vast, we are still under-informed as to whether the approach is useful for major target groups in the general population, such as people with relatively low incomes.

Much fuzziness stems from the questionable scientific rigour with which psychotherapy research is often approached (if not in its execution then in its interpretation), and from the way optimism threatens objectivity. All too often reviewers of research will draw positive conclusions about a particular psychotherapy, before only briefly alluding to the literature's preponderance of small samples sizes, inadequate control conditions, and null effects. As with other areas of psychology, the mantra that 'more high quality research is needed' receives greater airtime than the equally legitimate conclusion that 'our expectations are insufficiently supported by data'. Furthermore, a number of thought-provoking trends have emerged in the findings. For example, most psychotherapies (but not all) are supplemented by published treatment manuals, carefully prepared written guides as to how best to administer the therapy. This is often seen as useful for research in psychotherapy because it allows investigators to attach less weight to past studies in which a particular therapy has been poorly delivered. However, when adherence to treatment manuals has been tracked as a variable in its own right, it has been shown to be entirely unrelated to the success of therapy in achieving positive outcomes for those being treated (Webb, DeRubeis, & Barber, 2010). In other words, it appears as though the fidelity with which a therapy is administered makes no difference to its efficacy. In a similar vein, studies that take account of therapists' expertise in psychotherapy repeatedly find that their amount of expertise is unrelated to their effectiveness as therapists (Tracey, Wampold, Lichtenberg, & Goodyear, 2014). This finding is not new: way back in the 1980s, researchers showed that therapists with several years of experience were no more effective than therapists in the initial stages of their careers (Stein & Lambert, 1984). In one early study that would precipitate apoplexy amongst today's research ethics committees, a group of academic professors who were willing to *role-play* as psychotherapists were assigned to see some college student clients at a university counselling service (Strupp & Hadley, 1979). The professors were experienced teachers in areas such as history, mathematics, and philosophy, but none had ever been trained as a therapist or counsellor. The students they saw – for up to 25 hours in each case – presented with a range of formal diagnoses, including anxiety, obsessional trends, borderline tendencies, and (of course) neurosis. These students' outcomes were compared with those of a matched group who instead were sent to see real, properly trained, therapists, all of whom had over 20-years' experience of psychotherapy behind them. The researchers performed full psychiatric checks on the students at the start of the study, upon termination of therapy, and again after a year. When they compared the two groups, they found that the clients who saw the professors recovered just as well and just as quickly as those who underwent professional psychotherapy. Not only that, but they also maintained their recoveries just as well at the follow-up assessment one year later. In short, while psychotherapy is beneficial, there is ample evidence to suggest that it doesn't matter whether it is competently delivered, whether the therapist knows much about it, or even whether the therapist is an imposter with no formal training.

A related curiosity in psychotherapy research is known as the *Dodo bird effect* (Luborsky, Singer, & Luborsky, 1975). This is the common conjecture that all therapies, regardless of theoretical orientation, produce equivalent therapeutic outcomes. The allusion to the Dodo bird draws on imagery from Lewis Carroll's (1865) *Alice's Adventures in Wonderland*, in which the Dodo, wishing to equivocate as to the outcome of a rather badly organized footrace, declares that '*Everybody* has won, and *all* must have prizes'. In a similar way, debates about which psychotherapy is best can be said to be moot: all psychotherapies deserve a prize. Put empirically, the Dodo bird effect means that across all possible comparisons of any two standard therapies for a given condition, the average effect size of the difference in efficacy between them will be zero. Put *simply*, it means that all therapies seem equally good – look for differences between them, and they will typically be trivial at best. The effect was perhaps first comprehensively demonstrated by American psychotherapist and trained mathematician Bruce E. Wampold, who, along with colleagues, performed the necessary calculations on comparison studies published over a 25-year period up to 1995 (Wampold et al., 1997). While bona fide psychotherapies were found to be substantially beneficial when compared with placebos or no treatment, the average difference in benefit when any two psychotherapies were compared was, indeed, zero. The Dodo bird effect is apparent even though many psychotherapies are premised on theoretical assumptions that entirely contradict those of other psychotherapies. One interesting implication, therefore, is that these theoretical assumptions are essentially irrelevant. Psychotherapies work not because of the principles used to develop them, but because of other reasons. Cognitive therapies are effective for reasons unrelated to cognition. The benefits of behavioural treatments have nothing to do with behaviour. Psychodynamic psychotherapy is valuable notwithstanding the fact that Freud's ideas are irrelevant to the client lying on that couch (and, as a uniquely psychoanalytic prop, the couch too is neither here nor there).

The Dodo bird effect could be seen as implying that paradigmatic approaches to psychotherapy are pseudoscientific. What is said to go on in therapy – for example, the resolving of intrapsychic conflicts, the reframing of problematic thought patterns, or the achievement of self-as-context – is not what really goes on, or at least it is not what is really therapeutic. From this perspective, paradigm-laden views of psychotherapy involve rich tapestries of superfluous features, and so assuming these features to be integral to the therapeutic process lacks parsimony. In addition, the key ingredient of therapy is poorly defined and, if operationalized in terms of paradigmatic concepts, poorly measured. If a unique empirical evidence basis is not clearly there, then the standing of any one therapeutic approach will rely heavily on its reputation, a form of anecdotal evidence. In the case of some therapies, this reputation will be bolstered by endorsements from key guru-like authority figures, leading to what we might refer to as a suite of 'eminence-based' treatments. Allowing authority figures to influence therapist opinion in this way is far removed from the ethos of the scientific method. By this assessment, scientific examinations of psychotherapy should dispense with individual theoretical paradigms, and instead focus on those factors common to all therapeutic interventions. These factors include the therapeutic relationship, the client's sense of security, the exchange of information, and many other aspects of the therapeutic process that have become known as its 'nonspecific effects' (Perlman, 2001).

However, the Dodo bird perspective can itself be seen as hampered by empirical uncertainty. Most psychotherapists interpret the effect as a positive basis to declare all psychotherapies useful. This interpretation is arbitrary and, one might say, self-serving. Rather than infer that psychotherapies are pointlessly over-elaborated rituals, these commentators use the Dodo data to bolster claims that all psychotherapists are worth visiting. They focus on the part of the jar that is half-filled by nonspecific effects, rather than the part this is empty of psychologically refined concepts. Being thoroughgoing optimists, they conclude that every therapy has won the race, rather than that every therapy has *failed* to win. One eventual implication of this is that scientifically grounded therapies can be seen as no different from other therapies. If it is really true that individual therapies add no value to the therapeutic encounter, then psychotherapists may as well forget about rigour altogether. They can instead branch into pseudoscientific fields that are commercially attractive without requiring evidence bases to support them. If we are to accept this scenario as reasonable, then we need to be confident that the Dodo bird conjecture stacks up. The problem is that the Dodo approach involves an averaging out of distinctiveness across different psychotherapies. It may well be true that the average difference in outcomes between two therapies is zero, but this fails to take account of the fact that different therapies might be useful for achieving different outcomes in different situations. For example, as described above, the empirical literature on CBT appears to identify a number of specific use-cases in which CBT is particularly effective. Other psychotherapies can boast similar specialized applications where data imply strong therapeutic efficacy compared to that associated with other forms of therapy. Moreover, when Dodo bird proponents infer that relationship effects are the true active ingredients of psychotherapy, this claim is not itself beyond empirical testing. When relationship effects such as the quality of the therapeutic alliance have been scrutinized in research studies, their associations with therapeutic outcomes have been shown to be very inconsistent (Martin, Garske, & Davis, 2000). Essentially, the Dodo bird effect raises interesting questions about the impact of different psychotherapies, but it would be premature to respond to those questions by assuming that all therapies are the same. Indeed, when large multi-study datasets are examined from research *since* 1995 – research that better refines the use of different therapies for different outcomes – the Dodo bird effect becomes difficult to statistically replicate (Marcus, O'Connell, Norris, & Sawaqdeh, 2014).

In short, if we assume the Dodo bird effect to be universally and simply true, then paradigmatic approaches to psychotherapy may indeed appear to lack parsimony, to be vaguely defined, to rely on reputation, and so on. In other words, they may appear to possess some of the characteristics of pseudoscience. However, if we consider it in detail, then the Dodo bird effect might itself appear to possess such characteristics. It unparsimoniously assumes an influence for nonspecific effects (which, by definition, cannot be specified and so cannot be objectively corroborated), it is vague on what is meant by such central notions as the 'therapeutic relationship', and its reputation is bolstered by the seductiveness of its narrative. This narrative seductiveness is rooted in optimism: the reason 'all must have prizes' is that all acts of human helpfulness are assumed to be intrinsically worthwhile. In reality, as research on psychotherapy becomes more refined, the limits of the Dodo bird become more discernible (that being said, it should be welcomed that psychotherapy research has been refined *precisely because* of empirically reasoned concerns like the Dodo bird conjecture). It has been noted that clinical psychologists and psychotherapists often exhibit quite an ambivalent attitude towards the importance of science and scientific methods in their work (Baker, McFall, & Shoham, 2009). However, developing our knowledge of whether or which psychotherapy 'works' is best pursued by exploiting those features of the scientific approach that steer us away from confirmation bias – features such as parsimony, robust definitions of terms, the pursuit of accuracy in measurement, the eschewing of anecdotes, and so on. In other words, it benefits from more science rather than from less.

Being positive

Perhaps the most explicit way in which optimism informs psychology relates to a relatively new subfield known as *positive psychology*. This branch (or perhaps a better term might be 'brand') of psychology was explicitly proposed by University of Pennsylvania professor, Martin Seligman, shortly after he became president of the American Psychological Association in 1998 (Seligman, 1998). Previously, Seligman had become very well known for his work on learned helplessness. This included classic experiments in which dogs in cages were repeatedly exposed to unpredictable electric shocks to such an extent that they gave up trying to evade them. The obvious despair that gripped and then debilitated these poor animals, who ended up just lying there being electrocuted at random, was interpreted as a laboratory model of how depression might develop in humans. Being immersed (more than most) in the study of psychological adversity, it occurred to Seligman that his academic discipline was heavily focused on the negative. This was not an entirely new observation, with Maslow (see Chapter 8) among several figures to have earlier criticized psychology for ignoring the potential of human positivity, but Seligman was able to use his presidency of the APA to promote a new disciplinary agenda. He argued that psychology should study notions such as achievement, happiness, and optimism, rather than its conventional subjects of anxiety, depression, and other forms of misery. Such advocacy reflected claims that psychology had become arbitrarily reliant on medicalized thinking, leading it to focus on pathology rather than potency and on diagnosis and treatment instead of development and growth. To emphasize the point, Seligman suggested that psychologists should stop talking about 'the disorders' and instead discuss 'the "sanities"' (Seligman, 1998). In a similar vein, he later published an extensive taxonomy of character strengths and virtues, the CSV, intended to stand as a complement to the pathology-oriented DSM (Peterson & Seligman, 2004). This 800-page tome elaborated upon an extensive range of human attributes, structured within six higher-order categories (namely: *knowledge/wisdom*; *courage*; *humanity*; *justice*; *temperance*; and *transcendence*). As well as providing a refreshed vocabulary with which to describe the human condition, positive psychology also offered broad theories of the nature of life. These largely emphasized the importance of seeking a good or meaningful existence (which Seligman couched in the term 'authentic happiness'; Seligman, 2002) with the aid of such principles as *mindfulness* and *flow*. Positive psychologists assert that pursuing authentic happiness in this manner promotes tangible well-being in the form of enhanced physical and mental health, and improved productivity in life and labour.

Since its launch by Seligman in 1998, positive psychology has grown into a distinct movement within academic psychology, featuring in textbooks, conferences, and university courses, and having a number of its own learned societies. In some ways, positive psychology is a descendent of the self-help movement, a genre of mainstream advice-activism that first became popular in the late 19th century. It can also be seen as a sibling of modern-day 'pop' psychology, although positive psychologists would likely protest at such a comparison. According to them, pop psychology is comprised largely of jargon, judgement, and psychobabble, whereas positive psychology represents a rigorous attempt to formulate the empirical science of happiness. Positive psychologists will seek to bolster their theories with reference to formal empirical research data, often citing studies that link character strengths and virtues to indices of well-being, such as physical health, wealth, or satisfaction with life. For example, positive psychology research studies may examine links between positive self-statements and longevity (Danner, Snowdon, & Friesen, 2001), the effect of writing about gratitude on one's memory for positive life events (Watkins, Uhder, & Pichinevskiy, 2015), or the impact of positive psychology interventions on how patients cope with distressing physical diseases (Casellas-Grau, Font, & Vives, 2014).

It should be noted that positive psychology is less an alternative to the traditional sub-disciplines of psychology, and more a new way of doing psychology within these sub-disciplines. The positive psychology approach can be applied to the study of personality, cognition, social psychology, health psychology, educational psychology, and so on, and can be pursued using any existing methodology. In other words, while often heralded as a new branch of psychology, it might be more accurate to describe positive psychology as a novel perspective on all the old branches. In this way, positive psychology represents both a critique of, and a departure from, the original approach to psychology that had emerged throughout its development as a scientific field.

One point to bear in mind is that, unlike most developments in science, positive psychology did not evolve as a paradigm because it proved itself more useful and informative than those that preceded it. If anything, positive psychology achieved a platform because of high-profile advocates (most particularly Seligman), whose endorsement lent it the aura of a *cause célèbre*. This makes positive psychology quite an unusual field of psychology: its prominence results more from the value-system it represents (not least its optimism) than from the utility of its findings. It is, in fact, an open question whether the findings of positive psychology have tangibly advanced our understanding of the human condition or produced practical applications of meaningful worth. While there is now a multitude of studies linking character strengths and virtuous cognitions to positive states of being, several are correlational rather than causal, and in many the target variables are very vaguely measured. A large proportion, if not the majority, are of the *what-some-people-say-about-what-they-think-they-think* variety.

Accordingly, positive psychology attracts a degree of scepticism that would aggravate even its most Zen-like proponents. One prevalent criticism is that positive psychology over-sells its empirical offerings. It is said to do this by interpreting ambiguity with unwarranted certitude, downplaying measurement limitations, and allowing – if not cultivating – the proliferation of weakly designed studies. In particular, positive psychology stumbles over a problem commonly known as the *person-situation controversy* (Epstein & O'Brien, 1985). While it is always tempting to conclude that people's character traits influence their lives, this narrative distracts us from the often overwhelming influence exerted by situational circumstances that lie beyond an individual's control. Using mindfulness to transcend your daily hassles might be all very well if you have a good salary and live in prosperous times, but might be of very little relevance to someone on the breadline, an abuse victim, or a person facing discrimination at work. Whether or not our characters and virtues will lead to happiness or success will be difficult to predict if we ignore the contexts in which we find ourselves.

In one critical review of the positive psychology literature, processes such as optimistic expectation and kindness were found to have *adverse* effects in several contexts. When people were in situations where they frequently encountered interpersonal strife, these attributes were associated with *less* happiness rather than more (McNulty & Fincham, 2012). Likewise, positive thinking, trust, and forgiveness were shown to frequently backfire, especially when other people take advantage of them at your expense (McNulty & Fincham, 2012).

The more dire the circumstances, the more a belief in the power of positive psychology appears to verge on wishful thinking. For example, in the case of cancer care, researchers routinely claim that positive psychology interventions can aid longevity, impede cancer progression, and help patients to extract happiness from their experience (even when the illness is terminal) or achieve post-traumatic 'growth' (when it is not). However, these claims greatly exceed what the underlying empirical studies can sustain. When the relevant literature has been reviewed, it has been shown to have produced very ambiguous findings, to be based on biologically implausible assumptions, and to be methodologically inadequate in several important respects (Coyne & Tennen, 2010). Overall, whatever your views on the merit of trust, kindness, and optimism as human values, it is an empirical oversimplification to argue that these traits guarantee a return in the form of happiness, success, or health.

Other critics have accused positive psychology of resorting to pseudoscience. A common claim is that positive psychologists engage in circular reasoning. Whenever optimism is found to be associated with positive outcomes (such as hope and motivation), these are seen as integral to the very nature of optimism; but whenever optimism is found to be associated with negative outcomes (such as naïvety or the over-prediction of success), these correlates are dismissed as incidental side-effects (Miller, 2008). Overlapping measurement is also a problem: in some studies, optimism is seen as *comprising* hope and motivation, but in other studies it is described as *causing* them. In addition, positive psychology frequently locates its assertions within frameworks that rely on unproven assumptions. Examples include the assumption that cognitive life can be construed in terms of the setting and pursuit of goals, as well as the assumption that activities which reflect our personal strengths will necessarily make us happy (Miller, 2008). Insofar as these assumptions remain unproven, then research (or interventions) reliant upon them will, by definition, lack parsimony.

More clear-cut flirtations with pseudoscience emerge when positive psychology attempts to specify an arithmetic context for authentic happiness (Frawley, 2015). For example, Seligman (2002) himself attracted much criticism for conceiving of happiness as being subject to the following formula (where H, S, C, and V respectively represent actual happiness, the set range of possible happiness, circumstances, and voluntary control):

$$H = S + C + V$$

In presenting happiness as the arithmetic sum of three other factors, this formula implies that all the variables are measurable in interchangeable units, and that the three causal factors never interact with or affect each other (in other words, that the alternative formula $H = S \times C \times V$ would be wrong). In reality, these specificities are not covered by the theory underlying the formula at all, rendering the formula itself meaningless (Coyne, Tennen, & Ranchor, 2010) and attracting unflattering comparisons with pseudoscience.

An even more controversial case arose when positive psychologists claimed to have specified the optimal balance between positive and negative emotions (Fredrickson & Losada, 2005). According to the authors of this theory, a person's ideal positivity-negativity ratio will lie between 2.9013 and 11.6346, with no margins for error. In other words, if your positivity is more than 2.9013 times greater than your negativity *and* less than 11.6346 times greater, then you will 'flourish'. However, if your positivity is *less* than 2.9013 times greater than your negativity, or *more* than 11.6346 times greater, then you will 'languish' (i.e., not flourish). This was said to hold true regardless of your age, gender, cultural background, or socioeconomic status, and also regardless of whether you are an individual person, a couple, or an organization. The authors claimed to have derived these critical ratios using differential equations based on the principles of fluid dynamics, an approach ordinarily used by physicists and engineers to study fluctuations in air or water. Their paper, which appeared in one of the leading peer-reviewed journals of psychology, attracted hundreds of citations over several years. However, the popularity of the paper was not reflected in the robustness of its content. Although it took nearly a decade, the calculations underlying the so-called 'positivity ratio' were eventually shown not only to have been poorly described and incorrectly applied, but also to have been strewn with numerous mathematical errors, some of which were quite elementary (Brown, Sokal, & Friedman, 2013). It turned out that the paper's many complex equations and dramatic data visualizations were effectively meaningless. After this was revealed, the mathematical aspects of the paper were disowned by one of its authors (Fredrickson, 2013) and retracted by the journal concerned (Fredrickson & Losada, 2013). But perhaps the greater flaw with the positivity ratio was its conceptual approach. The idea that it was considered possible to measure 'positivity' and 'negativity' *at all* – never mind with such accuracy as to yield microscopically precise ratio-thresholds – seems, with hindsight at least, to have been deeply ill-conceived. How exactly were we supposed to measure these variables – in people, couples, or organizations – to four decimal places of accuracy? And what should we now make of all the hundreds of other published academic articles which cited the positivity ratio, as fact, without reservation? What should we make of the studies that purported to corroborate the ratio with their own data? Well, it seems that for an optimistic positive psychology researcher, there is always hope. Notwithstanding the exposure of the ratio as mathematically nonsensical, and thus the evidence for the ratio as empirically unsound, its authors continue to advocate for its validity *in principle*. It appears that, for some, the evaporation of evidence has only a marginal effect on the attractiveness of an idea.

It has frequently been pointed out that, by focusing so much on the contrast between 'positive' and 'negative' – in precisely those terms – positive psychology cannot but project an inherent value system. This then leads to a number of problems. Firstly, arguing for 'good' instead of 'bad' makes positive psychology very different to other scientific fields. It is not usually the job of science to tell people what they should or should not want, and doing so infringes upon the typical scientific aspiration for objectivity. Secondly, as any first-year philosophy student will tell you, determining what is good ('positive') or bad ('negative') is far from straightforward. For example, it may sometimes be a 'good' thing to allow a person to do 'bad' things, if that person so chooses. While the CSV declares *knowledge* and *courage* to be major character strengths and virtues, a particular individual might, for their own reasons, think otherwise and prefer to remain ignorant and cautious. Is positive psychology really saying that allowing an individual the freedom to exercise such a choice is authentically (or morally) 'negative', and represents a major character weakness – or even a *vice*?

Thirdly, and arising from the above, much of what positive psychology currently promotes as 'positive' can be directly questioned. Indeed, the very fact that human traits have unambiguously 'good' sides, and correspondingly alternative 'bad' sides, is itself questionable. Take *courage*, for example. Having lots of courage may mean that you lose the ability to fear negative outcomes. This would be a problem because our natural fear of risk is an important safeguard against recklessness. For sure, having too little courage would also be a handicap. Therefore, as with all human traits, the optimal range of functioning will be somewhere between the extremes of 'high' and 'low': a moderate level of courage would be the most adaptive. In fact, it is this principle that underlies the evolution and retention of variable traits in the human gene pool. Were 'high' courage always better than 'low' courage then, across time, only people with 'high' courage would thrive in the human environment; those with 'low' courage would gradually be selected out of the species. *Courage* itself would end up so homogenous across individuals – everyone would have the maximum amount of it – as to become unnoticeable as a human characteristic. Essentially, *courage* would cease to exist (as, soon enough, would humanity itself, as people blithely ignore ravines, predators, fire, and all the other hazards they would otherwise be afraid of).

A further problem with positive psychology's value system relates to its perspective on the field of psychology as a whole. By declaring itself 'positive' psychology, it effectively implies that the rest of psychology – including all that existed prior to 1998 – is 'negative'. By further implication, it dismisses as 'negative' those psychologists who criticize positive psychology, who are reticent about it, or who fail to see the point of it. By distinguishing positive psychologists from psychologists who are *not* positive psychologists, positive psychology conveys a generalized moral criticism of that majority of psychologists who work in the mainstream of the field. It is somewhat ironic that the enterprise of positive psychology should rest on such a misanthropic narrative.

Even when positive psychology seeks to avoid the pitfall of moral relativism by seeking objective validation for its perspective, it simply enters a different type of paradox. By empirically linking positive traits to tangible rewards such as enhanced physical health or business success, positive psychology seems to be attaching material enticements to principle, rather than arguing for principle on its own terms. In one of the most prominent critiques of the positive psychology movement, the American journalist and political activist Barbara Ehrenreich (2009) questioned the logic of incentivizing morality in this way:

> Would happiness stop being an appealing goal if it turned out to be associated with illness and failure?… Nothing underscores the lingering Calvinism of positive psychology more than this need to put happiness to work – as a means to health and achievement, or what positive thinkers call 'success'. (p. 159)

Ehrenreich also criticizes positive psychology for encouraging people to seek happiness in the status quo, thereby promoting conformity and discouraging dissent. Boosting people's satisfaction with life is an aspiration of ambiguous virtue if those lives are lived in a society (or world) where injustice and inequality are rampant.

In short, the mixing of value judgements with science creates very many problems for positive psychology. Some of these problems are empirical (concerning specificity, parsimony, and circularity), while others relate more to the broader viability of scientific research (such as the ability of positive psychology to be authentically objective). These problems are impossible for positive psychology to avoid because value judgements are automatic when one attempts to be definitive in separating 'positive' from 'negative'. By the very fact of orienting its worldview in these terms, positive psychology finds itself in a permanent scientific quagmire.

Neutrality is good

Optimism, positivity, and a policy of looking on the bright side might all seem very appealing when attempting to 'do' psychology, but it takes a substantial amount of all three to ignore the significant challenges they lead to. The fact that very many psychologists subscribe (through religions) to moral philosophies that encourage, if not insist upon, such orientations – and which encourage a form of epistemological compartmentalization in requiring adherents to accept truth without evidence – provides another layer of complexity. The difficulty here is not so much that these personal and human value systems exist. Rather, the problems arise when psychologists try to fold these noble values into the essence of psychology itself – such as when psychologists pick and choose when to require evidence, assume all helping to be meritorious, inflate the value of psychotherapy beyond what can be justified by data, or force a paradigm of arbitrary positivity as both normative framework and moral imperative.

Psychology is effectively a tool, akin to a hammer, a lawnmower, or a modem. For psychologists to view psychology as a mandate to do good in the world would be like plumbers claiming that their wrenches inspire them to fix taps. In reality, the motivation to do good resides with the person who becomes a psychologist, and not within psychology itself. As such, there is nothing inherent in psychology that necessitates altruism, and there is certainly nothing that makes research more likely to yield significant findings to the effect that human beings are naturally altruistic. The challenge for psychologists is to remember that their personal convictions about human nobility, helping, and character strengths are not necessarily rooted in empirical reality. The benefit of psychology, as done scientifically, is to interrogate such instinctive stereotypes, and to test them using objectively gathered and independently verifiable data.

None of this is to suggest that doing psychology well is without moral implications. Indeed, the call for scientific rigour in psychology, a call that is advanced throughout this book, is fundamentally a moral one. The wish is for psychology to be done correctly, because it is better that psychology be done right than it be done wrong. Doing psychology right will make things better. Likewise, the preference of science over pseudoscience is very clearly made with reference to values. It is aversive when people's heads are turned by pseudoscience, because getting sucked into it, while an understandable human failing, is hazardous. The point is that what makes psychology beneficial is not its psychological subject matter or the assumptions of its practitioners; it is its use of the scientific method and its adherence to scientific reasoning. Doing psychology scientifically is not only more logically coherent and productive than doing it pseudoscientifically, it is the *right* thing to do. Scientific psychology matters.

In our final chapter, we will look at why.

Chapter 10

Psychologists at the Threshold: Why Should We Care?

Being right or being wrong

The distinction between science and pseudoscience is essentially one between right and wrong. This is because the distinction is ultimately between fact and falsehood – whether or not the underlying epistemology of an effort to produce knowledge is effective in establishing truth. When grappling with such matters, aficionados of research methodology often like to refer to *reliability* and *validity*. *Reliability* is the extent to which a conclusion produced by a research method can be replicated. If the same conclusion is drawn every time a question is asked – and, especially, if independent observers produce the same conclusion when they research the matter for themselves – then the approaches being used are said to be 'reliable'. The term can be applied to the way a study is designed ('the experiment was reliable') or the way a variable is measured ('the instrument was reliable'). We can note that the conclusion need not actually be *correct*. It is its consistency across observations, and not its accuracy, that is 'reliable'. As such, in the historical past, when observers consistently arrived at the conclusion that the earth was flat rather than round, we can say that their approaches exhibited a high degree of reliability (even though their conclusion was wrong). This is why validity is important as well. *Validity* is the extent to which a method produces a conclusion that faithfully reflects matters as they are in actuality. In other words, if a research study or measurement technique provides a conclusion that is true – and is *truly* true – then it is said to be 'valid'. Therefore, in the historical present, when observers consistently now conclude that the earth is in fact round rather than flat, we say that their approaches exhibit validity as well as reliability.

Of course, we might worry about the fact that our understanding of the shape of the earth has changed from one historical period to another. How do we know that today's approach is 'valid' whereas historical approaches were not? Well, for one thing, our current understanding is informed by a much greater volume and breadth of relevant evidence than was available in the Middle Ages (such as satellite photography, projections of star constellations, data on daylight hours across different time zones, and so on). It is not so much that today's evidence is more important than that of the past, it is that putting together *all* the evidence – new *and* old – leads to our conclusion that the earth is round. Theoretically, it remains possible that scientific opinion could change again if new information arises; and were the evidence to change, then the scientific conclusion would have to change too. Hence it is true that a scientific consensus is always contingent on whatever evidence is available at a given time. However, it is also true that science is therefore undogmatic. In principle at least, confidence about a scientific explanation is drawn from the quality and quantity of the under-lying evidence, rather than from personal loyalty to a preferred scenario.

When considering the merits of a scientific claim, it is quite straightfor-ward to make a call concerning its reliability (in that it is easy to say whether multiple findings are consistent with one another, or not). Admittedly, it is more difficult to be definitive about its validity (in that evaluations of validity require a considered assessment of the overall weight of evidence). However, it is possible to gauge one's confidence on validity and to be transparent about one's reasons for this level of confidence. Overall then, *reliability* and *validity* are useful markers of how well knowledge is being produced: when conducting research, we aspire to maximize both.

As discussed throughout this book, the scientific approach to knowledge has devised a number of ways of pursuing this aim. For example, it seeks to maximize accuracy in measurement, to be specific in its definition of terms, and to be parsimonious in confining itself to that which is already known for sure. It prioritizes evidence-based assertions over the personal judgments of authority figures, and valorizes objectivity over subjectivity more generally. It seeks to steer evaluations towards hypotheses that are falsifiable in order to avoid confirmation bias. Where ignorance exists, the scientific method calls for modesty in recognizing it. And when findings are being disseminated, the scientific approach demands transparency, com-pleteness of disclosure, and independent peer-review.

In contrast, pseudoscientific approaches – practices that claim to be sci-entific but which fall short of all these rigours – are, by definition, *not* geared up to maximize reliability and validity. Reliability and validity can-not be assured if shortcuts are taken with the gathering and reporting of evidence. They are at risk if corners are cut through objectivity or peer-review. And they are unlikely to emerge if practitioners play fast and loose with poorly defined terms, unwarranted assumptions, or hypotheses that cannot be falsified. Reliability and validity will always be in short supply when pseudoscientific approaches are applied to esoteric fields like homoe-opathy, physiognomy, or the paranormal. But they will also be endangered when pseudoscience is used by behavioural scientists – even with honest intent – who work in the mainstream of academic and applied psychology.

Another way of phrasing the point is as follows. In being designed to boost reliability and validity, scientific approaches are more likely than pseudosciences to produce conclusions that are 'right' (assuming we take the term 'right' to mean 'in accordance with the facts insofar as they can ever be confirmed'). Correspondingly, failing to boost reliability and validity will make pseudoscientific approaches more likely to produce conclusions that are 'wrong'. Of course it is true that mistakes can be made in science and that scientists can sometimes be wrong. However, when this happens it is *despite* the structure of scientific reasoning, rather than *because of* it. Moreover, the continual rechecking required by science makes it somewhat self-correcting over time. On the other hand, when pseudoscientists go awry, we can reasonably blame this on the very fact that they are using pseudoscientific approaches. Pseudosciences are systematically unprotected against error, and will inevitably produce unwarranted or otherwise garbled conclusions. They do not advocate rechecking or self-correction. In short, pseudosciences are systematically wrong.

In Chapter 2, we noted that despite standing as sources of inaccurate information, pseudosciences can become extremely popular. In particular, we noted that the way they bypass the checks and balances of scientific epistemology makes them more adaptable, more agile, and ultimately more accessible to their audiences. In Chapter 4, we looked at the many ways in which human reasoning is uncomfortable with complexity, tolerant of error, and susceptible to social influence. All of these points help us to understand how bad ideas do not necessarily fail to prosper. In this final chapter, we will conclude our journey by considering two outstanding questions. Firstly, we will ask: why exactly are reasonable people so often attracted by pseudoscience and repelled by science? After all, while we know that pseudoscience is adaptable and that people are prone to cognitive shortcomings, these factors merely create the conditions for pseudoscience to exist; they do not guarantee that it will thrive. Many ideas are possible, but only some become popular or successful. Our second question is a related one. We will ask: why should we care about pseudoscience? Many people argue that pseudosciences do no real damage. They point out that human culture frequently grapples with trivia, banality, and pointless obsessions, without there being a looming disaster to concern us. Within psychology, subfields that flirt with the boundary between science and pseudoscience are often depicted as part of the field's rich tapestry of ideas. It is said that by adding methodological diversity and conceptual colour, they make psychology more stimulating. For sure, the costs are not always obvious. But they are certainly there.

(Why) is it fashionable to be wrong?

The capacity to learn from direct experience, to separate fact from nonfact, and to distinguish corroborated assertions from uncorroborated ones must surely be seen as beneficial. Therefore, the thriving of pseudoscience in human culture – a culture that is indisputably advancing in technological terms – is something of a paradox. Humanity's tendency to falter in the face of randomness and complexity, and to rely on cognitive shortcuts when grappling with logic and reason, has not prevented it from developing sophisticated medicines, building skyscrapers, or exploring the galaxy. In reality, the cognitive limitations that characterize *Homo sapiens* reflect more our ancient evolutionary history than our current predicament. While we are prone to error, we are also quite capable of taking stock and deliberating our way toward sensible conclusions when the stakes are sufficiently high. We can think clearly when we want to. This makes quite perplexing the fact that we very often seem to want to think *un*clearly. Indeed, the question of why people feel peculiarly *attracted by* unreliable thinking is so widely perplexing that it has been tackled by scholars and scientists from many different academic backgrounds. We will consider a number of these perspectives here.

Economists try to explain pseudoscience by considering how people naturally seek to pay the least possible cost for the highest possible well-being. We have already alluded to the notion that sloppy cognition is sometimes just quicker and easier than its logically sound alternative; this alone might account for its attractiveness. From the perspective of economic theory, the contrast is akin to that associated with a price incentive. Viewing *Homo sapiens* as *Homo economicus* – that symbolic figure used in economics to represent the human motivation to maximize utility – the procedures of scientific reasoning might appear like a tax on thinking. In economic terms, we can assume that consumers will be price-sensitive, and will seek to avoid having to pay taxes where they are not strictly necessary. American economist Anthony Downs (1957) captured the point when he posited the notion of *rational ignorance*, arguing that people will seldom educate themselves on a particular matter unless it clearly outweighs the costs of remaining uninformed. In reality, these so-called 'private error costs' are rarely overwhelming: it is usually much easier to guess or to rely on hearsay than it is to read up on the relevant science or to gather and analyse one's own evidence. More recently, another American economist, Bryan Caplan (2001) proposed the concept of *rational irrationality*, an effect where people find it most efficient to dispense with the accoutrements of logical reasoning altogether. According to Caplan, because of the very fact that private error costs associated with mental shortcuts are usually very low (or perceived to be such), the economic principle of the 'demand curve' comes into play. The demand curve refers to the association between price and demand: usually, the cheaper a commodity is, the more of it people wish to consume. By this logic, the very low cost of irrationality makes it a high-demand commodity. People buy it by the bucketload because it is so cheap, choosing to save the valuable currency that is their intellectual effort for other things. In Caplan's terms, in these situations, people rationally choose to think irrationally.

Anthropologists and social theorists try to account for the human fascination with pseudoscience by considering its social impact. Some studies have suggested that engagement with pseudoscientific practices (such as complementary and alternative medicine) may be linked to mysticism, increasing in popularity across time in order to fill the gap created by the decline of religion in secular societies (Hughes, 2006; Verheij, de Bakker, & Groenewegen, 1999). Others have argued that, by relying heavily on anecdotal support, pseudoscience attracts a form of social solidarity. Consumers find common cause in their attachment to pseudoscience, and feel gratified to be part of a group. The fact that some people criticize them for being scientifically unsophisticated is no discouragement at all; on the contrary, the shared persecution can serve to enhance the sense of group belongingness. The resulting emphasis on group loyalty and unity may be more rewarding than not.

One indication of this can be found in some intriguing archive research conducted by the American anthropologist Richard Sosis at the University of Connecticut. Sosis traced the history of 200 communes in the United States during the 19th century (Sosis, 2000). In that period of American history, it was reasonably common for groups of people to drop out of mainstream society and try to establish independent self-sustaining communities, typically located in woodlands, mountains, or other areas of previously uninhabited wilderness. The aim of these so-called utopian communities was to remain forever cut off from the official state, and to govern their own affairs unmolested in perpetuity. In reality, many of these communes didn't last very long. Some encountered problems with food security, others were hampered by health concerns, and others collapsed under the pressure of internecine conflict. Sosis wanted to examine the predictors of commune longevity. Noting that many communes were organized along religious grounds, he checked whether this made a difference to their success. While not directly pseudoscientific, religious communes were nonetheless committed to an ethos that was informed by mysticism: the basis for commune solidarity was a shared commitment to beliefs that rested on faith rather than evidence. In contrast, secular communes tended to be organized along more empirical lines. Their motivation to drop out of society was driven by logical argument (for example, grievances against the state) rather than religious fervour. Intriguingly, Sosis found a statistically significant difference between religious and secular communes: religious communes were far more likely to survive for longer than secular ones. For the 112 secular communes, the average length of survival was just over six years; in sharp contrast, for the 88 religious communes, the average length of survival was more than 25 years. Secular communes were more than four times more likely than religious ones to die out within their first two years of existence, and twice as likely die out during the subsequent three years. While there were many factors involved in determining the success of a commune in 19th-century America, the scale of this imbalance is consistent with the anthropological view that a shared commitment to faith-based, rather than empirical, reasoning systems helps to consolidate the social functioning of a group.

A different sociocultural perspective involves the notion of Luddism, a term used to describe people's reticence towards technology and industrialization. Historically, the Luddites were a formally organized protest movement in early 19th-century England, a period when the technological advances of the Industrial Revolution threatened to make many skilled labourers redundant. The Luddites were named after a possibly apocryphal character named Ned Ludd, who is said to have vandalized a number of labour-saving textile weaving machines. The Luddites organized a widespread campaign of similar vandalism and sabotage, ostensibly aimed at undermining the transition from artisan labour to machine-work, but also giving vent to a rising tide of working class dissent that was fuelled in part by the economic hardship of the time. Nowadays the term Luddism is used less formally to refer to popular scepticism towards technology, although its intertwining with social and economic circumstances remains. Various cultural theorists have argued that technophobia is extremely widespread, and very influential in shaping people's attitudes toward social change (Brosnan, 1998). Scepticism toward new technology emerges in every generation, and in every culture. While some consumers will queue around the block in order to get their hands on the latest smartphone, many others will bemoan its very existence and warn of negative effects on users' attention spans, social skills, or brains. Developments in energy production (such as nuclear power), food production (such as genetically modified crops), medicine (such as stem-cell engineering), or telecommunications (such as WiFi) are frequently met with levels of fear and hostility that seem to exceed what is justifiable in light of the available evidence. Often, this scepticism takes the form of a mass movement. However, what is most intriguing is that it frequently appears as though the concerns of protesters are only loosely connected to the targets of their complaints.

To illustrate the point, let us consider the anti-vaccination movement. Popular protests against vaccination are nearly as old as vaccination itself, with the earliest campaigns being launched soon after the introduction of human vaccines by Edward Jenner in the early 19th century. Even though today's circumstances, diseases, and technologies are all very different, the concerns of anti-vaccinationists have remained essentially unchanged over the past 150 years. For example, when medical researchers at Chicago's Northwestern University examined one anti-vaccination pamphlet published in England in 1878 (Wolfe & Sharp, 2002), they found that a number of distinct arguments against vaccination were articulated. Among these arguments were the following: that vaccines contain toxic additives; that vaccines cause rather than prevent illness; that vaccines provide only temporary immunity; that the promotion of vaccination distracts people from the benefits of natural living; that vaccines are promulgated solely for profit; and that mandatory vaccination amounts to a form of state totalitarianism.

The researchers then conducted a systematic thematic analysis of anti-vaccination literature produced in the late 20th century. Intriguingly, they found that the themes raised in the late 20th century mapped perfectly onto the same arguments as were identified in 1878 (Wolfe & Sharp, 2002).

In 1878 vaccines were a new technology aimed primarily at preventing smallpox, and were set against a relatively primitive medical understanding of disease pathogenesis. Furthermore, most of the people being inoculated lived in poverty and poor hygiene, and did not have access to either information or regular medicine. The context of modern vaccination is different in almost every respect. Today's vaccinations target a variety of pathogens – including flu, chicken pox, measles, mumps, polio, rubella, and human papillomavirus – and are developed and administered using approaches refined by a century of technological experience. Medical understanding of disease processes is much advanced and the typical recipient of vaccination, who is in good health, benefits from excellent hygiene and easy access to modern healthcare and information. The commercial aspect of pharmaceutical production is also entirely different today than it was in the 19th century, although the complaint of an unholy alliance between vaccine provider and paymaster is as intense as ever. In short, even though the context and circumstances have radically transformed, the grievance against technology has remained constant. It seems that strong sociocultural influences intrinsically stimulate a particular type of scepticism. Insofar as this scepticism reflects a distrust of science – and after all, technology is derived from science and scientists are its chief apologists – then we may conclude that the attraction of pseudoscience is at least partly driven by factors that lead societies to be tentative towards change in general.

Interest in explaining why people are attracted by pseudoscience is not confined to the social sciences. Biological scientists too have considered the question in various ways. Several neurological studies have suggested that people's brains are biologically pre-programmed to attach value to unfounded beliefs. As referred to in Chapter 6, brain imaging studies have revealed differences in brain function when religious believers are compared to non-believers (Harris et al., 2009). Of course, Chapter 6 also sets out several reasons why such imaging studies are limited in their explanatory power. Perhaps more informative are studies of brain damage. These studies provide a better perspective on cause-and-effect by allowing the pertinent sequence of brain events and cognition to be examined. For example, patients who experience temporal lobe epilepsy often exhibited intense spiritual feelings during seizures, as well as a subsequent heightening of interest in morality and religion. Some patients undergo radical alterations in their overall belief approach. The neuroscientist V. S. Ramachandran, who works at the University of California in San Diego, has described several patients who were lifelong atheists prior to their seizures but who became devoutly religious afterwards, as well as others who reported experiencing bouts of religious ecstasy in the midst of the seizure itself (Ramachandran & Blakeslee, 1998). Patients who suffered limbic system seizures also tended to show transformations in religiosity. Unlike brain imaging studies, the direction of causality here is conspicuous: the precipitating brain experiences cannot be said to have *resulted from* the changes in cognition that followed. These studies suggest that, at least in some respects, personal epistemology is a function of brain architecture. It could be that popular fascination with pseudoscience reflects a neurological default setting that, for whatever reason, has evolved to predispose human beings to favour mysticism over logic.

Other insights have come from zoologists working in the area of ethology. For example, the human tendency to attach credence to advice from authority (and, by extension, to anecdotal testimony of various kinds) seems very similar to principles of learning seen repeatedly throughout nature. Many species demonstrate filial imprinting, a phenomenon where newly born young acquire behavioural characteristics of their parents by observing them during a critical learning phase very early in their lives. Famously, when these baby animals are exposed to creatures who are *not* their parents – to human beings or members of other species, for example, or even to arbitrary shapes and objects – their imprinting is directed at these instead. The Austrian ethologist Konrad Lorenz showed how newly hatched goslings would treat him as their parent (and, as captured in several legendary photographs, devotedly follow him wherever he went) simply because he was the first creature they saw after they emerged from their incubators (Lorenz, 1952). Other forms of systematic vicarious learning, such as different types of imitation and observational conditioning, are seen in several species.

The point here is that, biologically, it makes perfect sense for newly born organisms to exhibit immediate attachment to whatever caregivers first present themselves and, from then on, to treat these caregivers as important role models and sources of reliable guidance. This natural instinct is an important survival skill. Indeed, in humans, it is difficult to see how children could possibly develop cognitively or socially were they not so utterly amenable to the advice and guidance of their elders. The inclination to place trust in the face-value of other people's advice is not only very important for ensuring successful development throughout childhood. In a global culture where it is simply impossible for people to independently fact-check everything they are told, adults too will benefit from being generally credulous by default. The problem of course is that such a predisposition leaves people open to acquiring *erroneous* beliefs when the information they receive is of poor quality; but this cost might just be outweighed by the evolutionary benefits. Overall, ethological principles of learning help explain why reasonable people are, in the main, quite trusting, and why they find science – which encourages them to be *less* so – to be counter-intuitive and thus unappealing.

A related theory from evolutionary biology concerns the phenomenon of costly signalling. This refers to the way certain animals engage in activities that are of little direct survival value in themselves, but which, through their symbolism, contribute indirectly – but powerfully – to evolutionary adaptedness. Perhaps the classic example in nature is that of the peacock's tail. As an appendage, the peacock's tail is extremely cumbersome in size and complexity. The volume of plumage accounted for by the tail requires a significant daily investment of calories, meaning that a peacock must spend large amounts of time foraging for food just to keep its tail healthy. Given that peacocks are largely terrestrial foragers, the hugeness of the tail presents the dual drawback of being both hard to conceal from predators and easy to snag on undergrowth. On simple survival-of-the-fittest grounds, it would seem that the average peacock would be far more successful in reaching reproductive adulthood had it no tail at all, instead of an iridescent fan many times larger than its body. However, peacocks are actually quite hardy, and typically live for more than twenty years in the wild. Many theorists argue that this is because the peacock's tail provides important signals to peahens. It is well established that the huger and more outrageous the tail, the more a typical peahen will find that peacock attractive. It is as if the tail's sheer wastefulness and conspicuousness convinces the peahen that this potential mate is able to make ends meet and look after itself.

Similar examples of costly signalling abound in nature. An important feature of all these signals is that they are hard, if not impossible, to fake. Some deer species have elaborate antlers, some insects have flamboyant colouring, and the claws and horns of various crustaceans and insects (such as horn beetles) often consume more calories than seems strictly necessary at first. Costly signalling also extends to behaviour patterns, such as the repeated on-the-spot jumping seen in some species of gazelle. Although this unsubtle behaviour certainly draws attention and consumes valuable energy, it also tells would-be predators how agile and energetic the gazelle is (and so how futile it would be to chase it).

Some biologists have suggested that costly signalling helps to account for various practices in human culture. The idea is that wasteful and pointless activities are not necessarily counterproductive in evolutionary terms, because they serve to demonstrate the robustness of individuals or groups. A community that invests time on frivolous things advertises itself as the type of community that *has* time to invest on frivolous things. Such communities are more appealing than ones pedantically concerned with ensuring a return on every investment of effort. For this reason, pseudoscience is attractive precisely because the benefits of one's efforts are not set out in advance, and so they allow people to luxuriate in uncertainty. In contrast, by insisting on uncertainty reduction, and by frowning on inefficiency, science comes across as somewhat po-faced and less culturally appealing.

Once again, we can refer to the work of Richard Sosis on communes (Sosis & Bressler, 2003). As part of his research, Sosis examined the various requirements set by the communes that established themselves in the United States during the 19th century. Many required residents to adhere to particular rules and to participate in mandatory activities. Sosis classified these requirements as either 'costly' or 'not costly' on the basis of whether they yielded a productive return to the individual or to the group. For example, communes that banned residents from owning photographs or jewellery, or which compelled residents to have particular hairstyles or wear particular hats, were deemed to have set 'costly' requirements. On the other hand, when communes obliged residents to help prepare meals or wash linen, these requirements were deemed to be 'not costly'. When these details were examined in conjunction with the data on commune longevity, yet another intriguing finding emerged. Sosis found that, for religious communes (which, as described above, tended to be the more robust), the number of costly requirements was positively correlated with longevity. In other words, the more pointless requirements the commune demanded of its residents, the *more* successful that commune was in maintaining its existence. This finding is consistent with the idea that frivolity serves as a signal of fitness: human groups which accommodate frivolity tend to thrive, whereas those that insist on rigour at all times tend to falter.

Each of these scholarly paradigms offers its own reasoned view as to why people might find pseudosciences attractive and sciences less so. Economists, anthropologists, and social theorists have suggested that human agents are affected by their perceptions in ways that make them wary or disregarding of science and correspondingly sympathetic to the alternative. Meanwhile, neurologists, ethologists, and evolutionary biologists have suggested that human beings might be naturally pre-programmed to be open to pseudoscientific ideas. In all these contexts, the views presented have concerned the core matter at hand – namely, the way people perceive and process sources of information in their world. However, there is another way in which diverse academic subject areas can become involved in the explanation of public scepticism toward science. This is when science becomes embroiled in the politics of academia, and the many fraught inter- (*and* intra-) disciplinary rivalries that tend to arise in ivory towers.

(Why) should psychologists care about pseudoscience?

One of the reasons to care about pseudoscience is that, far from being a banal distraction hovering at the margins of relevance, pseudoscience directly affects the well-being of science. Pseudoscience doesn't just coexist with science, it serves to undermine it; the pseudoscientific worldview is not ambivalent towards scientific principles, it is hostile towards them; and pseudoscientists don't just observe science, they attack it. As such, psychologists who retain sympathy for pseudoscientific notions – such as the idea that the human mind is special, that aspects of psychology lie beyond the reach of research, or that scientific approaches are intellectually impoverished and spiritually demeaning – are essentially harbouring ill-will towards their own discipline. Psychologists who are accidental pseudoscientists – those whose scientific literacy is so light-touch that they become blind to the faults in their own research paradigms, routinely embracing non-falsifiable hypotheses, unverifiable anecdotal data, theoretical circularity, and unwarranted reductionism – are perhaps less culpable. Nonetheless, their tolerant disposition aids the creep of pseudoscience into psychology and the dilution of its well-being as a science. Psychologists who actively *campaign* for non-scientific versions of psychology – by championing critical theory or qualitative interpretivism, for example – have an even more interesting case to answer. It is entirely reasonable to take the position that there are better ways than science with which to study human thoughts, feelings, and behaviours. However, this is akin to declaring that there are better ways than *psychology* with which to do so.

While scholars from many academic disciplines have attempted to explain negative attitudes towards science, some have sought to foster – if not *provoke* – negative attitudes towards science. For example, stemming originally from academic philosophy, a significant body of scholarly literature now actively promotes the idea that conventional science is of limited worth. A common criticism is that scientists are addled by unacknowledged and incorrigible bias. Philosophers such as Paul Feyerabend (see Chapter 8) have argued that scientists are so unable to suppress their subjectivity, they are no better than pseudoscientists when it comes to producing knowledge. Some of the most famous philosophers of the past century – figures such as Michel Foucault, Jacques Derrida, and Jean-François Lyotard – have been champions of postmodernism, according to whom all truth is relative and scientific findings little more than arbitrary social constructions. From the postmodernist perspective, no one claim to knowledge is more valid than another. Scientific descriptions are said not to reflect reasoned accounts of the world, but an agreed discourse shared within a particular narrative community. Many of these critics do not stop at bemoaning science as a source of distorted knowledge; they also argue that, because of its dominance in human affairs, science stands as a real and present threat to human welfare. According to the received postmodernist view, conventional science is contaminated by the vested interests of industrialized Western society and its various hegemonic powers. As a result, it perpetuates the bourgeois privileges of the capitalist order, leaving in its wake the many victims of global socioeconomic oppression. In postmodernism, mainstream science is not merely critiqued. It is roundly castigated.

There is no doubt that many postmodernist philosophers of science hold their views extremely sincerely. For others, it may be relevant that they cohabit universities with scientists and, in that arena, compete with them for resources and prestige. Whatever the origins of the dynamics, it is certainly the case that many philosophers – as well as many literary theorists, cultural scholars, and political scientists – are very influenced by postmodernist ideas regarding relativism in research and the social construction of knowledge. Indeed, much of their criticism has been so vituperative that the resulting inter-academic conflicts have themselves become an object of scholarship (e.g., Ashman & Barringer, 2001; Gross & Levitt, 1994; Rehg, 2009).

The postmodernist critique of science is not all-encompassing and it falls foul of a number of limitations. For one thing, any appeal to truth-relativism is hampered by the inherent paradox at its core – an assertion to the effect that 'there is no such thing as a reliable assertion' is self-contradictory and thus automatically nonsensical. As pointed out in Chapter 5, there is also the matter of conventional science's track record of productivity. The fact that aeronautics provides knowledge that can enable a 300-tonne machine to fly through the air suggests that it is far more than just an agreed discourse shared within a particular narrative community (Hughes, 2012).

Many of the strongest criticisms of postmodernism come from within philosophy itself. Nonetheless, it is certainly the case that there now exist large movements within academia offering perspectives that are intended to undermine public confidence in science. Postmodernist philosophy is a densely complex pursuit, and its literature can be difficult for even a specialist to follow, but it may be this appearance of sheer hyper-scholarship in the absence of genuine popular scrutiny that allows it to influence public opinion (Hughes, 2008b). Popular scepticism toward science (and sympathy toward pseudoscience) may in part result from this permeation of postmodernist ideas outside the academy.

Throughout this book we have considered the interconnectedness of psychology, science, and pseudoscience. Psychology is a field that strives to apply science to its study of human thoughts, feelings, and behaviours; science is the use of empirical methods to resolve uncertainties in our understanding of nature; and pseudosciences are activities that purport to be scientific, but which lack the rigours required of true empiricism. Psychology is particularly pertinent to any discussion of science and pseudoscience because its research explicitly tries to explain how people make sense of their world. The very existence of science as a formal epistemological method reflects the fact that spontaneous human reasoning is simply not a good way to achieve reliable and valid knowledge. Psychology research has allowed us to better appreciate the limitations of human cognition and its susceptibility to social influence.

But, as mentioned previously, there is another reason why psychology is relevant to this discussion. Psychology, as a field of study in its own right, has struggled to distinguish itself as a science. While it is easy to detect the pseudoscientific mien in fringe topics like telepathy, physiognomy, and quantum consciousness, it can be much more difficult – analytically *and* politically – to broach that which is dubious within the mainstream. Nonetheless, camouflaged threats warrant greater vigilance than those that are easily identified. Right across the spectrum of mainstream psychology are assumptions, methods, and theoretical conventions that stray toward the boundary with pseudoscience. The risks arise even in psychology's most superficially 'scientific' domains (those activities invoking biological frames of reference), as well as those which rely on people's personally perceived perspectives to provide insights on the human predicament. Objectivity is a significant scientific challenge: psychologists must struggle to withstand both conscious and unconscious biases, including those stemming from personal value systems and aspirations to create a better world. Psychology faces a near-constant struggle to do justice to scientific epistemology. But then nobody said that science was easy.

As set out in Chapter 3, psychology does not just aspire to scientific epistemology, it squarely *is* a science. Claims that it isn't tend to be grounded in misapprehension: simply put, most scientists (including most psychologists) have a poor understanding of what scientific epistemology is. Those who do understand may be reticent for other reasons. After all, as discussed above, humankind seems to have much warmer feelings for pseudoscience than for science. When psychology emphasizes its scientific standing, it elevates the risk that ordinary people (that is to say, non-psychologists) will become wary of it. As such, psychologists find themselves standing at something of a threshold, between scientific literacy on one side and pseudoscientific sympathy on the other. Some stand facing pseudoscience and ushering passers-by in that direction; a gratifyingly larger group face the other way, encouraging people to choose scientific literacy instead. However, perhaps the majority of psychologists have their backs turned, and are unaware that they stand at a threshold at all.

There are several grounds to be concerned by such neutrality. Pseudoscientific approaches are not just inert or pointless – they mislead and misinform. And there are many costs associated with misinformation. Misinformation undermines our ability to solve problems, and misinformed human cultures are unlikely to address significant social challenges in meaningful ways. Misinformation leads to nihilism, because claims that something is worth striving for cannot ever be properly substantiated. Misinformation also leads to cynicism, because its deviation from the stable roots of reality guarantees a lack of a consistency; and when the information people get is consistently inconsistent, it inevitably fosters mistrust. Contemporary history is replete with examples of the costs wrought by poor scientific literacy. Take, for example, arguments about whether human activity contributes to climate change. These are mired in a near-permanent confusion that could substantially be remedied not by better knowledge of meteorology, but by popular awareness of the rudiments of scientific epistemology. People do not need to better understand wind flows and barometric pressure, they need to understand the difference between a falsifiable hypothesis and a non-falsifiable one. Both advocates of the view that anthropogenic global warming is a real problem (of whom there are several) as well as their opponents (of whom there are relatively few) accuse the other side of talking pseudoscience. If the population at large were adept at making the distinction for themselves, then the entire political discourse on the matter would be fundamentally altered.

In today's media-saturated culture, consensus denialism can appear almost inescapable. The idea that most of what everyone tells you is false is as seductive as it is recurrent. It comes in an almost endless number of forms – HIV/AIDS denialism, the 9/11 truth movement, Holocaust denial, evolution denial, belief in conspiracy theories concerning political assassinations or plane crashes, claims that water fluoridation is damaging to public health, claims that evidence of safety of GM foods has been fabricated, claims that pharmaceutical companies have developed cures for cancer but withhold them in order to sell palliative medications, and so on. Virtually all such movements rely on limitations in people's ability to think coherently about empirical evidence. Moreover, their cultural currency perpetuates this diminution by reinforcing perverse attitudes towards evidence. They popularize the view that most of what is presented as evidence is inherently untrustworthy – and, paradoxically, what really matters is what those in the contrarian *minority* feel is important. Such controversies would be significantly disrupted were we better able to appreciate the differences between science and pseudoscience. Even our participation in the economy and in politics would be improved. Our understanding of global economic forces, currency markets, growth rates, property bubbles, and credit crunches would be greatly enhanced if we could easily separate that which is evidence-based from that which is anecdotal, or which lacks parsimony, or which smacks of confirmation bias. The same could be said for our understanding of the effects of immigration, of the causes of crime, and of the distinct track records of different political parties.

Psychologists would not only benefit as citizens, but as professionals too. Many of the most pressing controversies surrounding mental health and psychological development are propelled by pseudoscientific energy. One of the most conspicuous examples is the campaign to implicate MMR vaccines in the aetiology of autism (a form of modern Luddism, as discussed above). This campaign represents much of what is costly about popular misinformation. Not only are parents of children with autism lumbered with unwarranted guilt about their vaccination choices, but the boycott-led diminution in herd immunity has resulted in marked increases in unnecessary illnesses and deaths due to vaccine-preventable infections (Piccirilli et al., 2015). A wider appreciation of what distinguishes scientific evidence from pseudoscientific rationalization would undoubtedly make a difference here of the most constructive kind.

For psychologists, the distinction between science and pseudoscience is ethical as well as epistemological. As referred to in Chapter 3, many professional psychology associations require an adherence to science as part of their codes of ethics. The British Psychological Society's (2009) *Code of Ethics and Conduct* begins by defining psychology as 'the scientific study of behaviour' (p. 6), and goes on to require that psychologists remain abreast of the latest scientific research (p. 16), that they seek assistance when the demands of circumstances exceed their scientific knowledge (p. 16), and even that they assume general responsibility for the scientific standards of their colleagues (p. 18). Moreover, the *Code* has an entire section on the ethical principles relating to informed consent. These require that psychologists properly inform clients and stakeholders before engaging their participation. In therapeutic contexts, for example, it would mean informing clients of the evidence base for interventions. It is difficult to envisage how this could be ethically done were a psychologist to have a poor grasp on the difference between science and pseudoscience, or an ambivalent attitude to the scientific literacy of others. In summary, psychologists should care about pseudoscience because of the threats posed by pseudoscience to their discipline, both epistemological *and* ethical.

Finally, perhaps the most important reason psychologists should advocate for scientific literacy – and why they should oppose the inward creep of pseudoscience – relates to dignity. There is a dignity inherent in being conscientiously complete in our understanding of human behaviour, a dignity we lack when we are casual about accuracy. In short, there is a dignity in being right instead of wrong. Scientific approaches might be pedantic and slow – they might even at times be boring – but the knowledge they lead to is far more likely to be not only real and reliable, but profound and inspiring too.

Modern democracies, with their spin-doctored political jousting and marketplace-of-ideas culture, certainly need citizens who can reasonably weigh up claims against counter-claims and evaluate the heft of a manifesto. But as citizens of the world, we are surely also enhanced when we can confidently dispense with myths – false claims about the risks posed by other people, for example, or misconceptions about groups in society that are held to be inherently superior. It is true that parents benefit from practical competence in separating constructive child-rearing advice from misguided dogma, not to mention an ability to forecast children's likely developmental trajectories. But there is also an intrinsic benefit in simply appreciating what is known empirically about our children's feelings and experiences, insights far more enriching than uncorroborated opinions or self-serving social prejudices. And as cultural agents, there is no doubt that humans can prosper by being aware of skills and strategies relevant to their relationships and workplaces, and knowing what advice can be discarded as spurious. But we will also garner dignity simply from being aware of the known scope of human nature – the range of human abilities, the similarities and differences between people, our biological mortality and relatedness to other species in nature, and the fact that there are simply many things about humans that we don't yet understand. Pseudoscience does not just erode judgement competence in a way that undermines people's ability to appreciate the practical merits of psychology. It also erodes our ability to participate meaningfully in the human world.

Psychology is one of the few movements in human culture that itself seeks to explain humanity. We can surely declare that the enterprise is intended to provide truths rather than falsehoods. Distinguishing science from pseudoscience is not just a skill to be commended to psychologists. It is fundamental to the comprehension of humanity, and to our dignity as a self-conscious species.

References

AAAS. (2013). *About AAAS: Organization/Governance – Sections, Psychology (J)*. Retrieved October 26, 2013, from American Academy for the Advancement of Science: http://www.aaas.org/aboutaaas/organization/sections/psych.shtml

Adler, L. L., & Mukherji, B. R. (1995). *Spirit versus scalpel: Traditional healing and modern psychotherapy*. Westport, CT: Bergin & Garvey.

Ajzen, I. (1985). From intentions to actions: A theory of planned behavior. In J. Kuhl, & J. Beckman (Eds.), *Action-control: From cognition to behavior* (pp. 11–39). Heidelberg: Springer.

Ajzen, I. (1991). The theory of planned behavior. *Organizational Behavior and Human Decision Processes, 50*, 179–211.

Ajzen, I. (2011). The theory of planned behaviour: Reactions and reflections. *Psychology & Health, 26*(9), 1113–1127.

Ajzen, I., & Fishbein, M. (2004). Questions raised by a reasoned action approach: Comment on Ogden (2003). *Health Psychology, 23*(4), 431–434.

Alcock, J. E. (2010). The parapsychologist's lament. In S. C. Krippner, & H. L. Friedman (Eds.), *Mysterious minds: The neurobiology of psychics, mediums, and other extraordinary people* (pp. 35–44). Santa Barbara, CA: Praeger.

Allen, C. N. (1927). Studies in sex differences. *Psychological Bulletin, 24*(5), 294–304.

Allen, W. D. (2007). The reporting and underreporting of rape. *Southern Economic Journal, 73*(3), 623–641.

American Psychiatric Association. (2013). *Diagnostic and statistical manual of mental disorders* (5th ed.). Washington, DC: American Psychiatric Association.

American Psychological Association. (2012, June 11). *APA on Children Raised by Gay and Lesbian Parents: How Do These Children Fare?* Retrieved October 24, 2014, from http://www.apa.org/news/press/response/gay-parents.aspx

American Psychological Association. (2013). *About APA: Definition of "Psychology"*. Retrieved October 26, 2013, from American Psychological Association: http://www.apa.org/about/index.aspx

Arbuthnot, J. (1710). An argument for divine providence, taken from the constant regularity observed in the births of both sexes. *Philosophical Transactions of the Royal Society of London, 27*, 186–190.

Ashman, K., & Barringer, P. (2001). *After the science wars: Science and the study of science*. London: Routledge.

Ashmore, M. (1989). *The reflexive thesis: Wrighting sociology of scientific knowledge*. Chicago: University of Chicago Press.

Baker, T. B., McFall, R. M., & Shoham, V. (2009). Current status and future prospects of clinical psychology: Toward a scientifically principled approach to mental and behavioral health care. *Psychological Science in the Public Interest, 9*(2), 67–103.

Bala, A. (2012). *Asia, Europe, and the emergence of modern science: Knowledge crossing boundaries.* New York: Palgrave Macmillan.

Bandolier. (2007). *Causes of Death in the USA.* Retrieved March 24, 2014, from http://www.medicine.ox.ac.uk/bandolier/booth/risk/top15usa.html

Bannister, D. (1966). Psychology as an exercise in paradox. *Bulletin of the British Psychological Society, 19*(63), 21–26.

Barlow, D. H., Bullis, J. R., Comer, J. S., & Ametaj, A. A. (2013). Evidence-based psychological treatments: An update and a way forward. *Annual Review of Clinical Psychology, 9*, 1–27.

Baron-Cohen, S. (2004). *The essential difference: Men, women and the extreme male brain.* London: Penguin.

Barrera, M. J. (1986). Distinctions between social support concepts, measures, and models. *American Journal of Community Psychology, 14*, 413–445.

Bartz, J. A., Zaki, J., Ochsner, K. N., Bolger, N., Kolevzon, A., Ludwig, N., et al. (2010). Effects of oxytocin on recollections of maternal care and closeness. *Proceedings of the National Academy of Sciences, 107*(50), 21371–21375.

Bassman, L. E., & Uellendahl, G. (2003). Complementary/alternative medicine: Ethical, professional, and practical challenges for psychologists. *Professional Psychology: Research and Practice, 34*, 264–270.

Beard, G. M. (1881). *American nervousness: Its causes and consequences – a supplement to nervous exhaustion (neurasthenia).* New York: Putnam.

Bem, D. J. (2011). Feeling the future: Experimental evidence for anomalous retroactive influences on cognition and affect. *Journal of Personality and Social Psychology, 100*, 407–425.

Bennett, D. (1998). *Randomness.* Cambridge, MA: Harvard University Press.

Berezow, A. B. (2012, July 13). Why psychology isn't science. *Los Angeles Times.* Retrieved January 10, 2013, from http://articles.latimes.com/2012/jul/13/news/la-ol-blowback-pscyhology-science-20120713

Bering, J. (2011). *The belief instinct: The psychology of souls, destiny, and the meaning of life.* New York: Norton.

Bering, J. M. (2006). The folk psychology of souls. *Behavioral and Brain Sciences, 29*, 453–498.

Bhati, K. S., Hoyt, W. T., & Huffman, K. L. (2014). Integration or assimilation? Locating qualitative research in psychology. *Qualitative Research in Psychology, 11*, 98–114.

Boring, E. G. (1929). *A history of experimental psychology.* New York: Century.

Bösch, H., Steinkamp, F., & Boller, E. (2006). Examining psychokinesis: The interaction of human intention with random number generators – A meta-analysis. *Psychological Bulletin, 132*(4), 497–523.

Bouchard Jr, T. J., & McGue, M. (2003). Genetic and environmental influences on human psychological differences. *Journal of Neurobiology, 54*(1), 4–45.

Briggs, L. (2000). The race of hysteria: "Overcivilization" and the "savage" woman in late nineteenth-century obstetrics and gynecology. *American Quarterly, 52*(2), 246–273.

British Psychological Society. (2009). *Code of ethics and conduct: Guidance published by the Ethics Committee of the British Psychological Society.* Leicester: British Psychological Society.

British Psychological Society. (2013). *'Accreditation through partnership' handbook: Guidance for undergraduate and conversion programmes in psychology, September 2013.* Leicester: British Psychological Society.

British Psychological Society. (2013). *The British Psychological Society: Promoting Excellence in Psychology.* Retrieved October 26, 2013, from British Psychological Society: http://www.bps.org.uk/

British Psychological Society. (2015). Retrieved January 30, 2015, from Find a Chartered Psychologist specialising in psychotherapy: http://www.bps.org.uk/psychology-public/find-psychologist/psychotherapy-register/find-chartered-psychologist-specialising-

Brizendine, L. (2007). *The female brain.* London: Bantam.

Brizendine, L. (2010). *The male brain.* London: Bantam.

Broadcasters' Audience Research Board. (2014). *Viewing Data: Average Weekly Viewing.* Retrieved March 24, 2014, from http://www.barb.co.uk/viewing/trend-graph-average-weekly-viewing

Brosnan, M. J. (1998). *Technophobia: The psychological impact of information technology.* London: Routledge.

Brown, J. W. (1969). *The rise of biblical criticism in America, 1800–1870: The New England scholars.* Middletown, CT: Wesleyan University Press.

Brown, N. J., Sokal, A. D., & Friedman, H. L. (2013). The complex dynamics of wishful thinking: The critical positivity ratio. *American Psychologist, 68*(9), 801–813.

Buchanan, M. (2007). Conviction by numbers. *Nature, 445,* 254–255.

Buller, D. J. (2005). *Adapting minds: Evolutionary psychology and the persistent quest for human nature.* Cambridge, MA: MIT Press.

Burt, C. (1962). The concept of consciousness. *British Journal of Psychology, 53,* 229–242.

Buss, D. (2013). *Evolutionary psychology: The new science of the mind* (4th ed.). Harlow, Essex: Pearson Education.

Bussey, K., & Bandura, A. (1999). Social cognitive theory of gender development and differentiation. *Psychological Review, 106,* 676–713.

Button, K. S., Ioannidis, J. P., Mokrysz, C., Nosek, B. A., Flint, J., Robinson, E. S., et al. (2013). Power failure: Why small sample size undermines the reliability of neuroscience. *Nature Reviews: Neuroscience, 14,* 365–376.

Buunk, B. P., & Hoorens, V. (1992). Social support and stress: The role of social comparison and social exchange processes. *British Journal of Clinical Psychology, 31,* 445–457.

Byrne, R. (2006). *The secret.* New York: Atria.

Byrne, R. (2010). *The power.* New York: Atria.

Byrne, R. (2012). *The magic.* New York: Atria.

Campbell, D. (2013, September 16). NHS spells out breast cancer screening risk. *The Guardian.*

Cannon, W. (1932). *Wisdom of the body.* New York: Norton.

Caplan, B. (2001). Rational ignorance versus rational irrationality. *Kyklos, 54*(1), 3–26.

Carothers, B. J., & Reis, H. T. (2013). Men and women are from Earth: Examining the latent structure of gender. *Journal of Personality and Social Psychology, 104,* 385–407.

Carroll, L. (1865). *Alice's Adventures in Wonderland.* London: Macmillan.

Carter, C. S. (2014). Oxytocin pathways and the evolution of human behavior. *Annual Review of Psychology, 65,* 17–39.

Casciani, D. (2014, June 5). *Measures Aimed at Addressing Drop in Rape Convictions Launched.* Retrieved October 1, 2014, from BBC News: http://www.bbc.com/news/uk-27726280

Casellas-Grau, A., Font, A., & Vives, J. (2014). Positive psychology interventions in breast cancer: A systematic review. *Psychooncology, 23*(1), 9–19.

Casscells, W., Schoenberger, A., & Graboys, T. (1978). Interpretation by physicians of clinical laboratory results. *New England Journal of Medicine, 299*(18), 999–1001.

Ceci, S. J., & Williams, W. M. (2011). Understanding current causes of women's underrepresentation in science. *Proceedings of the National Academy of Sciences of the United States of America, 108*(8), 3157–3162.

CFP. (2005). *Código de Ética Profissional do Psicólogo.* Brasília: Conselho Federal de Psicologia.

Chalabi, M. (2013, October 8). Do we spend more time online or watching TV? *The Guardian.*

Chambers, D. W. (1983). Stereotypic images of the scientist: The Draw-A-Scientist Test. *Science Education, 67*(2), 255–265.

Chopra, D. (1989). *Quantum healing: Exploring the frontiers of mind/body medicine.* New York: Bantam.

Chopra, D. (1993). *Ageless body, timeless mind: The quantum alternative to growing old.* New York: Three Rivers Press.

Christenfeld, N., Gerin, W., Linden, W., Sanders, M., Mathur, J., Deich, J. D., et al. (1997). Social support effects on cardiovascular reactivity: Is a stranger as effective as a friend? *Psychosomatic Medicine, 59,* 388–398.

Cohen, J. (1962). The statistical power of abnormal-social psychological research: A review. *Journal of Abnormal and Social Psychology, 65,* 145–153.

Cohen, J. (1992). A power primer. *Psychological Bulletin, 112*(1), 155–159.

Cohen, M. R., & Doner, K. (2006). *The Chinese way to healing: Many paths to wholeness.* Lincoln, NE: Author's Choice.

Cohn, M. A., Fredrickson, B. L., Brown, S. L., Mikels, J. A., & Conway, A. M. (2009). Happiness unpacked: Positive emotions increase life satisfaction by building resilience. *Emotion, 9*(3), 361–368.

Collins, A. F. (1999). The enduring appeal of physiognomy: Physical appearance as a sign of temperament, character, and intelligence. *History of Psychology, 2*(4), 251–276.

Complementary & Natural Healthcare Council. (2014). *Welcome to the Complementary and Natural Healthcare Council (CNHC).* Retrieved May 20, 2014, from http://www.cnhc.org.uk/

Conlisk, J. (2011). Professor Zak's empirical studies on trust and oxytocin. *Journal of Economic Behavior & Organization, 78,* 160–166.

Connellan, J., Baron-Cohen, S., Wheelwright, S., Batki, A., & Ahluwalia, J. (2000). Sex differences in human neonatal social perception. *Infant Behavior & Development, 23,* 113–118.

Coon, D. J. (1992). Testing the limits of sense and science: American experimental pscyhologists combat spiritualism, 1880–1920. *American Psychologist, 47*(2), 143–151.

Costa, P. T., Terracciano, A., & McCrae, R. R. (2001). Gender differences in personality traits across cultures: Robust and surprising findings. *Journal of Personality and Social Psychology, 81*, 322–331.

Coyne, J. C., & Tennen, H. (2010). Positive psychology in cancer care: Bad science, exaggerated claims, and unproven medicine. *Annals of Behavioral Medicine, 39*, 16–26.

Coyne, J. C., Tennen, H., & Ranchor, A. V. (2010). Positive psychology in cancer care: A story line resistant to evidence. *Annals of Behavioral Medicine, 39*(1), 35–42.

Crompton, J. (2013). *Unbelievable! The bizarre world of coincidences.* London: Michael O'Mara.

Cuddy, A. J., Fiske, S. T., Kwan, V. S., Glick, P., Demoulin, S., Leyens, J.-P., et al. (2009). Stereotype content model across cultures: Towards universal similarities and some differences. *British Journal of Social Psychology, 48*, 1–33.

Cunningham, J. A., & Selby, P. L. (2007). Implications of the normative fallacy in young adult smokers aged 19–24 years. *American Journal of Public Health, 97*(8), 1399–1400.

Cuvier, G. (1830). Lectures on the history of the natural sciences, lecture ninth: Theophrastus. *Edinburgh New Philosophical Journal, 18*, 76–83.

Dahl, G. B., & Moretti, E. (2008). The demand for sons. *Review of Economic Studies, 75*, 1085–1120.

Danner, D. D., Snowdon, D. A., & Friesen, W. V. (2001). Positive emotions in early life and longevity: Findings from the nun study. *Journal of Personality and Social Psychology, 80*(5), 804–813.

Davies, J. B. (2004). Bring on the physics revolution. *The Psychologist, 17*(12), 692–693.

Davies, N. (2009). *Flat earth news: An award-winning reporter exposes falsehood, distortion and propaganda in the global media.* London: Vintage.

Davison, H. K., & Burke, M. (2000). Sex discrimination in simulated employment contexts: A meta-analytic investigation. *Journal of Vocational Behavior, 56*, 225–248.

Dawkins, R. (2006). *The god delusion.* London: Bantam.

de Gobineau, A. (1915). *The inequality of human races.* New York: Putnam.

Dean, M. E. (2001). Homeopathy and "the progress of science". *History of Science, 39*, 255–283.

Declerck, C. H., Boone, C., & Kiyonari, T. (2010). Oxytocin and cooperation under conditions of uncertainty: The modulating role of incentives and social information. *Hormones and Behavior, 57*(3), 368–374.

DeLorme, D. E., Huh, J., & Reid, L. N. (2007). "Others are influenced, but not me": Older adults' perceptions of DTC prescription drug advertising effects. *Journal of Aging Studies, 21*, 135–151.

Dennett, D. C. (1991). *Consciousness explained.* London: Penguin.

Derksen, T., & Meijsing, M. (2009). The fabrication of facts: The lure of the incredible coincidence. In H. Kaptein, H. Prakken, & B. Verheij (Eds.), *Legal evidence and proof* (pp. 39–70). Farnham, Surrey: Ashgate.

Derry, G. N. (1999). *What Science Is and How It Works*. Princeton, NJ: Princeton University Press.

Dethlefsen, L., Huse, S., Sogin, M. L., & Relman, D. A. (2008). The pervasive effects of an antibiotic on the human gut microbiota, as revealed by deep 16S rRNA sequencing. *PLOS Biology, 6*(11), e280.

DeWitt, R. (2004). *Worldviews: An introduction to the history and philosophy of science*. Malden, MA: Blackwell.

DGP. (2013). *Was ist die DGPs?* Retrieved October 26, 2013, from Deutsche Gesellschaft für Psychologie: http://www.dgps.de/index.php?id=83

Diener, E., Emmons, R. A., Larson, R. J., & Griffin, S. (1985). The satisfaction with life scale: A measure of life satisfaction. *Journal of Personality Assessment, 49*, 1–5.

'Dogs walked by men are more aggressive.' (2011, November 3). Retrieved November 7, 2011, from Discovery.com: http://news.discovery.com/animals/zoo-animals/dog-walking-behavior-111103.htm

Downs, A. (1957). *An economic theory of democracy*. New York: Harper.

Duarte, J. L., Crawford, J. T., Stern, C., Haidt, J., Jussim, L., & Tetlock, P. E. (2015). Political diversity will improve social psychological science. *Behavioral and Brain Sciences, 38*(e130), 1-13.

Dunbar, R. (2012). *The science of love and betrayal*. London: Faber and Faber.

Eagly, A. H., & Crowley, M. (1986). Gender and helping behavior: A meta-analytic review of the social psychological literature. *Psychological Bulletin, 100*, 283–308.

Eagly, A. H., & Wood, W. (1999). The origins of sex differences in human behavior: Evolved dispositions versus social roles. *American Psychologist, 54*, 408–423.

Earp, B. D., & Trafimow, D. (2015). Replication, falsification, and the crisis of confidence in social psychology. *Frontiers in Psychology, 6*, 621.

Ecklund, E. H., & Scheitle, C. P. (2007). Religion among academic scientists: Distinctions, disciplines, and demographics. *Social Problems, 54*, 289–307.

Ehrenreich, B. (2009). *Smile or die: How positive thinking fooled America and the world*. London: Granta.

Einstein, A. (1936). Physik und Realität. *Journal of The Franklin Institute, 221*(3), 313–347.

Encyclopaedia Britannica. (2013). *Psychology*. Retrieved October 26, 2013, from Britannica Academic Edition: http://www.britannica.com/EBchecked/topic/481700/psychology

Englich, B., Mussweiler, T., & Strack, F. (2006). Playing dice with criminal sentences: The influence of irrelevant anchors on experts' judicial decision making. *Personality and Social Psychology Bulletin, 32*(2), 188–200.

Epstein, S., & O'Brien, E. J. (1985). The person-situation debate in historical and current perspective. *Psychological Bulletin, 98*(3), 513–537.

Ernst, E., & Canter, P. H. (2006). A systematic review of systematic reviews of spinal manipulation. *Journal of the Royal Society of Medicine, 99*, 192–196.

Ernst, E., Lee, M. S., & Choi, T. Y. (2011). Acupuncture: Does it alleviate pain and are there serious risks? A review of reviews. *Pain, 152*, 755–764.

Ernst, E., Pittler, M. H., Stevinson, C., & White, A. R. (2006). Craniosacral therapy. In E. Ernst, M. H. Pittler, & B. Wider (Eds.), *The desktop guide to complementary and alternative medicine: An evidence-based approach* (2nd ed., pp. 317–319). Philadelphia, PA: Mosby Elsevier.

European Commission. (2012). *Special Eurobarometer 393: Discrimination in the EU in 2012*. Retrieved December 20, 2014, from http://ec.europa.eu/public_opinion/archives/ebs/ebs_393_en.pdf

Evans, H. E. (1983). Tales from the outback: The discovery of Aha Ha (Sphecidae, Miscophini). *Sphecos, 7,* 14.

Eysenck, H. J. (1952). The effects of psychotherapy: An evaluation. *Journal of Consulting Psychology, 16,* 319–324.

Fact Stream. (2011, April 21). Of white blood cells and stars. Retrieved January 6, 2015, from http://factstream.blogspot.ie/2011/04/of-white-blood-cells-and-stars.html

Fanelli, D. (2012). Negative results are disappearing from most disciplines and countries. *Scientometrics, 90*(3), 891–904.

Fanelli, D., & Ioannidis, J. P. (2013). US studies may overestimate effect sizes in softer research. *Proceedings of the National Academy of Sciences of the United States of America, 110*(37), 15031–15036.

Feyerabend, P. K. (1975). *Against method: Outline of an anarchistic theory of knowledge.* London: New Left Books.

Fine, C. (2010). *Delusions of gender: The real science behind sex differences.* London: Icon.

Fischer, M., Broeckel, U., Holmer, S., Baessler, A., Hengstenberg, C., Mayer, B., et al. (2005). Distinct heritable patterns of angiographic coronary artery disease in families with myocardial infarction. *Circulation, 111*(7), 855–862.

Fleming, J., & Darley, J. M. (1986). *Perceiving intention in constrained behaviour: The role of purposeful and constrained action cues in correspondence bias effects.* Unpublished manuscript, Princeton, NJ: Princeton University Press.

Forer, B. R. (1949). The fallacy of personal validation: A classroom demonstration of gullibility. *Journal of Abnormal and Social Psychology, 44*(1), 118–123.

Frawley, A. (2015). Happiness research: A review of critiques. *Sociology Compass, 9*(1), 62–77.

Frazier, K. (2003). Are science and religion conflicting or complementary? Some thoughts about boundaries. In P. Kurtz (Ed.), *Science and religion: Are they compatible?* (pp. 25–28). New York: Prometheus.

Freberg, L. (2009). *Discovering biological psychology, international edition.* Belmont, CA: Wadsworth.

Fredrickson, B. L. (2013). Updated thinking on positivity ratios. *American Psychologist, 68*(9), 814–822.

Fredrickson, B. L., & Cohn, M. A. (2008). Open hearts build lives: Positive emotions, induced through loving-kindness meditation, build consequential personal resources. *Journal of Personality and Social Psychology, 95*(5), 1045–1062.

Fredrickson, B. L., & Losada, M. F. (2005). Positive affect and the complex dynamics of human flourishing. *American Psychologist, 60,* 678–686.

Fredrickson, B. L., & Losada, M. F. (2013). Correction to Fredrickson and Losada (2005). *American Psychologist, 68*(9), 822.

French, C. (2012, March 15). *Precognition studies and the curse of the failed replications.* Retrieved on November 21, 2013, from The Guardian: http://www.theguardian.com/science/2012/mar/15/precognition-studies-curse-failed-replications

Fu, T. S.-T., Koutstaal, W., Poon, L., & Cleare, A. J. (2012). Confidence judgment in depression and dysphoria: The depressive realism vs. negativity hypotheses. *Journal of Behavior Therapy and Experimental Psychiatry, 43*, 699–704.

Fuchs, A. H. (2000). Contributions of American mental philosophers to psychology in the United States. *History of Psychology, 3*(1), 3–19.

Furnham, A. (2004). Belief in a just world: research progress over the past decade. *Personality and Individual Differences, 34*, 795–817.

Gallup. (2014, May 28). *Conservative lead on social and economic ideology shrinking.* Retrieved November 10, 2014, from http://www.gallup.com/poll/170741/conservative-lead-social-economic-ideology-shrinking.aspx

Gallup International. (2014, December 30). *WIN/Gallup International's annual global end of year survey shows that happiness is on the rise.* Retrieved January 19, 2015, from http://www.wingia.com/en/services/end_of_year_survey_2014/global_results/8/45/

Galton, F. (1883). *Inquiries into human faculty and its development.* London: Macmillan.

Galton, F. (1962). *Hereditary genius: An inquiry into its laws and consequences.* London: Fontana [Original publication 1869].

Ganske, K. H., & Hebl, M. R. (2001). Once upon a time there was a math contest: Gender stereotyping and memory. *Teaching of Psychology, 28*(4), 266–268.

Garver-Apgar, C. E., Gangestad, S. W., Thornhill, R., Miller, R. D., & Olp, J. J. (2006). Major histocompatibility complex alleles, sexual responsivity, and unfaithfulness in romantic couples. *Psychological Science, 17*(10), 830–835.

Gazzaniga, M. S. (1967). The split brain in man. *Scientific American, 217*, 24–29.

Geary, D. C., & Flinn, M. V. (2002). Sex differences in behavioral and hormonal response to social threat: Comment on Taylor et al. (2000). *Psychological Review, 109*(4), 745–750.

Gelman, A., & Weakliem, D. (2009). Of beauty, sex, and power: Statistical challenges in estimating small effects. *American Scientist, 97*, 310–316.

Gergen, K. J. (2001). Psychological science in a postmodern context. *American Psychologist, 56*, 803–813.

Gergen, K. J. (2008). On the very idea of social psychology. *Social Psychology Quarterly, 71*(4), 331–337.

Gettier, E. L. (1963). Is justified true belief knowledge? *Analysis, 23*, 121–123.

Gigerenzer, G., & Gaissmaier, W. (2011). Heuristic decision making. *Annual Review of Psychology, 62*, 451–482.

Gilbert, A. (2008). *What the nose knows: The science of scent in everyday life.* New York: Crown Publishers.

Giles, K. (1994). The Biblical argument for slavery: Can the Bible mislead? A case study in hermeneutics. *Evangelical Quarterly, 66*(1), 3–17.

Gill, L. (1987, July 23). Health: Putting fear to flight – Fear of flying. *The Times* (62828).

Gill, R. D., Groeneboom, P., & de Jong, P. (2010). *Elementary statistics on trial (the case of Lucia de Berk).* Leiden: Mathematical Institute, Leiden University.

Gilmour, R. (2008). Raymond Domenech looks to the stars. *The Telegraph.* Retrieved July 28, 2014, from http://www.telegraph.co.uk/sport/football/teams/england/2295526/Raymond-Domenech-looks-to-the-stars.html.

Gilovich, T. (1990). Differential construal and the false-consensus effect. *Journal of Personality and Social Psychology, 59*(4), 623–634.

Gilovich, T. (1991). *How we know what isn't so: The fallibility of human reason in everyday life.* New York: Free Press.

Global Industry Analysts. (2012). *Global alternative medicine industry.* San Jose, CA: Global Industry Analysts, Inc.

Goertzen, J. R. (2008). On the possibility of unification: The reality and nature of the crisis in psychology. *Theory & Psychology, 18*(6), 829–852.

Goldenberg, J. L., Pyszczynski, T., Greenberg, J., & Solomon, S. (2000). Fleeing the body: A terror management perspective on the problem of human corporeality. *Personality and Social Psychology Review, 4*(3), 200–218.

Goodman, S., & Greenland, S. (2007). *Assessing the unreliability of the medical literature: A response to "Why most published research findings are false".* Department of Biostatistics. Baltimore, MD: Johns Hopkins University Press.

Gøtzsche, P. C., & Jørgensen, K. J. (2013). Screening for breast cancer with mammography (Article no. CD001877). *Cochrane Database of Systematic Reviews (Issue 6).*

Gould, S. J. (1978). Morton's ranking of races by cranial capacity. *Science, 200,* 503–509.

Gould, S. J. (1996). *The mismeasure of man.* London: Penguin.

Gould, S. J. (1999). *Rocks of ages: Science and religion in the fullness of life.* New York: Ballantine.

Graham, J., Haidt, J., & Nosek, B. A. (2009). Liberals and conservatives rely on different sets of moral foundations. *Journal of Personality and Social Psychology, 96*(5), 1029–1046.

Gramer, M., & Reitbauer, C. (2010). The influence of social support on cardiovascular responses during stressor anticipation and active coping. *Biological Psychology, 85,* 268–274.

Greenaway, K. H., Louis, W. R., Hornsey, M. J., & Jones, J. M. (2014). Perceived control qualifies the effects of threat on prejudice. *British Journal of Social Psychology, 53,* 422–443.

Greenwald, A. G., McGhee, D. E., & Schwartz, J. L. (1998). Measuring individual differences in implicit cognition: The implicit association test. *Journal of Personality and Social Psychology, 74*(6), 1464–1480.

Grether, N. (2014, June 6). *Men's rights activist: Feminists have used rape 'as a scam'.* Retrieved October 18, 2014, from Al Jazeera America: http://america.aljazeera.com/watch/shows/america-tonight/articles/2014/6/6/mena-s-rights-activistfeministshaveusedrapeaasascama.html

Grinin, L. (2010). The role of an individual in history: A reconsideration. *Social Evolution & History, 9*(2), 116–117.

Groombridge, B., & Jenkins, M. D. (2002). *World atlas of biodiversity: Earth's living resources in the 21st century.* Oakland, CA: University of California Press.

Gross, P. R., & Levitt, N. (1994). *Higher superstition: The academic left and its quarrels with science.* Baltimore, MD: Johns Hopkins University Press.

Gross, R. (2009). *Themes, issues and debates in psychology* (3rd ed.). Abingdon: Hodder.

Guinot, B., & Schneder, B. (2011, August 18). *Notre fédération est partie intégrante.* Retrieved October 26, 2013, from Collège des Psychologues Cliniciens Spécialisés en Neuropsychologie: http://www.cpcn.fr/attachments/article/626/27._FFPP_CNFPS_adherents_18_08_11.pdf

Gustafsson, E., Levréro, F., Reby, D., & Mathevon, N. (2013). Fathers are just as good as mothers at recognizing the cries of their baby. *Nature Communications, 4,* 1698.

Guthrie, R. V. (2004). *Even the rat was white: A historical view of psychology.* Boston, MA: Pearson.

Haapala, E. A., Poikkeus, A.-M., Kukkonen-Harjula, K., Tompuri, T., Lintu, N., Väistö, J., et al. (2014). Associations of physical activity and sedentary behavior with academic skills: A follow-up study among primary school children. *PLoS ONE, 9*(9), e107031.

Halpern, D. (1993). Minorities and mental health. *Social Science & Medicine, 36*(5), 597–607.

Halpern, D. F. (2007). Choosing the sex of one's child. In J. Brockman (Ed.), *What is your dangerous idea?* (pp. 97–100). New York: HarperCollins.

Ham, K. (2012). *The true account of Adam and Eve.* Green Forest, AR: Master Books.

Haraway, D. (1989). *Primate visions: Gender, race, and nature in the world of modern science.* New York: Routledge.

Harding, S. (1986). *The science question in feminism.* Ithaca, NY: Cornell University Press.

Harris, S. (2011). *The moral landscape: How science can determine human values.* London: Bantam.

Harris, S., Kaplan, J. T., Curiel, A., Bookheimer, S. Y., Iacoboni, M., & Cohen, M. S. (2009). The neural correlates of religious and nonreligious belief. *PLoS ONE, 4*(10), e0007272.

Hart, W., Albarracín, D., Eagly, A. H., Brechan, I., Lindberg, M. J., & Merrill, L. (2009). Feeling validated versus being correct: A meta-analysis of selective exposure to information. *Psychological Bulletin, 135*(4), 555–588.

Haslam, S. A., O'Brien, A., Jetten, J., Vormedal, K., & Penna, S. (2005). Taking the strain: Social identity, social support, and the experience of stress. *British Journal of Social Psychology, 44,* 355–370.

Haub, C. (2002). How many people have ever lived on Earth? *Population Today, 30*(8), 3–4.

Hawking, S. (1988). *A brief history of time: From the big bang to black holes.* New York: Bantam Dell.

Hay, L. L. (1984). *You can heal your life.* Carlsbad, CA: Hay House.

Haynes, R. (2003). From alchemy to artificial intelligence: Stereotypes of the scientist in Western literature. *Public Understanding of Science, 12,* 243–253.

Hehman, E., Leitner, J. B., Deegan, M. P., & Gaertner, S. L. (2013). Facial structure is indicative of explicit support for prejudicial beliefs. *Psychological Science, 24*(3), 289–296.

Heisenberg, W. (1927). Über den anschaulichen Inhalt der quantentheoretischen Kinematik und Mechanik. *Zeitschrift für Physik, 43*(3–4), 172–198.

Hickman, L. (2011, June 8). Do the police use psychics to help them? *The Guardian.*

Higher Education Statistics Agency. (2014). *2012/13 Students by Subject*. Retrieved on March 31, 2015, from http://www.hesa.ac.uk/dox/dataTables/studentsAndQualifiers/download/Subject1213.xlsx

Hill, R. (2004). Multiple sudden infant deaths: Coincidence or beyond coincidence. *Pediatric and Perinatal Epidemiology, 18*, 320–326.

Hines, T. (2003). *Pseudoscience and the paranormal* (2nd ed.). Amherst, NY: Prometheus.

Hofmann, S. G., Asnaani, A., Vonk, I. J., Sawyer, A. T., & Fang, A. (2012). The efficacy of cognitive behavioral therapy: A review of meta-analyses. *Cognitive Therapy and Research, 36*(5), 427–440.

Hough, A. (2010, May 14). Frano Selak: 'World's luckiest man gives away his lottery fortune'. *The Telegraph*.

House, J. (1981). *Work stress and social support*. Reading, MA: Addison-Wesley.

Howe, K. R. (1988). Against the quantitative-qualitative incompatibility thesis: Or dogmas die hard. *Educational Researcher, 17*(8), 10–16.

Hughes, B. M. (2006). Regional patterns of religious affiliation and availability of complementary and alternative medicine. *Journal of Religion and Health, 45*(4), 549–557.

Hughes, B. M. (2007). Self-esteem, performance feedback, and cardiovascular stress reactivity. *Anxiety, Stress, & Coping, 20*(3), 239–252.

Hughes, B. M. (2008a). Evidence-based helping: Dispositional, situational, and temporal parameters of social support. In P. Buchwald, T. Ringeisen, & M. W. Eysenck (Eds.), *Stress and anxiety: Application to lifespan development and health promotion* (pp. 121–132). Berlin: Logos Verlag.

Hughes, B. M. (2008b). How should clinical psychologists approach complementary and alternative medicine? Empirical, epistemological, and ethical considerations. *Clinical Psychology Review, 28*, 657–675.

Hughes, B. M. (2011, November 7). *Who let the pseudoscientists out?* Retrieved November 7, 2011, from The Science Bit: http://thesciencebit.net/2011/11/07/who-let-the-pseudoscientists-out/

Hughes, B. M. (2012). *Conceptual and historical issues in psychology*. Harlow: Prentice Hall.

Hughes, B. M. (2013). Blood pressure reactivity or responses. In M. D. Gellman, & J. R. Turner (Eds.), *Encyclopedia of behavioral medicine* (Vol. II, pp. 235–239). New York: Springer.

Hughes, B. M. (2015). Complementary and alternative therapies for psychological problems. In R. Cautin, & S. O. Lilienfeld (Eds.), *Encyclopedia of clinical psychology*. Hoboken, NJ: Wiley-Blackwell.

Hughes, B. M., & Creaven, A.-M. (2009). Achieving greater theoretical sophistication in research on socially supportive interactions and health. In A. T. Heatherton, & V. A. Walcott (Eds.), *Handbook of social interactions in the 21st century* (pp. 125–135). New York: Nova Science.

Hwang, Y. (2010). Selective exposure and selective perception of anti-tobacco campaign messages: The impacts of campaign exposure on selective perception. *Health Communication, 25*(2), 182–190.

Hyde, J. S. (2005). The gender similarities hypothesis. *American Psychologist, 60*(6), 581–592.

Hyde, J. S. (2014). Gender similarities and differences. *Annual Review of Psychology, 65*, 373–398.

Hyde, J. S., Fennema, E., & Lamon, S. (1990). Gender differences in mathematics performance: A meta-analysis. *Psychological Bulletin, 107*, 139–155.

Hyde, J. S., Lindberg, S. M., Linn, M. C., Ellis, A., & Williams, C. (2008). Gender similarities characterize math performance. *Science, 321*, 494–495.

Inbar, Y., & Lammers, J. (2012). Political diversity in social and personality psychology. *Perspectives on Psychological Science, 7*(5), 496–503.

Innes, E. (2013, April 17). Is maternal instinct a myth? Mothers and fathers are EQUALLY good at recognising their baby's cry. *Daily Mail.*

Innoplexus. (2013). Face reading app. Retrieved from https://play.google.com/store/apps/details?id=com.innoplexus.facereading

Institute of Psychology. (2012, September 17). *Brief Introduction.* Retrieved October 26, 2013, from Institute of Psychology, Chinese Academy of Sciences: http://english.psych.cas.cn/au/

Ioannidis, J. P. (2005). Why most published research findings are false. *PLoS Medicine, 2*(8), e124.

Ipsos. (2011, September 12). *Canadians Split On Whether Religion Does More Harm in the World than Good.* Retrieved December 20, 2014, from ipsos-na.com: http://www.ipsos-na.com/news-polls/pressrelease.aspx?id=5328

Ipsos MORI. (2013). *Perceptions are not reality: The top 10 we get wrong.* Retrieved on July 13, 2013, from http://www.ipsos-mori.com/researchpublications/research archive/3188/Perceptions-are-not-reality-the-top-10-we-get-wrong.aspx

Irigaray, L. (1982). Le sujet de la science est-il sexué? [Is the subject of science sexual?]. *Les Tempes Modernes, 39*, 960–974.

Isaacson, W. (2011). *Steve Jobs.* New York: Simon & Schuster.

Jaccard, J., McDonald, R., Wan, C. K., Dittus, P. J., & Quinlan, S. (2002). The accuracy of self-reports of condom use and sexual behavior. *Journal of Applied Social Psychology, 32*, 1863–1905.

James, W. (1986). *Essays in psychical research (The works of William James).* Cambridge, MA: Harvard University Press.

Jastrow, J. (1889). The problems of "psychic research". *Harper's New Monthly Magazine, 79*, 76–82.

Jenkinson, J. (1997). Face facts: A history of physiognomy from ancient Mesopotamia to the end of the 19th century. *Journal of Biocommunication, 24*(3), 2–7.

Jessop, V. (1997). *Titanic survivor: The memoirs of Violet Jessop, stewardess.* New York: Sheridan House.

Jin, P. (1992). Efficacy of Tai Chi, brisk walking, meditation, and reading in reducing mental and emotional stress. *Journal of Psychosomatic Research, 36*(4), 361–370.

Jinha, A. E. (2010). Article 50 million: An estimate of the number of scholarly articles in existence. *Learned Publishing, 23*(3), 258–263.

Johnson, V. E. (2013). Revised standards for statistical evidence. *PNAS: Proceedings of the National Academy of Sciences, 110*(48), 19313-19317.

Johnson-Laird, P. N., Legrenzi, P., & Sonino Legrenzi, M. (1972). Reasoning and a sense of reality. *British Journal of Psychology, 63*(3), 395–400.

Jussim, L. (2012). Liberal privilege in academic psychology and the social sciences: Commentary on Inbar & Lammers (2012). *Perspectives on Psychological Science, 7*(5), 504–507.

Kahneman, D. (2012). *Thinking, fast and slow.* London: Penguin.

Kalauokalani, D., Sherman, K. J., & Cherkin, D. C. (2001). Acupuncture for chronic low back pain: Diagnosis and treatment patterns among acupuncturists evaluating the same patient. *Southern Medical Journal, 94,* 486–492.

Keller, E. F. (1983). The force of the pacemaker concept in theories of aggregation in cellular slime mold. *Perspectives in Biology and Medicine, 26,* 515–521.

Keller, E. F. (1985). *Reflections on gender and science.* New Haven, CT: Yale University Press.

Killen, M., & Smetana, J. G. (Eds.). (2006). *Handbook of moral development.* Mahwah, NJ: Lawrence Erlbaum Associates.

Kosfeld, M., Heinrichs, M., Zak, P. J., Fischbacher, U., & Fehr, E. (2005). Oxytocin increases trust in humans. *Nature,* 673–676.

Koyré, A. (1939). *Études Galiléennes.* Paris: Hermann.

Kruger, J., & Dunning, D. (1999). Unskilled and unaware of it: How difficulties in recognizing one's own incompetence lead to inflated self-assessments. *Journal of Personality and Social Psychology, 77*(6), 1121–1134.

Lambert, T. A., Kahn, A. S., & Apple, K. J. (2003). Pluralistic ignorance and hooking up. *Journal of Sex Research, 40*(2), 129–133.

Landsburg, S. E. (2003, October 2). *Oh, no: It's a girl!* Retrieved October 18, 2014, from Slate: http://www.slate.com/articles/arts/everyday_economics/2003/10/oh_no_its_a_girl.html

Lang, R., O'Reilly, M., Healy, O., Rispoli, M., Lydon, H., Streusand, W., et al. (2012). Sensory integration therapy for autism spectrum disorders: A systematic review. *Research in Autism Spectrum Disorders, 6*(3), 1004–1018.

Lannin, P., & Ek, V. (2011, October 5). *Ridiculed crystal work wins Nobel for Israeli.* Retrieved February 1, 2013, from http://www.reuters.com/article/2011/10/05/nobel-chemistry-idUSL5E7L51U620111005

Laughland, O. (2012, April 7). The insider's guide to cancer prevention. *The Guardian.*

Lavater, J. C. (1840). *Essays on physiognomy: Designed to promote the knowledge and love of mankind* (3rd ed.). London: Thomas Tegg [Original publication 1775–1778].

Lee, J. J. (2013, May 19). *Six women scientists who were snubbed due to sexism.* Retrieved October 20, 2014, from National Geographic: http://news.national geographic.com/news/2013/13/130519-women-scientists-overlooked-dna-history-science/

Lee, M. (2011, June 22). *Cognitive distortions with Eeyore.* Retrieved January 19, 2015, from HyphenMagazine.com: http://www.hyphenmagazine.com/blog/archive/2011/06/cognitive-distortions-eeyore

Leeb, R. T., & Rejskind, F. G. (2004). Here's looking at you, kid! A longitudinal study of perceived gender differences in mutual gaze behavior in young infants. *Sex Roles, 50*(1/2), 1–14.

Leggett, N. C., Thomas, N. A., Loetscher, T., & Nicholls, M. E. (2013). The life of p: 'Just significant' results are on the rise. *Quarterly Journal of Experimental Psychology, 66*(12), 2303–2309.

Lemay Jr, E. P., Clark, M. S., & Greenberg, A. (2010). What is beautiful is good because what is beautiful is desired: Physical attractiveness stereotyping as projection of interpersonal goals. *Personality and Social Psychology Bulletin, 36*(3), 339–353.

Lennard, N. (2012, September 24). Thousands of Wall Street traders follow astrology. *Salon.*

Leydesdorff, L., & Rafols, I. (2009). A global map of science based on the ISI subject categories. *Journal of the American Society for Information Science and Technology, 60,* 348–362.

Liberman, M. (2006, August 6). *Sex-linked lexical budgets.* Retrieved August 16, 2014, from Language Log: http://itre.cis.upenn.edu/~myl/languagelog/archives/003420.html

Lin, F., Zhou, Y., Du, Y., Qin, L., Zhao, Z., Xu, J., et al. (2012). Abnormal white matter integrity in adolescents with Internet Addiction Disorder: A tract-based spatial statistics study. *PLoS ONE, 7*(1), e30253.

Lindberg, S. M., Hyde, J. S., Petersen, J., & Linn, M. C. (2010). New trends in gender and mathematics performance: A meta-analysis. *Psychological Bulletin, 136,* 1123–1135.

Loftus, E. F., & Palmer, J. C. (1974). Reconstruction of automobile destruction: An example of the interaction between language and memory. *Journal of Verbal Learning and Verbal Behavior, 13,* 585–589.

Loma Linda University. (2015). *Loma Linda University mission statement.* Retrieved January 10, 2015, from http://www.llu.edu/central/mission.page?

Longino, H. E. (1987). Can there be a feminist science? *Hypatia, 2*(3), 51–64.

Longino, H. E. (1990). *Science as social knowledge: Values and objectivity in scientific inquiry.* Princeton, NJ: Princeton University Press.

Longino, H. E. (2004). How values can be good for science. In P. Machamer, & G. Walters (Eds.), *Science, values, and objectivity* (pp. 127–142). Pittsburgh, PA: University of Pittsburgh Press.

Lorenz, K. (1952). *King Solomon's ring: New light on animal ways.* London: Methuen.

Lorenzi-Cioldi, F., Chatard, A., Marques, J. M., Selimbegovic, L., Konan, P., & Faniko, K. (2011). What do drawings reveal about people's attitudes toward countries and their citizens? *Social Psychology, 42*(3), 231–240.

Luborsky, L., Singer, B., & Luborsky, L. (1975). Comparative studies of psychotherapies: Is it true that "everyone has won and all must have prizes"? *Archives of General Psychiatry, 32,* 995–1008.

Mackay, A. L. (1991). *A dictionary of scientific quotations.* London: Institute of Physics Publishing.

Madueme, H., & Reeves, M. (Eds.). (2014). *Adam, the fall, and original sin: Theological, biblical, and scientific perspectives.* Grand Rapids, MI: Baker Academic.

Maeda, Y., & Yoon, S. Y. (2013). A meta-analysis on gender differences in mental rotation ability measued by the Purdue Spatial Visualization Tests: Visualization of rotations (PSVT:R). *Educational Psychology Review, 25,* 69–94.

Makel, M. C., Plucker, J. A., & Hegarty, B. (2012). Replications in psychology research: How often do they really occur? *Perspectives on Psychological Science, 7*(6), 537–542.

Marcus, A., & Oransky, I. (2013). Retrieved from Retraction Watch: http://retractionwatch.com/

Marcus, D. K., O'Connell, D., Norris, A. L., & Sawaqdeh, A. (2014). Is the Dodo bird endangered in the 21st century? A meta-analysis of treatment comparison studies. *Clinical Psychology Review, 34*, 519–530.

Martin, D. J., Garske, J. P., & Davis, M. K. (2000). Relation of the therapeutic alliance with outcome and other variables: A meta-analytic review. *Journal of Consulting and Clinical Psychology, 68*, 438–450.

Masalu, J. R., & Astrom, A. N. (2001). Predicting intended and self-perceived sugar restriction among Tanzanian students using the theory of planned behaviour. *Journal of Health Psychology, 6*, 435–445.

Maslow, A. H. (1954). *Motivation and personality.* New York: Harper & Row.

Masters, K. S., Spielmans, G. I., & Goodson, J. T. (2006). Are there demonstrable effects of distant intercessory prayer? A meta-analytic review. *Annals of Behavioral Medicine, 32*(1), 21–26.

Matchar, E. (2014, February 26). *"Men's rights" activists are trying to redefine the meaning of rape.* Retrieved October 18, 2014, from New Republic: http://www.newrepublic.com/article/116768/latest-target-mens-rights-movement-definition-rape

May, P. W. (2008). *Molecules with silly or unusual names.* London: Imperial College Press.

McCabe, D. P., & Castel, A. D. (2008). Seeing is believing: The effect of brain images on judgments of scientific reasoning. *Cognition, 107*, 343–352.

McClean, S. (2013). The role of performance in enhancing the effectiveness of crystal and spiritual healing. *Medical Anthropology, 32*, 61–74.

McDonald, H. (2011, September 23). Spontaneous combustion killed Irish pensioner, inquest rules. *The Guardian.* Retrieved on July 13, 2013, from http://www.guardian.co.uk/world/2011/sep/23/irish-pensioner-killed-spontaneous-combustion

McDougall, W. (1923). *Outline of psychology.* New York: Scribner's.

McGrath, A., & Collicut McGrath, J. (2007). *The Dawkins delusion? Atheist fundamentalism and the denial of the divine.* London: Society for Promoting Christian Knowledge.

McLemore, J., & Hallengren, A. L. (2010). X-ray appearance of subcutaneous gemstones as part of alternative/holistic medicine: A case report and review of the literature. *Clinical Imaging, 34*, 316–318.

McNulty, J. K., & Fincham, F. D. (2012). Beyond positive psychology? Toward a contextual view of psychological processes and well-being. *American Psychologist, 67*(2), 101–110.

Mendrick, H., & Francis, B. (2012). Boffin and geek identities: Abject or privileged? *Gender and Education, 24*(1), 15–24.

Merriam-Webster Online. (2006). Word of the year 2006. Retrieved June 20, 2013, from http://www.merriam-webster.com/info/06words.htm

Mezulis, A. H., Abramson, L. Y., Hyde, J. S., & Hankin, B. L. (2004). Is there a universal positivity bias in attributions? A meta-analytic review of individual, developmental, and cultural differences in the self-serving attributional bias. *Psychological Bulletin, 130*(5), 711–747.

Miller, A. (2008). A critique of positive psychology: Or 'the new science of happiness'. *Journal of Philosophy of Education, 42*(3–4), 591–608.

Miller, D. T., & McFarland, C. (1987). Pluralistic ignorance: When similarity is interpreted as dissimilarity. *Journal of Personality and Social Psychology, 53,* 298–305.

Milne, A. A. (1926). *Winnie-The-Pooh.* London: Methuen.

Milton, J., & Wiseman, R. (1999). Does psi exist? Lack of replication of an anomalous process of information transfer. *Psychological Bulletin, 125*(4), 387–391.

Minsky, M. (1991). Conscious machines. In *Machinery of Consciousness,* Proceedings of the National Research Council of Canada, 75th Anniversary Symposium on Science in Society, Ottawa, June 1991.

Missinne, S., & Bracke, P. (2012). Depressive symptoms among immigrants and ethnic minorities: Population based study in 23 European countries. *Social Psychiatry and Psychiatric Epidemiology, 47*(1), 97–109.

Möbius, J. S. (1901). The physiological mental weakness of woman. *Alienist and Neurologist, 22,* 624–642.

Moore, S. J. (1914). The articulation of the concepts of normal and abnormal psychology. *American Journal of Psychology, 25*(2), 283–287.

Mora, C., Tittensor, D. P., Adl, S., Simpson, A. G., & Worm, B. (2011). How many species are there on Earth and in the ocean? *PLoS Biology, 9*(8), e1001127.

Morgan, M. (1998). Qualitative research: Science or pseudo-science? *The Psychologist, 11*(10), 481–483.

Moss-Racusin, C. A., Dovidio, J. F., Brescoll, V. L., Graham, M. J., & Handelsman, J. (2012). Science faculty's subtle gender biases favor male students. *PNAS: Proceedings of the National Academy of Sciences, 109,* 41.

Motzkin, J. C., Baskin-Sommers, A., Newman, J. P., Kiehl, K. A., & Koenigs, M. (2014). Neural correlates of substance abuse: Reduced functional connectivity between areas underlying reward and cognitive control. *Human Brain Mapping, 35*(9), 4282–4292.

Myers, P. Z. (2014, August 31). *Homosexuality and evolution.* Retrieved August 2014, from ScienceBlogs: Pharyngula: http://scienceblogs.com/pharyngula/2014/08/31/homosexuality-and-evolution/

Nakagaki, T. (2001). Smart behavior of true slime mold in a labyrinth. *Research in Microbiology, 152*(9), 767–770.

NAOP. (2013). *Objectives of NAOP, India.* Retrieved October 26, 2013, from National Academy of Psychology India: http://www.naopindia.org/info/objectives

NASA. (2013). *Dark energy, dark matter.* Retrieved November 11, 2013, from http://science.nasa.gov/astrophysics/focus-areas/what-is-dark-energy/

Nash, A., & Grossi, G. (2007). Picking Barbie's brain: Inherent sex differences in scientific ability? *Journal of Interdisciplinary Feminist Thought, 2*(1), 5.

National Board for Certification in Occupational Therapy. (2004). A practice analysis study for entry-level occupational therapist registered and certified occupational therapy assistant practice. *Occupational Therapy Journal of Research, 24,* 1–31.

National Center for Complementary and Alternative Medicine. (2012). *What is complementary and alternative medicine? (NCCAM Publication No. D347).* Washington, DC: US Government Printing Office.

National Center for Complementary and Intergrative Health. (2015). *Complementary, Alternative, or Integrative Health: What's In a Name?* Retrieved March 30, 2015, from https://nccih.nih.gov/health/whatiscam

National Transportation Safety Board. (2014). *Review of accident data: 2012 aviation statistics.* Retrieved March 24, 2014, from http://www.ntsb.gov/data/aviation_stats.html

Nickerson, R. S. (1988). Confirmation bias: A ubiquitous phenomenon in many guises. *Review of General Psychology, 2*(2), 175–220.

Nieuwenhuis, S., Forstmann, B. U., & Wagenmakers, E.-J. (2011). Erroneous analyses of interactions in neuroscience: A problem of significance. *Nature Neuroscience, 14*(9), 1105–1107.

O'Keeffe, C., & Wiseman, R. (2005). Testing alleged mediumship: Methods and results. *British Journal of Psychology, 96*, 165–179.

Offit, P. A. (2013). *Do you believe in magic? The sense and nonsense of alternative medicine.* New York: HarperCollins.

Ogden, J. (2003). Some problems with social cognition models: A pragmatic and conceptual analysis. *Health Psychology, 22*(4), 424–428.

Orehek, E., Sasota, J. A., Kruglanski, A. W., Dechesne, M., & Ridgeway, L. (2014). Interdependent self-construals mitigate the fear of death and augment the willingness to become a martyr. *Journal of Personality and Social Psychology, 107*(2), 265–275.

Oxford Dictionaries. (2012). *Definition of pneumonoultramicroscopicsilicovolcano-coniosis.* Retrieved November 11, 2012, from Oxford Dictionaries: http://oxforddictionaries.com/definition/english/pneumonoultramicroscopicsilico volcanoconiosis

Oxford English Dictionary. (2013). *Psychology.* Retrieved October 26, 2013, from Oxford Dictionaries Pro: http://english.oxforddictionaries.com/definition/psychology

Paley, J. (2005). Error and objectivity: Cognitive illusions and qualitative research. *Nursing Philosophy, 6*, 196–209.

Panico, R., Richer, J., & Powell, W. H. (1993). *A guide to IUPAC nomenclature of organic compounds.* Hoboken, NJ: Wiley-Blackwell.

Pashler, H., & Harris, C. R. (2012). Is the replicability crisis overblown? Three arguments examined. *Perspectives on Psychological Research, 7*(6), 531–536.

Penrose, R. (1997). *The large, the small and the human mind.* Cambridge: Cambridge University Press.

Perlman, L. M. (2001). Nonspecific, unintended, and serendipitous effects in psychotherapy. *Professional Psychology: Research and Practice, 32*(3), 283–288.

Peschek-Bohmer, F., & Schreiber, G. (2004). *Healing crystals and gemstones: From amethyst to zircon.* Old Saybrook, CT: Konecky & Konecky.

Peterson, C., & Seligman, M. E. (2004). *Character strengths and virtues: A handbook and classification.* Oxford: Oxford University Press.

Peterson, G. R. (2003). Demarcation and the scientistic fallacy. *Zygon: Journal of Religion and Science, 38*, 751–761.

Pew Research Center. (2014). *Worldwide, many see belief in god as essential to morality: Richer nations are exception.* Washington, DC: Pew Research Center.

Phillips, A. C., Ginty, A. T., & Hughes, B. M. (2013). The other side of the coin: Blunted cardiovascular and cortisol reactivity are associated with negative health outcomes. *International Journal of Psychophysiology, 90,* 1–7.

Piccirilli, G., Lazzarotto, T., Chiereghin, A., Serra, L., Gabrielli, L., & Lanari, M. (2015). Spotlight on measles in Italy: Why outbreaks of a vaccine-preventable infection continue in the 21st century. *Expert Review of Anti-infective Therapy, 13*(3), 355–362.

Pielke Jr, R. A. (2007). *The honest broker: Making sense of science in policy and politics.* Cambridge: Cambridge University Press.

Pinker, S. (1997). *How the mind works.* New York: Norton.

Pinker, S. (2011). *The better angels of our nature: The decline of violence in history and its causes.* London: Allen Lane.

Pirsig, R. M. (1974). *Zen and the art of motorcycle maintenance: An inquiry into values.* New York: William Morrow & Company.

Plain English Campaign. (2012, July 24). *Before and after.* Retrieved November 11, 2012, from http://www.plainenglish.co.uk/examples/before-and-after.html

Popper, K. (1934). *Logik der Forschung: Zur Erkenntnistheorie der Modernen Naturwissenschaft.* Vienna: Verlag Von Julius Springer.

Porter, E. H. (1913). *Pollyanna.* Boston, MA: L C Page.

Porto, M. D., & Romano, M. (2013). Newspaper metaphors: Reusing metaphors across media genres. *Metaphor and Symbol, 28,* 60–73.

Pronin, E., Gilovich, T., & Ross, L. (2004). Objectivity in the eye of the beholder: Divergent perceptions of bias in self versus others. *Psychological Review, 111*(3), 781–799.

PsySSA. (2013). Retrieved October 26, 2013, from Psychological Society of South Africa: http://www.psyssa.com/

Putnam, H. (1995). Review of 'Shadows of the Mind'. *Bulletin of the American Mathematical Society, 32,* 370–373.

Quine, W. V. (1951). Two dogmas of empiricism. *Philosophical Review, 60*(1), 20–43.

Ramachandran, V. S., & Blakeslee, S. (1998). *Phantoms in the brain: Probing the mysteries of the human mind.* New York: William Morrow.

Rapaport, M. H., Nierenberg, A. A., Howland, R., Dording, C., Schettler, P. J., & Mischoulon, D. (2011). The treatment of minor depression with St. John's wort or citalopram: Failure to show benefit over placebo. *Journal of Psychiatric Research, 45,* 931–941.

Raven, H. (2005). *Heal yourself with crystals: Crystal medicine for body, emotions and spirit.* London: Godsfield Press.

Redding, R. E. (2001). Sociopolitical diversity in psychology: The case for pluralism. *American Psychologist, 56*(3), 205–215.

Redding, R. E. (2012). Likes attract: The sociopolitical groupthink of (social) psychologists. *Perspectives on Psychological Science, 7*(5), 512–515.

Redding, R. E. (2013). Politicized science. *Society, 50,* 439–446.

Rehg, W. (2009). *Cogent science in context: The science wars, argumentation theory, and Habermas.* Cambridge, MA: MIT Press.

Reid, L. D. (2011). Students' estimates of others' mental health demonstrate a cognitive bias. *Psychology, 2*(5), 433–439.

Rennels, J. L., & Langlois, J. H. (2014). Children's attractiveness, gender, and race biases: A comparison of their strength and generality. *Child Development, 85*(4), 1401–1418.

Reuben, J. A. (1996). *The making of the modern university: Intellectual transformation and the marginalization of morality.* Chicago: University of Chicago Press.

Reyna, V. F., & Lloyd, F. J. (2006). Physician decision making and cardiac risk: Effects of knowledge, risk perception, risk tolerance, and fuzzy processing. *Journal of Experimental Psychology: Applied, 12*(3), 179–195.

Řezáč, P., Viziová, P., Dobešová, M., Havlíček, Z., & Pospíšilová, D. (2011). Factors affecting dog–dog interactions on walks with their owners. *Applied Animal Behaviour Science, 134*(3–4), 170–176.

Rice, C. (2012, June 29). *Science: It's a girl thing! A viral fiasco.* Retrieved January 13, 2014, from The Guardian: http://www.theguardian.com/science/blog/2012/jun/29/science-girl-thing-viral-fiasco

Richards, G. (1997). *'Race', racism and psychology: Towards a reflexive history.* London: Routledge.

Riley, S., Frith, H., Archer, L., & Veseley, L. (2006). Institutional sexism in academia. *The Psychologist, 19*(2), 94–97.

Risman, B. J. (1987). Intimate relationships from a microstructural perspective: Men who mother. *Gender & Society, 1*(1), 6–32.

Roese, N. J., & Vohs, K. D. (2012). Hindsight bias. *Perspectives on Psychological Science, 7*(5), 411–426.

Roll, W. G., & Williams, B. J. (2010). Quantum theory, neurobiology, and parapsychology. In S. Krippner, & H. L. Friedman (Eds.), *Mysterious minds: The neurobiology of psychics, mediums, and other extraordinary people* (pp. 1–33). Santa Barbara, CA: Praeger.

Ronson, J. (2004). *The men who stare at goats.* London: Picador.

Rooney, P. (1991). Gendered reason: Sex metaphor and conceptions of reason. *Hypatia, 6*, 77–103.

Rosenzweig, S. (1936). Some implicit common factors in diverse methods in psychotherapy. *American Journal of Orthopsychiatry, 6*, 412–415.

Rossi, E., Bartoli, P., Bianchi, A., Endrizzi, C., & Da Frè, M. (2012). Homeopathic aggravation with Quinquagintamillesimal potencies. *Homeopathy, 101*, 112–120.

Rossiter, M. W. (1982). *Women scientists in America: Struggles and strategies to 1940.* Baltimore, MD: Johns Hopkins University Press.

RPS. (2013). *The Russian Psychologist's Oath.* Retrieved October 26, 2013, from Official Website of the Russian Psychological Society: http://www.psyrus.ru/en/documents/oath.php

Russell, B. (1952). Is there a god? In J. G. Slater, & P. Köllner (Eds.), *The collected papers of Bertrand Russell, Volume 11: Last philosophical testament, 1943–68* (pp. 543–548). London: Routledge, 1997.

Russell, N. (2010). *Communicating science: Professional, popular, literacy.* Cambridge: Cambridge University Press.

Russell, P. (1998). The palaeolithic mother-goddess: Fact or fiction? In K. Hays-Gilpin, & D. S. Whitley (Eds.), *Reader in gender archaeology* (pp. 261–268). London: Routledge.

Russett, C. E. (1989). *Sexual science: The Victorian construction of womanhood.* Cambridge, MA: Harvard University Press.

Sarason, I. G., Sarason, B. R., & Pierce, G. R. (1990). Social support: The search for theory. *Journal of Social and Clinical Psychology, 9,* 133–147.

Scarf, M. (2005). *Secrets, lies, betrayals: How the body holds the secrets of a life, and how to unlock them.* New York: Ballantine Books.

Schick Jr., T., & Vaughn, L. (2008). *How to think about weird things: Critical thinking for a new age* (5th ed.). Boston, MA: McGraw-Hill.

Schmidt, I. W., Berg, I. J., & Deelman, B. G. (1999). Illusory superiority in self-reported memory of older adults. *Aging, Neuropsychology, and Cognition, 6*(4), 299–301.

Science Council. (2013). *About Us.* Retrieved October 26, 2013, from Science Council: http://www.sciencecouncil.org/our-members

Seligman, M. E. (1998). President's column: What is the 'good life'? *APA Monitor, 29*(10), 1.

Seligman, M. E. (2002). *Authentic happiness: Using the new positive psychology to realize your potential for lasting fulfillment.* New York: Free Press.

Shanahan, J., & Good, J. (2000). Heat and hot air: Influence of local temperature on journalists' coverage of global warming. *Public Understanding of Science, 9,* 285–295.

Shang, A., Huwiler-Müntener, K., Nartey, L., Jüni, P., Dörig, S., Sterne, J. A., et al. (2005). Are the clinical effects of homoeopathy placebo effects? Comparative study of placebo-controlled trials of homoeopathy and allopathy. *Lancet, 366,* 726–732.

Shechtman, D., Blech, I., Gratias, D., & Cahn, J. W. (1984). Metallic phase with long-range orientational order and no translational symmetry. *Physical Review Letters, 50*(20), 1951–1954.

Shields, S. A. (1975). Functionalism, Darwinism, and the psychology of women: A study in social myth. *American Psychologist, 30,* 739–754.

Shiozaki, M., Hirai, K., Koyama, A., & Inui, H. (2011). Negative support of significant others affects psychological adjustment in breast cancer patients. *Psychology & Health, 26*(11), 1540–1551.

Shiraishi, N., Nishida, A., Shimodera, S., Sasaki, T., Oshima, N., Watanabe, N., et al. (2014). Relationship between violent behavior and repeated weight-loss dieting among female adolescents in Japan. *PLoS ONE, 9*(9), e107744.

Silventoinen, K., Magnusson, P. K., Tynelius, P., Kaprio, J., & Rasmussen, F. (2008). Heritability of body size and muscle strength in young adulthood: A study of one million Swedish men. *Genetic Epidemiology, 32*(4), 341–349.

Simmons, J. P., Nelson, L. D., & Simonsohn, U. (2011). False-positive psychology: Undisclosed flexibility in data collection and analysis allows presenting anything as significant. *Psychological Science, 22*(11), 1359–1366.

Sivin, N. (1995). *Science in Ancient China: Researches and reflections.* Brookfield, VT: Ashgate.

Skaggs, E. B. (1933). The meaning of the term "abnormality" in psychology. *Journal of Abnormal and Social Psychology, 28*(2), 113–118.

Sniehotta, F. F., Presseau, J., & Araújo-Soares, V. (2014). Time to retire the theory of planned behaviour. *Health Psychology Review, 8*(1), 1–7.

Sosis, R. (2000). Religion and intragroup cooperation: Preliminary results of a comparative analysis of utopian communities. *Cross-Cultural Research, 34*(1), 70–87.

Sosis, R., & Bressler, E. R. (2003). Cooperation and commune longevity: A test of the costly signaling theory of religion. *Cross-Cultural Research, 37*(2), 211–239.

Spicer, J. I. (2006). *Biodiversity.* New York: Rosen Publishing Group.

Spyros, A. (1980). Gene cloning by press conference. *New England Journal of Medicine, 302,* 743–746.

Statistics Canada. (2005). *Population by religion, by province and territory (2001 Census).* Retrieved December 20, 2014, from http://www.statcan.gc.ca/tables-tableaux/sum-som/l01/cst01/demo30a-eng.htm

Stein, D. M., & Lambert, M. L. (1984). On the relationship between therapist experience and psychotherapy outcome. *Clinical Psychology Review, 4,* 127–142.

Stenger, V. (1992). The myth of quantum consciousness. *The Humanist, 53*(3), 13–15.

Stephens, L. F. (2005). News narratives about nano S&T in major U.S. and non-U.S. newspapers. *Science Communication, 27*(2), 175–199.

Storey, A. E., Walsh, C. J., Quinton, R. L., & Wynne-Edwards, K. E. (2000). Hormonal correlates of paternal responsiveness in new and expectant fathers. *Evolution & Human Behavior, 21*(2), 79–95.

Stove, D. (1991). *The Plato cult and other philosophical follies.* Oxford: Blackwell.

Strack, F., & Mussweiler, T. (1997). Explaining the enigmatic anchoring effect: Mechanisms of selective accessibility. *Journal of Personality and Social Psychology, 73,* 437–446.

Strevens, M. (2003). The role of the priority rule in science. *Journal of Philosophy, 100*(2), 55–79.

Strupp, H. H. (1963). The outcome problem in psychotherapy revisited. *Psychotherapy: Theory, Research and Practice, 1,* 1–13.

Strupp, H. H., & Hadley, S. W. (1979). Specific vs nonspecific factors in psychotherapy: A controlled study of outcome. *Archives of General Psychiatry, 36,* 1125–1136.

Sullivan, A. (2012). *The art of asking questions about religion.* London: Centre for Longitudinal Studies, Institute of Education, University of London.

Svenson, O. (1981). Are we all less risky and more skillful than our fellow drivers? *Acta Psychologica, 47,* 143–148.

Swift, J. J., Johnson, J. A., Morton, T. D., Crepp, J. R., Montet, B. T., Fabrycky, D. C., et al. (2013). Characterizing the cool KOIs IV: Kepler-32 as a prototype for the formation of compact planetary systems throughout the galaxy. *Astrophysical Journal, 764,* 105.

Tappe, D., Kern, P., Frosh, M., & Kern, P. (2010). A hundred years of controversy about the taxonomic status of Echinococcus species. *Acta Tropica, 115*(3), 167–174.

Tavris, C. (1993). *The mismeasure of woman.* New York: Touchstone.

Taylor, J. G. (1975). *Superminds: An enquiry into the paranormal.* London: Macmillan.

Taylor, L. D., Bell, R. A., & Kravitz, R. L. (2011). Third-person effects and direct-to-consumer advertisements for antidepressants. *Depression and Anxiety, 28,* 160–165.

Taylor, S. E., & Brown, J. D. (1988). Illusion and well-being: A social psychological perspective on mental health. *Psychological Bulletin*, *103*(2), 193–210.

Taylor, S. E., & Master, S. L. (2011). Social responses to stress: The tend-and-befriend model. In R. Contrada, & A. Baum (Eds.), *The handbook of stress science: Biology, psychology, and health* (pp. 101–109). New York: Springer.

'Teen boys losing virginity earlier and earlier, report teen boys'. (2014). Retrieved August 17, 2014, from The Onion: http://www.theonion.com/video/teen-boys-losing-virginity-earlier-and-earlier-rep,35906/

Tegmark, M. (2000). Importance of quantum decoherence in brain processes. *Physical Review E*, *61*(4), 4194–4206.

The Dictionary Project. (2011). *A student's dictonary & gazetteer* (19th ed.). Sullivan's Island, SC: The Dictionary Project, Inc.

Tracey, T. J., Wampold, B. E., Lichtenberg, J. W., & Goodyear, R. K. (2014). Expertise in psychotherapy: An elusive goal? *American Psychologist*, *69*(3), 218–229.

Trafimow, D. (2014). Considering quantitative and qualitative issues together. *Qualitative Research in Psychology*, *11*, 15–24.

Tudor, A. (1989). Seeing the worst side of science. *Nature*, *340*, 589–592.

Tugade, M. M., & Fredrickson, B. L. (2004). Resilient individuals use positive emotions to bounce back from negative emotional experiences. *Journal of Personality and Social Psychology*, *86*(2), 320–333.

Tversky, A., & Kahneman, D. (1974). Judgment under uncertainty: Heuristics and biases. *Science*, *185*(4157), 1124–1131.

Tversky, A., & Kahneman, D. (1982). Judgments of and by representativeness. In D. Kahneman, P. Slovic, & A. Tversky (Eds.), *Judgment under uncertainty: Heuristics and biases* (pp. 84–98). Cambridge: Cambridge University Press.

Tversky, A., & Kahneman, D. (1983). Extensional versus intuitive reasoning: The conjunction fallacy in probability judgment. *Psychological Review*, *90*(4), 293–315.

Tversky, A., & Koehler, D. J. (1994). Support theory: A nonextentional representation of subjective probability. *Psychological Review*, *101*, 547–567.

Twitmyer, E. B. (1905). Knee jerks without simulation of the patellar tendon. *Psychological Bulletin*, *2*, 43.

Tyson, P. J., Jones, D., & Elcock, J. (2011). *Psychology in social context: Issues and debates*. Chichester: BPS Blackwell.

UNESCO. (1988). *Proposed international standard nomenclature for fields of science and technology*. Paris: UNESCO.

Ungar, S. (2000). Knowledge, ignorance and the popular culture: Climate change versus the ozone hole. *Public Understanding of Science*, *9*, 297–312.

Urbach, P. (1987). *Francis Bacon's philosophy of science: An account and a reappraisal*. Chicago: Open Court Publishing.

US Bureau of Labor Statistics. (2014). *American Time Use Survey*. Retrieved March 24, 2014, from http://www.bls.gov/tus/data.htm

Uttal, W. R. (2001). *The new phrenology: The limits of localizing cognitive processes in the brain*. Boston, MA: Massachusetts Institute of Technology.

Valentine, E. R. (1992). *Conceptual issues in psychology* (2nd ed.). London: Routledge.

Verheij, R. A., de Bakker, D. H., & Groenewegen, P. P. (1999). Is there a geography of alternative medical treatment in The Netherlands? *Health & Place, 5*(1), 83–97.

Verplanken, B. (2006). Beyond frequency: Habit as mental construct. *British Journal of Social Psychology, 45,* 639–656.

Wagenmakers, E.-J., Wetzels, R., Borsboom, D., & van der Maas, H. L. (2011). Why psychologists must change the way they analyze their data: The case of psi: Comment on Bem (2011). *Journal of Personality and Social Psychology, 100*(3), 426–432.

Wampold, B. E. (2001). *The great psychotherapy debate: Model, methods, and findings.* Mahwah, NJ: Lawrence Erlbaum Associates.

Wampold, B. E. (2013). The good, the bad, and the ugly: A 50-year perspective on the outcome problem. *Psychotherapy, 50*(1), 16–24.

Wampold, B. E., Mondin, G. W., Moody, M., Stich, F., Benson, K., & Ahn, H.-n. (1997). A meta-analysis of outcome studies comparing bona fide psychotherapies: Empirically, "all must have prizes". *Psychological Bulletin, 122*(3), 203–215.

Wang, C., Bannuru, R., Ramel, J., Kupelnick, B., Scott, T., & Schmid, C. H. (2010). Tai Chi on psychological well-being: Systematic review and meta-analysis. *BMC Complementary & Alternative Medicine, 10,* 23.

Wansink, B., Kent, R. J., & Hoch, S. J. (1998). An anchoring and adjustment model of purchase quantity decisions. *Journal of Marketing Research, 35,* 71–81.

Watkins, P. C., Uhder, J., & Pichinevskiy, S. (2015). Grateful recounting enhances subjective well-being: The importance of grateful processing. *Journal of Positive Psychology, 10*(2), 91–98.

Webb, C. A., DeRubeis, R. J., & Barber, J. P. (2010). Therapist adherence/competence and treatment outcome: A meta-analytic review. *Journal of Consulting and Clinical Psychology, 78,* 200–211.

Webster, R. (2012). *Face reading quick & easy.* Woodbury, MN: Llewellyn Publications.

Wegenstein, B., & Ruck, N. (2011). Physiognomy, reality television and the cosmetic gaze. *Body & Society, 17*(4), 27–54.

Wegrocki, H. J. (1938). A critique of cultural and statistical concepts of abnormality. *Journal of Abnormal and Social Psychology, 34*(2), 166–178.

Weinstein, N. D. (1980). Unrealistic optimism about future life events. *Journal of Personality and Social Psychology, 39,* 806–820.

Weinstein, N. D. (2007). Misleading tests of health behavior theories. *Annals of Behavioral Medicine, 33*(1), 1–10.

Weisstein, N. (1971). Psychology constructs the female. *Journal of Social Education, 35,* 362–373.

Wheen, F. (2004). *How mumbo-jumbo conquered the world: A short history of modern delusions.* London: Fourth Estate.

White, A. R., Rampes, H., Liu, J. P., Stead, L. F., & Campbell, J. (2011). Acupuncture and related interventions for smoking cessation (Article no. CD000009). *Cochrane Database of Systematic Reviews (Issue 1).*

Whorton, J. C. (2002). *Nature cures: The history of alternative medicine in America.* Oxford: Oxford University Press.

Wills, T. A. (1985). Supportive functions of interpersonal relationships. In S. Cohen, & S. L. Syme (Eds.), *Social support and health.* Orlando: Academic Press.

Wills, T. A. (1998). Social support. In E. A. Blechman, & K. D. Brownell (Eds.), *Behavioral medicine and women: A comprehensive handbook* (pp. 118–128). New York: Guilford Press.

Wilson, T. D. (2012, July 12). Stop bullying the 'soft' sciences. *Los Angeles Times*. Retrieved January 10, 2013, from http://articles.latimes.com/2012/jul/12/opinion/la-oe-wilson-social-sciences-20120712

Wolfe, R. M., & Sharp, L. K. (2002). Anti-vaccinationists past and present. *BMJ*, *325*, 430–432.

Woolley, H. T. (1910). Psychological literature: A review of the recent literature on the psychology of sex. *Psychological Bulletin*, *7*, 355–342.

Woolley, H. T. (1914). The psychology of sex. *Psychological Bulletin*, *11*, 353–379.

Wu, J., Yeung, A. S., Schnyer, R., Wang, Y., & Mischoulon, D. (2012). Acupuncture for depression: A review of clinical applications. *Canadian Journal of Psychiatry*, *57*, 397–405.

Xu, X., Aron, A., Brown, L., Cao, G., Feng, T., & Weng, X. (2011). Reward and motivation systems: A brain mapping study of early-stage intense romantic love in Chinese participants. *Human Brain Mapping*, *32*, 249–257.

Yong, E. (2012). Bad copy. *Nature*, *485*, 298–300.

Yong, E. (2012, July 17). *One molecule for love, morality, and prosperity?* Retrieved September 1, 2013, from Slate.com: http://www.slate.com/articles/health_and_science/medical_examiner/2012/07/oxytocin_is_not_a_love_drug_don_t_give_it_to_kids_with_autism_.single.html

Young, I. M. (1990). *Throwing like a girl and other essays in feminist political theory*. Bloomington, IN: Indiana University Press.

Zak, P. J. (2008). Values and value: Moral economics. In P. J. Zak (Ed.), *Moral markets* (pp. 259–279). Princeton, NJ: Princeton University Press.

Zak, P. J. (2012). *The moral molecule: The new science of what makes us good or evil*. London: Bantam.

Zak, P. J., & Fakhar, A. (2006). Neuroactive hormones and interpersonal trust: International evidence. *Economics and Human Biology*, *4*, 412–429.

Zebrowitz, L. A., & Franklin, R. G. (2014). The attractiveness halo effect and the babyface stereotype in older and younger adults: Similarities, own-age accentuation, and older adult positivity effects. *Experimental Aging Research*, *40*, 375–393.

Zebrowitz, L. A., Montepare, J. M., & Lee, H. K. (1993). They don't all look alike: Individuated impressions of other racial groups. *Journal of Personality and Social Psychology*, *65*, 85–101.

Zebrowitz, L. A., Wang, R., Bronstad, P. M., Eisenberg, D., Undurraga, E., Reyes-García, V., et al. (2012). First impressions from faces among US and culturally isolated Tsimane' people in the Bolivian rainforest. *Journal of Cross-Cultural Psychology*, *43*, 119–134.

Index